Plant Diseases and Vectors:
Ecology and Epidemiology

Plant Diseases and Vectors: Ecology and Epidemiology

EDITED BY

KARL MARAMOROSCH
Waksman Institute of Microbiology
Rutgers University
New Brunswick, New Jersey

KERRY F. HARRIS
Department of Entomology
Texas A&M University
College Station, Texas

1981

ACADEMIC PRESS

A Subsidiary of Harcourt Brace Jovanovich, Publishers

New York   London   Toronto   Sydney   San Francisco

ACADEMIC PRESS, INC.
111 Fifth Avenue, New York, New York 10003

United Kingdom Edition published by
ACADEMIC PRESS, INC. (LONDON) LTD.
24/28 Oval Road, London NW1 7DX

Library of Congress Cataloging in Publication Data
Main entry under title:

Plant diseases and vectors.

   Includes bibliographical references and index.
   1. Plant diseases. 2. Virus diseases of plants.
3. Insects as carriers of plant disease. I. Maramorosch,
Karl. II. Harris, Kerry F.
S3731.P65   632'.3   80-28311
ISBN 0-12-470240-6

PRINTED IN THE UNITED STATES OF AMERICA

81 82 83 84   9 8 7 6 5 4 3 2 1

CONTENTS

Chapter   6.   Ecology and Control of Soybean Mosaic Virus

Michael E. Irwin and Robert M. Goodman

Chapter   7.   Early Events in Plant Virus Infection

G. A. de Zoeten

Chapter   8.   Virus Transmission through Seed and Pollen

C. L. Mandahar

Chapter   9.   Seedborne Viruses: Virus–Host Interactions

Thomas W. Carroll

**Chapter   10.   Man–Made Epidemiological Hazards in Major
Crops of Developing Countries**

L. Chiarappa

CONTRIBUTORS

Numbers in parentheses indicate the pages on which the authors' contributions begin.

Moshe Bar-Joseph (35), Virus Laboratory, Agricultural Research Organization, The Volcani Center, P. O. Box 6, Bet-Dagan 50-20, Israel.

Luite Bos (1), Research Institute for Plant Protection, Binnenhaven 12, Wageningen, P. O. Box 42, The Netherlands.

Lloyd N. Chiykowski (105), Chemistry and Biology Research Institute, Agriculture Canada, Ottawa, Ontario K1A 0C6, Canada.

Thomas W. Carroll (293), Department of Plant Pathology, Montana State University, Bozeman, Montana 59717.

Luigi Chiarappa (319), Plant Protection Service, Food and Agricultural Organization of the United Nations, Via delle Terme di Caracalla, 00100– Rome, Italy.

Gustaaf A. de Zoeten (221), Department of Plant Pathology, University of Wisconsin, 1630 Linden Drive, Madison, Wisconsin 53706.

James E. Duffus (161), United States Department of Agriculture, Science and Education Administration, 1636 E. Alisal Street, P. O. Box 5098, Salinas, California 93915.

Bryce W. Falk (161), Department of Plant Pathology, University of California, Riverside, California 92521.

Steven M. Garnsey (35), Horticultural Research Laboratory, United States Department of Agriculture, Science and Education Administration, 2120 Camdem Road, Orlando, Florida 32803.

Robert M. Goodman (181), Department of Plant Pathology and International Soybean Program (INTSOY), University of Illinois, Urbana, Illinois 61801.

Michael E. Irwin (181), Office of Agricultural Entomology, Department of Plant Pathology, and International Soybean Program (INTSOY),

University of Illinois and Illinois Natural History Survey, Urbana, Illinois 61801.

C. L. Mandahar (241), Department of Botany, Panjab University, Chandigarh– 160014, India.

Pătru G. Ploaie (61), Laboratory of Phytopathology, Plant Protection Institute, Academy of Agricultural and Silvicultural Sciences, Bucharest 71592, Rumania.

PREFACE

This is the fourth in a five-volume series of books on vectors of plant disease agents. The first three volumes, *Aphids as Virus Vectors, Leafhopper Vectors and Plant Disease Agents,* and *Vectors of Plant Pathogens,* are up-to-date treatises on the interactions of plant pathogens such as viruses, mycoplasmalike organisms, spiroplasmas, fungi, bacteria, and rickettsialike organisms with their plant hosts and insect, nematode, mite, or fungal vectors. Such interactions form the bases of vector-dependent, pathogen transmission cycles and, hence, become integral parts of any studies aimed at elucidating the dynamics of disease spread. In this sense, disease epidemiology was certainly not ignored in earlier books in the series, but neither was it the topic of primary emphasis, as is the case in this volume.

Successful transmission of vector–dependent pathogens requires that the transmission participants, vector, pathogen, and host, interact in a manner compatible with pathogen acquisition, carry-over, and inoculation by the vector and subsequent infection site development in the host. The newly emerging science that asks how various biotic and abiotic components of the environment affect pathogen–vector–host compatibility as measured by pathogen spread or vector transmission efficiency might be referred to as *transmission ecology*. Considering the complexity of any one transmission system and the number and types of vectors and pathogens covered in the three previous volumes, it would of course be impossible to include all of the information available on transmission ecology and disease epidemiology in a single volume. Therefore, we have attempted to choose timely topics that illustrate some of the incipient overriding principles in this challenging field of research.

The book is comprised of ten chapters representing the expertise of 13 outstanding scientists from a total of seven different countries. The occurrence of disease epidemics or, to state it differently, the extent to which pathogens are dispersed by arthropod vectors, is basically a reflection of the degree of spatial and temporal overlap between vectors and pathogen sources. Appropriately, therefore, we begin our book with a chapter on the ecological involvement of wild plants in plant virus pathosystems. This is followed by an equally timely discussion of the principles and applications of enzyme-linked immunosorbent

assay (ELISA) in diagnosing plant viruses and monitoring their movement in the environment. Chapters three and four give detailed accounts of
the epidemiologies of diseases caused by leafhopper-borne viruses, mollicutes,
and rickettsialike organisms. Chapter five brings us abreast of the latest developments in understanding the importance of helper agents to the transmission ecol-
ogies of many aphid-borne plant viruses. And, in Chapter six, the aphid-borne
and seed transmitted soybean mosaic virus serves as an excellent model for pre-
senting an in-depth coverage of the many and varied factors that can contribute to
the epidemiology and control of a disease affecting a major agricultural crop of
the world. Chapter seven, ''Early Events in Plant Virus Infection,'' focuses
attention on the ecological importance of virus-plant interactions and the development of disease at the organismic level, including virus genome release and
cell ingress, host–virus recognition and virus replication in the first cell entered,
and transport of the infectious entity from cell to cell and over long distances
within the plant. A vector of plant viruses not covered in earlier volumes of the
series—the host plant itself—is dealt with in Chapters eight and nine, with discussions of the ecological and epidemiological significance of virus transmission
from one sporophyte generation to the next via virus infected or infested seed or
pollen. And, finally, man himself, yet another participant in and manipulator of
the transmission cycle, is the focus of attention in Chapter ten; especially with
respect to the epidemiological hazards he has created in major crops of developing countries around the world.

It is the editors' hope that this volume furthers our knowledge of transmission
ecology and disease epidemiology, not only by serving as a valuable supplemen-
tal textbook, reference work and bibliographical source but also by catalyzing
novel syntheses of present thinking and stimulating further research in the area.
If we have succeeded in these regards, it is thanks to our authors for their schol-
arly contributions and to the staff of Academic Press for their continuing encour-
agement and assistance in producing these volumes.

Chapter 1

WILD PLANTS IN THE ECOLOGY OF VIRUS DISEASES

L. Bos

Research Institute for Plant Protection (IPO),
Wageningen,
The Netherlands

1.1 INTRODUCTION

In recent years, crops have been increasingly studied as *ecosystems*. No matter how artificial they may seem—as suggested by the designation 'crop industry'—crops remain highly complex natural systems, which comprise a multitude of interrelated factors. Several of these act from the natural environment at large.

Among these factors, viruses play an intriguing part and their involvement can be described as an *ecosystem* or *virus pathosystem*. I have recently discussed its general principles and outlined a scheme to survey the various factors and to clarify their multicausal and dynamic interplay (Bos, 1980). It is also meant to comprehend ways of managing crop ecosystems (interfering with virus ecology) to control virus diseases.

The present book on virus ecology goes into much greater detail and I have been asked to write on the role of wild plants. Wild plants have long been presumed to contribute to virus spread and disease incidence. For earlier discussions and general information see Hein (1953), Schwarz (1959), Brcák and Polák (1966), Heathcote (1970), Murant (1970), Duffus (1971) and Schmidt (1977). Only recently though has wild plant involvement been described in ecological terms. This chapter is an elaboration of a paper presented during the 3rd International Congress of Plant Pathology in Munich, West Germany, September 1978 (Bos, 1978). It will analyze the various aspects, and illustrate these with examples from the literature.

1.2   THE ECOLOGICAL MODEL

Figure 1 summarizes the groups of factors involved in virus ecology and surveys the dynamic interrelationships within the virus ecosystem as developing with *time*. It allows a first orientation as to the involvement of wild plants and will be used as a basis for further elaboration of the subject.

From a cultural point of view, man's chief concern is the *crops*. These have to be protected, for instance from various *viruses*. Some of the viruses of plants have been characterized, but several are still unidentified. In addition to form and size, viruses differ in ecological relationships, so they may require completely different ways of control.

Besides the crops and the viruses come the *sources of infection*. These may be (1) similar crops, or closely or remotely related crops, or even a single infected plant, nearby or at a distant place, (2) individual plants within the crop to be protected, introduced by infected seed or planting stock, or (3) uncultivated plants in or near the crop, or far away. That is where wild plants enter the picture.

The next group of factors is *vectors* or vehicles of spread. Plants themselves, including wild ones, may 'actively' participate through contact or by their natural propagation material including true seeds. Man himself, although long unaware, has been actively engaged in spreading viruses with crop propagation material and in germplasm, part of which is derived from wild crop relatives often from distant parts of the world.

Growing conditions greatly influence susceptibility and sensitivity of crops and wild plants to virus infection, as well as vector behavior. Crop micro-climate is considerably affected by the presence of wild plants within the crop territory. Nutrition and even infection with viruses may have a direct bearing on host pal-

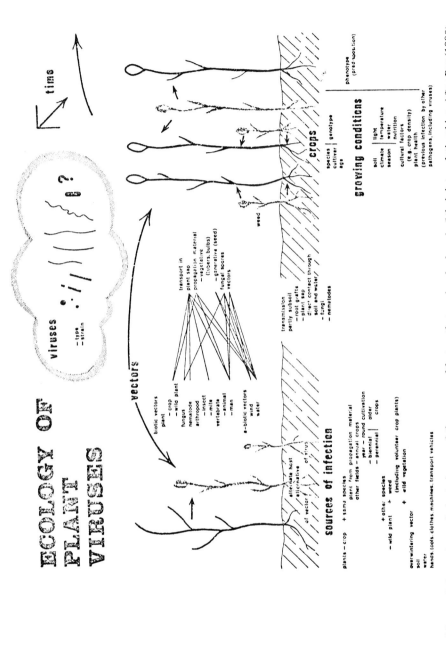

FIG. 1. Scheme of the ecology of plant viruses, of the groups of factors that are involved, and of their interrelationships. After Bos (1980).

atability to insect and other vectors and even on vector behavior. Thus there are a multitude of interdependences and I will later return to some of these in so far as they bear on the role of wild plants.

1.3 DEFINITION OF WILD PLANTS

Webster (1976) defines 'wild' as "living in a state of nature, inhabiting natural haunts, not tamed or domesticated, growing or produced without the aid and care of man, not living near or associated with man." Hence in plain words, *wild plants* are uncultivated plants, plants that develop without human aid, for instance in natural habitats. They have long been ignored by agronomists and left to the interest of biologists. However agricultural interest in wild plants has rapidly increased because wild plants, still growing in their natural habitat or assembled in botanic gardens, may include species or types that offer prospects for cultivation, either directly or for provision of genes in crop diversification. Wild plants also include the ancestors of present-day crops. Interbreeding with related cultivars may broaden genetic variation. Collections of wild germplasm are now thriving. Wild plants include also crop species or their descendants that have gone astray and run wild, such as groundkeepers and volunteer plants, and crop plants in abandoned fields. Wild plants usually grow in places where they present no direct threat to crop development, as in wild vegetation and on uncultivated land.

Weeds also are uncultivated plants, and Webster's definition at first sight includes them, except that weeds live near or in association with man. They usually grow on cultivated land and may in certain instances depend on man. Thus, Webster (1976) defines a weed as a "plant growing in ground that is or has been in cultivation usually to the detriment of the crop or to the disfigurement of the place." Duffus (1971), in referring to literature, describes weeds as "plants with harmful or objectionable characteristics that grow where they are unwanted." They directly interfere with crops and compete with crop plants for space, food, water and light.

Hence, the terms weed and wild plant are not exactly equivalent but there is much overlap. Weeds are wild plants that grow on cultivated land, but no sharp limit exists. Likewise, there is no clear distinction of wild plants or weeds from cultivated plants, since weeds and volunteer plants developing in a crop may later be used for grazing cattle in a stubble. The term weed is emotive; it has a derogatory connotation as something threatening and thus requiring destruction. Wild plants, even when distant from crops but harboring pathogens including viruses that are harmful to crops, would be weeds in the sense of Duffus (1971), and so should be destroyed. Calling something a weed has direct ethical consequences. This is even more apparent with the equivalents in German, 'Unkraut', and French, 'mauvaise herbe.' There is now a tendency to speak more neutrally of 'wild plants.' For their role in the environment, these plants are now enjoying more goodwill.

1.4 VIRUS INFECTIONS OF WILD PLANTS

Infections of uncultivated plants, including virus infections, have long escaped attention. Phytopathologists had no interest in diseases of such plants (Wilson, 1969). Virus infections were first detected in cultivated plants by clear symptoms (tulip mosaic, tobacco mosaic, cucumber mosaic, beet curly top, maize streak), by often heavy (epidemic) incidence due to uniformity of the crop, and by their impact on yield and quality. The last makes virus infections of crops economically important. So crops are closely watched by their growers down to the level of individual plants, and weights (yields) are determined. Thus, even 'latent' viruses may attract attention if they only affect plant or crop weight. Plants that have been killed because of infection may still betray their earlier presence by gaps in otherwise even stands. This now readily explains why most virus infections in mixed wild vegetations escaped attention.

Wild species are usually genetically very variable. No two plants react identically to infection (Fig. 2). High natural competition then rapidly leads towards *selection for resistance and latency* (symptomless infection): sensitive host plants and virulent virus strains disappear. For this reason, *Stellaria media* plants from Britain, North America and Australia were least affected by the strain of cucumber mosaic virus from their country of origin, and showed more severe reactions when infected with two alien strains (Tomlinson and Walker, 1973). The place of wild plants that have succumbed in the natural habitat is soon taken by more vital individuals of the same or totally different species and their absence is not noticed. The same has been reported for other pathogens in weeds (Wilson, 1969).

FIG. 2. Beet western yellows virus in *Capsella bursa-pastoris*, six weeks after inoculation with aphids. Reactions range from highly sensitive (left) to tolerant (right). After Ashby *et al.* (1979).

Moreover the presence in natural vegetations of totally different kinds of plants together prevents or reduces direct contact between susceptible plants and thus reduces the chance of rapid epidemic infection that would be noticed.

However it was soon realized that the mere presence of viruses in wild plants may have a bearing on the health of nearby crops.

In the early 1920s when an increasing number of crop virus diseases were studied and certain viruses were found to infect different crops, Doolittle and Walker (1925) were among the first to substantiate the overwintering of cucumber mosaic virus in wild cucumber (*Echinocystis lobata*) and some other uncultivated species in Wisconsin. In Florida, *Commelina nudiflora* was soon found to be the most important source of infection for the virus of celery crops (Doolittle and Wellman, 1934). The virus has since been reported from various other weeds (e.g. Hein, 1959; Schwarz, 1959). Bruckart and Lorbeer (1976) more recently detected it in 12 of 66 species of weed collected near affected lettuce and celery fields in the State of New York, five of these being newly recognized hosts. Cucumber mosaic virus is now known to be infectious to 242 genera of 64 families of flowering plants. There are various strains of the virus and its multipartite genome may add to its polyphagous nature (Quiot *et al.*, 1976).

In Indonesia, Thung (1932) detected the transmission of tobacco leaf curl by the whitefly *Bemisia tabaci* and 'overwintering' of the virus in a number of weeds, such as *Ageratum conyzoides*, *Synedrella nodiflora* and *Vernonia cinerea* in ruderals near villages. Another classic example is of beet curly-top virus. Severin (1934) mentioned 39 weed species belonging to 11 families in which the virus overwintered in California. Later, Bennett (1971) listed more than 300 species of 44 families as susceptible to the virus and infection in many was symptomless.

Although information on virus infections of wild plant species is accumulating, data on incidence of virus infection in the natural vegetation are still scarce. MacClement and Richards' paper (1956) is still unique. They systematically collected samples at six different biotopes in 700 ha of natural terrain of the Royal Botanic Gardens of Hamilton, Ontario. They did so every two weeks during four growing seasons, and tested the samples (2,193 in total) for sap-transmissible viruses. Some aquatic species were included. On average, 10% of the samples turned out to be infected. During one year, all 21 species then tested were found once or more often to be infected. Of certain species over 50% of the samples contained virus and several single plants were infected with more than one!

Widely diverse wild plants such as cacti and ferns may contain viruses. Chessin and Lesemann (1972) sampled 340 flat-padded prickly pears (*Opuntia*) in Arizona, Nevada and Utah and found that about half contained the TMV-like Sammon's *Opuntia* virus and a flexible rod resembling cactus virus X or *Zygocactus* virus throughout the collecting area. Nienhaus *et al.* (1974) isolated a 785-nm elongate plant virus from diseased plants of the ferns *Polypodium vulgare* and *Dryopteris filix-mas* in forests in West Germany.

I will not endeavor to list all information on virus infections of wild plants, which is already overwhelming. For a survey of those infections that had been reported up to and including 1973 to cause natural diseases of wild plants in Europe see Schmidt (1977). A number of viruses have now been described exclusively or primarily from wild hosts.

1.5 VIRUS 'POLYPHAGISM'

Some of the viruses detected early in history of plant virology to infect wild and cultivated species were soon found to have many natural hosts. During recent decades with increasing ease of diagnosis, in addition to cucumber mosaic virus and beet curly-top virus several viruses became known for their wide artificial and natural host ranges including several wild species.

For alfalfa mosaic virus, 430 species in 51 families could be infected and about 150 species of these in 22 families were found to be susceptible to natural infection, often without symptoms (Schmelzer et al., 1973). For broad bean wilt virus, 67 species of 27 families of flowering plants could be infected (Schmelzer et al., 1975).

Soil-borne nematode-transmitted viruses, particularly nepoviruses, have attracted special attention and the importance of wild plants in their ecology is well documented (Murant, 1970). Noordam (1956) isolated tobacco rattle virus from the roots of 24 of 35 plant species including various weeds and the Gramineae *Apera spica-venti*, another weed, and *Secale cereale*. In some species, infection was restricted to the roots. The virus is now known to infect more than 400 species of more than 50 families of flowering plants (Schmelzer, 1957). With soil-borne viruses infection is often symptomless and may thus easily escape notice.

The more recently discovered beet western yellows virus is a persistent aphid-borne luteovirus with a very wide range of susceptible plants comprising over 150 species in 23 dicotyledonous families (Duffus, 1973).

Other viruses that were long assumed to be restricted to certain species or families are now increasingly found to have wider host ranges. Bean common mosaic virus causes disease in different *Phaseolus* spp., not only *P. vulgaris* but also mungbean (*P. aureus, P. radiatus*; Kaiser and Mossahebi, 1974), and tepary bean (*P. acutifolius* var. *latifolius*; Provvidenti and Cobb, 1975). It also affects other legumes such as phasemy bean (*Macroptilium lathyroides*; Provvidenti and Braverman, 1976). It has even been detected in the weed legume *Rhynchosia minima*, which is common near bean fields in tropical areas (Meiners et al., 1978). It can infect some other legumes (Quantz, 1961) and non-legumes such as *Nicotiana clevelandii* (Drijfhout and Bos, 1977) and *N. benthamiana* (Christie and Crawford, 1978). Although in practice the virus seems to rely on seed transmission in bean, wild alternative hosts might well play a role in its perennation under certain conditions.

Bean yellow mosaic virus used also to be considered a typical legume virus, with gladiolus (McWhorter et al., 1947) and freesia (van Koot et al., 1954) as exceptional non-legume hosts. Now the virus has been isolated from naturally infected plants of *Cheopodium album, Cirsium arvense, Papaver somniferum*, and *Sesamum indicum* (Kovachevski, 1972), *Proboscidea jussieui* (Provvidenti and Schroeder, 1972) and *Viola odorata* (Provvidenti and Granett, 1973) and was found to cause widespread disease in summer squash (*Cucurbita pepo*) in the northeastern United States (Provvidenti and Uyemoto, 1973).

These few examples demonstrate how viruses described from crop plants are much more 'polyphagous' than used to be supposed. Hence, most viruses are bound to have wild species among their natural hosts, even though they do not infect plants indiscriminately and though some are highly specific. In view of these facts crop pathologists are tending to include wild plants in their field of operation.

1.6   IMPACT OF WILD PLANTS ON CROP VIRUSES

The ecological scheme (Fig. 1) has already indicated that wild plants act also in other ways than as sources of infection of the virus. They have a much more intricate and diverse impact on virus ecology, and a number of aspects should be distinguished, although some are interrelated. Wild plants allow or greatly assist in virus survival through adverse periods. (Section 1.6.1), but may also be the main origin of crop viruses (Section 1.6.2). They may also play a role in virus spread from one field to another and even to geographically distant areas (Section 1.6.3), and in virus establishment after introduction into new regions (Section 1.6.4). They also indirectly act as refuges and sources of virus vectors, even when themselves exempt from infection by the viruses concerned (Section 1.6.5). Finally they are of great help in virus control through plant breeding as sources of crop resistance to viruses, although this coin has two sides (Section 1.6.6).

1.6.1   Virus Survival

Several crops are short-lived and absent from the field during winter or dry summer. They may also be absent for long periods of time in crop rotations, and intervening crops may be immune. Then, wild alternative hosts may be essential for virus survival (overwintering, oversummering, perennation). Several weeds in crops and wild plants around fields, along roadsides or in natural habitats may be short-lived as well, but may have growing periods that overlap those of crops (spring annuals, biennials) or may be long-lived (perennials). Once infected with virus, plants usually remain infected throughout life, and latent infections do not usually curtail their life span.

Early observations by Doolittle and Walker (1925) showed that in the north-central United States cucumber mosaic virus occurs mainly in weeds near cucumber fields. It was found to overwinter in roots of *Asclepias syriaca* (milkweed), *Phytolacca decandra* (pokeweed), *Nepeta cataria* (catnip) and certain perennial

TABLE I. Relationships between Vegetable Crops, Viruses and Weed Hosts in South Florida*

Cause of major virus disease	Major susceptible crop	Major and (in parentheses) minor weed host
potato virus Y	pepper, potato, tomato	black nightshade[2], groundcherry[6]
tobacco etch virus	pepper, tomato	black nightshade[2], groundcherry[6]
tobacco mosaic virus	pepper, tomato	black nightshade[2], groundcherry[6]
celery mosaic virus	celery	dayflower[5], (balsam apple[1])
cucumber mosaic virus	celery	(mock bishopweed[9])
watermelon mosaic virus	cucumber, summer squash, watermelon	creeping cucumber[3], (balsam apple[1])
Bidens mottle virus	endive, escarole, lettuce	hairy beggarticks[7], Virginia pepperweed[10], (cressleaf groundsel[4])
lettuce mosaic virus	endive, escarole, lettuce	(lambsquarters[8])

*Taken from compilation of data of Orsenigo and Zitter (1971).

[1] *Momordica charantia* L., [2] *Solanum nigrum* L., [3] *Melothria pendula* L., [4] *Senecio glabellus* Poir., [5] *Commelina* spp., [6] *Physalis* sp., [7] *Bidens pilosa* L., [8] *Chenopodium album* L., [9] *Ptilimnium capillaceum* (Michx.) Raf., [10] *Lepidium virginicum* L.

species of *Physalis*. In Florida, *Commelina nudiflora* was soon reported to be the main oversummering host of cucumber mosaic virus from celery crops (Doolittle and Wellman, 1934).

Between about 1950 and 1970, much research was done in South Florida by a number of workers as summarized by Orsenigo and Zitter (1971) on the relationships between vegetable crops, viruses and weed hosts (Table I). Most of the viruses were non-persistently transmitted by aphids, and infected weed-host populations occurred especially along field ditches, laterals, canals, fencerows, noncrop and waste areas, and cull piles and dumps.

Ecologically, perennial weeds are of special importance. Severin (1934) showed that 10 out of 39 wild plant species found naturally infected with beet curly-top virus in California were perennials.

Woody plants are of special interest, notably with soil-borne nepoviruses. In parts of Britain, both *Arabis* mosaic virus and its nematode vector *Xiphinema diversicaudatum* are restricted to woods and hedgerows containing woody plants (Fig. 3), perhaps because the vector is adversely affected when soil is disturbed by cultivation (Harrison and Winslow, 1961; Pitcher and Jha, 1961).

Viruses may also perennate in annual weeds if the viruses can pass through the seeds of such hosts. Seed transmission was already suspected by Doolittle and Gilbert (1919) for cucumber mosaic virus in *Echinocystis* (*Micrampelis*) *lobata* (wild cucumber) near cucumber fields. Such transmission was later detected to be prevalent in Europe in *Stellaria media* (chickweed) (first detected by Noordam et al., 1965; Häni et al., 1970: 10%; Tomlinson and Carter, 1970: 21-40%) and in a number of other annual weeds, such as *Cerastium holosteoïdes* (2%), *Lamium purpureum* (4%) and *Spergula arvensis* (2%) (Tomlinson and Carter, 1970). Tomlinson and Walker (1973) found that the virus survived at least 21 months in buried chickweed seeds. It is likely that viruses that infect embryos remain infective in seeds as long as seeds remain viable and that infection may

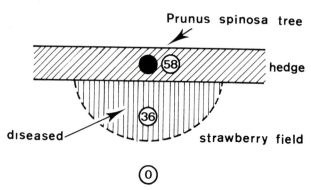

FIG. 3. Patchy outbreak (D) of *Arabis* mosaic virus in strawberry field in relation to numbers of nematode vector (*Xiphinema diversicaudatum*) in 250 ml of soil at three sampling sites, and both apparently related to the presence of a *Prunus spinosa* tree (P) in adjacent hedge (H). After Harrison and Winslow (1961).

not reduce seed viability and vitality (reviewed by Bos, 1977). For example, Pierce and Hungerford (1929) isolated bean common mosaic virus from a bean seedling grown from seed after 30 years of storage.

Soil-borne nepoviruses are notorious for their high rates of seed transmission in several of their hosts, including weeds, and their perennation in weed seeds has been studied in considerable detail (Lister and Murant, 1967; Murant and Lister, 1967; Murant, 1970).

In a recent review (Bos, 1977), after reference to a compilation of data by Phatak (1974), I concluded that seed transmission is much more prevalent than supposed earlier. Most research on seed transmission has been in cultivated plants, but a few examples suggest that such transmission is no less prevalent in wild species, which often have seeds that remain long dormant in the soil. Cultivated land may contain enormous numbers of weed seeds. The top 15 cm of the soil of a vegetable field has been reported to contain 500,000 viable seeds per acre (124 per square meter) of *Capsella bursa-pastoris* and 440,000 seeds per acre (109 per square meter) of *Stellaria media*. Single plants of these species can produce 3,500 to 4,000 and 2,200 to 2,700 seeds, respectively, and these may lie dormant for 35 and 60 years, respectively. Such seeds may also pass unharmed through the gut of animals (for literature see Heathcote, 1970). Where a virus is transmitted by weed seeds it has a tremendous potential to survive in the soil, even for extended periods of time. Virus-infected seedlings of weeds form major in-crop reservoirs of infection for efficient short-distance spread to crop and other plants by vectors such as aphids (in the non-persistent manner) and free-living nematodes. For the direct contribution of virus transmission through seed to virus spread see the section 1.6.3.

Virus infection in wild hosts is often symptomless and such hosts may not themselves suffer at all. Hence the role of weeds may remain obscure but for the ultimate effect of infection on the crop. Thus, the wild sources of infection may be hard to detect. These may not only be wild and one or a few species of the wide range within or outside the crop but they may be remote taxonomically and in vegetative habit from the crop species. Soil-borne viruses of a variety of crops may overwinter in woody plants. Viruses primarily known from woody crops may have wild herbaceous hosts. The aphid-transmitted plum pox (Sharka) virus affects *Prunus* species only, but is artifically transmissible to a range of herbaceous plants (reviewed by van Oosten, 1971); it has also been isolated from 10 herbaceous hosts growing in the open, most of them perennial weeds (Kröll, 1975). Tobacco rattle virus has even been isolated from the roots of the weed grass *Apera spica-venti* (Noordam, 1956). With such viruses, infection may remain restricted to the roots (Noordam, 1956).

Special mention should be made of virus retention in cultivated plants or their offspring that have run wild, for instance groundkeepers and volunteers. With increasing mechanical harvesting of potatoes, over 100,000 small potato tubers per hectare (10 per square meter) may be left on and in the top soil (van der Zweep, 1976). With a series of mild winters in continental Western Europe, the resulting volunteer plants have recently caused much concern as a weed of following crops and as a major reservoir of potato nematodes and viruses. In Britain with mild winters as a rule, Doncaster and Gregory (1948) had already shown that such potato volunteers could still be found after five years of successive cereal crops. A potato crop grown on land with groundkeepers from a crop grown two years before contained 96% of virus-infected plants at the end of the season, whereas part of the same crop without groundkeepers contained only 9% of infected plants. Howell and Mink (1977) found that in the Columbia Basin no natural weed hosts but volunteer carrots together with carrots grown for seed and raised for root processing formed a continuous yearly cycle of hosts which perpetuated the viruses of carrot thin leaf and motley dwarf. In Washington State, sugar-beets that escape the digger and are covered with soil may survive the winter as do seedlings from seeds produced by bolting plants, and the resulting volunteer plants may serve as an overwintering source of beet western yellows virus (Wallis, 1967). In Britain, virus-infected sugar-beet plants can sometimes be found on roadside grass where they have fallen from lorries carrying beet to the processing factory, and may act as sources of virus (Heathcote, 1973).

1.6.2 Origin of Crop Viruses

Since crop viruses often survive crop-free periods in wild plants, since wild plants in their natural habitats may show high incidence of virus infection, and since crop plants have developed from wild ancestors, most crop viruses probably have originated from wild plants. An exception to this hypothesis might be found in the leafhopper-borne viruses that multiply in their insect vectors and often pass from generation to generation of vectors vertically. These viruses probably originated as insect viruses (Maramorosch, 1972). In an interesting contribution on "origin and distribution of new or little known virus diseases," Bennett (1952) was among the first to conclude that this is what happened in past ages. During early agriculture with slow developments and little transport of plants and their products, the cultivated and uncultivated species were likely to achieve a state of equilibrium with their viruses and to acquire a high resistance or tolerance.

Bennett (1952) listed some 'new' virus diseases showing up in crops after their introduction into new agricultural regions where they had not been grown before. The viruses were later found to be indigenous to native plants in such regions. Soon after its introduction into the western United States, sugar-beet was severely attacked by curly-top virus, which nearly ruined the sugar-beet industry there. In Africa, corn soon after its introduction severely suffered from maize streak, which was later discovered also to occur on sugarcane and over 22 species of grasses in various parts of Africa (Storey and McClean, 1930), in which it may have existed before cultivation of cane (Storey, 1925) and maize.

Among more recent examples are the following. Lettuce big vein virus, transmitted in the soil by the fungus *Olpidium brassicae*, is regarded as a widespread indigenous but unnoticed component of the flora although possibly confined to Compositae (Campbell, 1965). The original hosts of rice hoja blanca virus seem to be *Echinochloa* spp. (jungle grasses) and the leafhopper vector *Sogota cubanus*, which transmits the virus through eggs. The virus has recently come into prominence in rice in the Americas where it is very destructive. Rice and the vector *S. oryzicola* are probably adapted hosts (Everett and Lamey, 1969). Common sow-thistle (*Sonchus oleraceus*) appears to be the principal, and probably the only natural source of lettuce necrotic yellows virus, causing marginal but destructive infection in lettuce fields in Australia after spread by aphids (Stubbs *et al.*, 1963).

In Scotland, soil-borne nematode-transmitted viruses are now primarily re-garded as pathogens of wild plants, taking their toll when infested land is taken into cultivation. A sensitive crop then acts as a 'developer' to reveal a situation that had existed unobserved in the indigenous plant species (Murant, 1970). Sim-ilarly in the Netherlands, *Arabis* mosaic virus may damage lettuce crops in new greenhouses on ploughed grassland (van Dorst and van Hoof, 1965). In England, the virus often occurs in crops near hedges (Harrison and Winslow, 1961).

Meantime, Bennett's (1952) conclusion still holds true: "no estimate of the number of potentially destructive viruses that remain hidden in the wild plants of various parts of the world may be made." Such viruses may show up in newly introduced crops or new cultivars that had no opportunity of adapting naturally.

1.6.3   Virus Spread

Virus spread by wild plants themselves is obvious for those species that pro-duce propagules which are easily spread. Theoretically, vegetative propagules could most easily play a role because they are systemically infected if produced by virus-infected plants. But their spread, as with flowing water, seems minor. Where viruses do pass through seed to the offspring of infected plants, spread in seeds of wild plants, for instance by wind, can do much to spread virus locally, from region to region, or even country to country.

Virus spread in seeds of wild plants has especially been studied for soil-borne viruses, which move very little within the soil because of low mobility of their nematode vectors (Lister and Murant, 1967). Murant and Lister (1967) suggested that tomato black-ring virus and raspberry ringspot virus have reached a large proportion of the populations of their vector, *Longidorus elongatus*, in eastern Scotland, through seed dispersal. For such viruses, spread through seed of wild plants may be the main means above ground. To mention one out of many ex-amples, cherry leafroll virus may be disseminated in pollen and seed of birch (*Betula* spp.), where up to 22% of seed collected from open-pollinated and nat-urally infected trees in southern England has been found to carry the virus (Cooper, 1976). With viruses that have other means of aerial spread, dissemina-tion in seed of wild hosts may help in distant spread, though not essential.

Spread of seeds with viruses through *wind* is especially likely in wild hosts such as *Betula* spp. and in weeds such as groundsel (*Senecio vulgaris*) and other Compositae. *S. vulgaris* may carry several nepoviruses in its seed, as well as lettuce mosaic virus (Ainsworth and Ogilvie, 1939; Phatak, 1974).

Virus spread in seeds may also be brought about mechanically by animals, especially birds, although none of these have been documented. Viable seed may be retained for long periods in the digestive tract of birds and pass out undamaged (Proctor, 1968; Heathcote, 1970). Of course, weed seeds are also spread short distances on wheels of farm machinery, in straw, dung and even on clothes.

Man's role in spread over short distances is greatly exceeded by that in the exploration of the world's wild gene centres, exchange and transport of wild germplasm from country to country and its maintenance in and distribution from gene banks. This is now general policy and practice in order to diversify crops and to improve resistance to diseases and pests (Section 1.6.6). Transfer of germplasm is in vegetative propagation material as well as in seed.

The risks of virus spread with such material have been clearly demonstrated by Kahn and Monroe (1970). They held 551 samples of imported vegetative wild and cultivated *Solanum* spp. in quarantine at the United States Department of Agriculture's Plant Introduction Station, Glenn Dale, Maryland, and tested them for a period of ten years. Of the 445 samples of cultivated material 73% proved to be infected with virus and of the 106 samples of wild material 39%. Thus, on average two out of every five samples of wild material contained virus, though the material had been sampled by experienced 'plant explorers,' avoiding visibly diseased plants. Kahn and Sowell (1970) detected virus in 11% of the samples of vegetative wild material of *Arachis* sp.

Besides spreading viruses, such wild material may even be the original source of several crop viruses (Section 1.6.2). The Andes region of South America, the centre of origin of the potato, is now also considered the centre of dissemination of the most common potato viruses (Silberschmidt, 1954). The International Potato Centre (CIP) in Lima, Peru, has an extensive germplasm collection including much wild material. When testing 930 tubers from such wild material in West Germany, Bode (1977) recently detected only 4.6% to be free of virus and 28.3% to contain one virus only. Of those infected, 18.7% contained Andean potato latent virus, 9.5 Andean potato mottle virus, 5.8% potato leafroll virus, 49% potato virus Y, 61% virus X, 3.9% virus M, and 40.2% virus S. Koenig and Bode (1977) stressed the risks that germplasm would introduce the Andean potato latent and mottle viruses, as yet unknown in Europe. The first virus has a misleading name since it may cause severe symptoms under certain conditions, and it is readily spread by simple contact (Jones and Fribourg, 1978).

Use of true seed may reduce but not prevent the risk of virus dissemination in germplasm. For example, true seed of *Solanum* spp. may be infected with Andean potato latent virus (Jones and Fribourg, 1977), potato virus T (Salazar and Harrison, 1978), and potato spindle tuber viroid (Diener and Raymer, 1971). These risks are now fully recognized by the International Board for Plant Genetic Resources in the Crop Ecology and Genetic Resources Unit of FAO, Rome, as indicated by a recent extensive report on "plant health and quarantine in international transfer of genetic (including wild) resources" (Hewitt and Chiarappa, 1977).

The actual, though little studied, role of botanic gardens with diverse collections of wild species from all over the world and regular international exchange of material is worth mention. Their collections often serve as intermediate stations or repositories of wild plants on their way to domestication. Blattný and Schmelzer (1976) detected viruses in 9 out of 13 perennial alpine plant species grown in the botanic garden of the Charles University, Prague, Czechoslovakia. One of the viruses was possibly new. Of course, most of the infections could have been contracted while in the botanic garden. If so, then that would be an additional illustration of the risks of distributing material from such collections, especially since only one of the species they examined showed symptoms. I have also often isolated viruses from material from botanic gardens.

These few data indicate the danger of spreading viruses over long distances, even between continents, in wild plants through botanic gardens, germplasm collections and breeding nurseries. The same risks apply also for other pathogens, but for viruses the danger is greater because of the prevalence of latent infection and because of lack of knowledge of viruses among those in charge of botanical collections.

1.6.4 Virus Establishment

Besides the small amounts of wild and other germplasm, international exchange of genetic material includes a heavy traffic in large amounts of commercial propagation material. This is to achieve national and international division of labor, to improve agricultural productivity and to exploit to the full climatic conditions of a place, or the skill and expertise of a country. However there are the attendant risks. In my opinion, even the strictest quarantine cannot absolutely prevent introduction of new viruses. A diverse natural vegetation at the site of introduction greatly enhances the chances of newly introduced alien viruses or virus strains getting established.

Prerequisite are virus transmissibility, availability of a vector, and absence of host specificity. 'Polyphagism' of many, if not most, plant viruses has already been discussed (Section 1.5). This characteristic holds especially for viruses carried by aphids in a non-persistent manner, thus by a wide range of aphids, and for viruses carried by nematodes. But certain persistently transmitted aphid-borne viruses, although more exclusively transmitted by certain aphid-species, may also have wide host ranges, as mentioned for beet western yellows virus.

Once a trace of virus has been introduced, there may be a slow buildup of infection in the wild vegetation, where the virus may remain hidden. However buildup of inoculum potential in the entire environment may also be rapid with crops getting reinfected from the wild vegetation. Rapid multiplication in susceptible crops may then continuously or, with crop rotation, intermittently further charge the wild vegetation with inoculum, and eventually lead to epidemic development.

Simons *et al.* (1956) and Simons (1959) have presented an interesting example of the establishment of potato virus Y in wild vegetation, mainly nightshade (*Solanum gracile*), in Florida with the advent of commercial potato production. Peppers and tomatoes were severely damaged only in those areas where potatoes had been grown in previous years (Fig. 4).

Another example may be celery mosaic virus, which has recently made cultivation of celeriac (*Apium graveolens* var. *rapifera*) virtually impossible in the original center of celeriac growing in the southwest of the Netherlands (Bos, un-

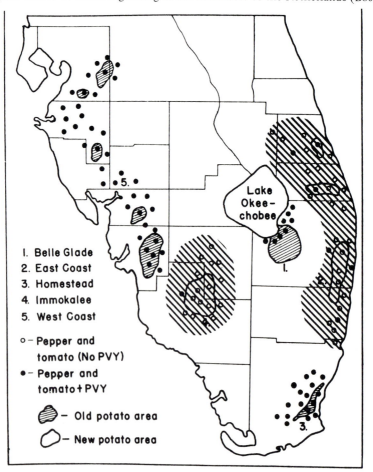

1. Belle Glade
2. East Coast
3. Homestead
4. Immokalee
5. West Coast

o – Pepper and tomato (No PVY)
● – Pepper and tomato + PVY

◎ – Old potato area
◯ – New potato area

FIG. 4. Potato virus Y distribution in pepper and tomato before 1956 (solid dots represent areas with infested fields) in areas where potatoes had been produced commercially. Broken lines indicate new areas of occurrence as observed during 1959, where at the time of potato introduction pepper and tomato were still free from the virus. After Simons (1959).

published data). The virus is known to infect various wild umbellifers (e.g., Walkey and Cooper, 1971). In the Netherlands, celeriac, grown as a summer crop, is apparently the only crop that becomes infected. So epidemic development depends on wild hosts. With lack of acceptable virus resistance and other means of control, the crop is now moving away from the original center of cultivation. But the disease is following it, apparently gradually building up inoculum in the wild vegetation. The epidemic has not been systematically investigated and we have not yet traced the wild sources of infection. Possibly latent infection may assist the wild hosts in escaping eradication by the growers.

The mutual and cumulative effect of wild plants and crops on inoculum potential has also been demonstrated by Häni (1971). He showed that rates of infection of *Stellaria media* by cucumber mosaic virus were higher in areas in Switzerland with continuous cultivation of tobacco than in regions with regular crop rotation, but that infection of tobacco each year mainly originates from *S. media* (in which the virus is seed-transmitted) (Fig. 5). In a field where tobacco was grown for the first time, the degree of infection of *S. media* rapidly increased during the season.

1.6.5 Refuges and Sources of Virus Vectors

Besides harboring crop viruses and other pathogens, wild plants act as important refuges and sources of insects, mites, nematodes, and fungi that may directly damage crops and may be essential in the ecology of vectors. Certain wild species may be indispensable to a vector as its alternate host. Even though itself not sus-

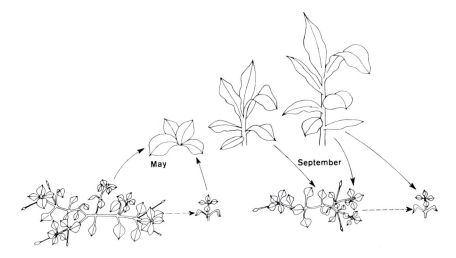

FIG. 5. Infection cycle of chickweed (*Stellaria media*) and cultivated tobacco. Solid lines, aphid transmission; interrupted lines, seed transmission. After Häni (1971).

ceptible to infection by the virus, such a species is then an essential intermediary in the ecology of the virus.

Arthropod vectors often have ranges of diverse food plants including several wild ones. Despite or even because of definite host preference, winged aphids in search of proper food plants probe many different species on which they happen to alight. Nonfood plants, either cultivated or wild, may then act as a reservoir of virus or may be infected by stylet-borne (nonpersistent) viruses. The vectors may do so during spring migration, in summer when moving around after crowding, and again when returning to winter hosts, on which they have to seek refuge if their summer food plant is an annual, as often.

Host-alternating aphid species are usually obliged to overwinter on special woody plant species, which are often wild (Table II). If such aphids act as specific vectors of certain viruses, the ecology of these completely depends upon the particular wild plants, even though they may not harbor the viruses. These plants may occur in natural habitats at great distances from the crops attacked in spring. The same, of course, holds for all wild sources of vectors. The leafhopper vector, *Circulifer tenellus*, of beet curly-top virus has been reported to fly 200-300 miles (300-400 km) or more from natural breeding grounds to areas growing sugarbeets in the western United States (review by Bennett, 1971).

Wild plants and weeds may also serve as reservoirs of both virus and vector. Then the virus can spread rapidly, as does beet western yellows virus in the western United States. The virus is transmitted especially by *Myzus persicae* and in circulative (persistent) manner. In the Salinas Valley of California, the disease has developed into a 'yellow plague' and various vegetable crops show almost complete infection (Duffus, 1977). In the Pacific northwest, wild plants that grow in protected areas such as drainage ditches provide a major means of overwintering to both the green peach aphid and beet western yellows virus (Wallis, 1967).

In many instances, winter hosts of vector and virus differ as with *Cavariella aegopodii*, overwintering on willow (*Salix* spp.) and spreading a number of viruses among Umbelliferae in summer. Spring flights start free from virus and may first colonize cow parsley (*Anthriscus sylvestris*), whence they may spread the viruses causing motley dwarf to carrot crops (Dunn and Kirkley, 1966). However when the winter is mild in Britain, the vector can anholocyclically overwinter on umbelliferous weeds, whence it may carry the motley dwarf virus to carrot crops early in spring and cause severe losses (Watson and Serjeant, 1963; Dunn, 1965). Both viruses are transmitted in persistent manner and are thus not spread by probing. Woody plants may harbor enormous numbers of overwintering eggs. In Germany, Heinze (1939) recorded up to 20,000 eggs of *M. persicae* per tree of peach (*Prunus persicae*), 4,000 being a good average. Further details on aphids in the ecology of plant viruses are reviewed by Swenson (1968) and Harris and Maramorosch (1977) and details on leafhoppers by Bennett (1967). Much information on insects in relation to plant diseases is also presented by Carter (1973). There is much other literature on these subjects.

TABLE II. Some Host-Alternating Virus-Transmitting Aphid Species*

Name	Primary host	Secondary host (summer host)	Number of transmitted viruses
Aphis fabae black bean aphid	*Euonymus europaeus* (spindle-tree) *Viburnum opulus* (european cranberry bush) *Philadelphus coronarius* (mock-orange)	numerous crops	40
Aphis nasturtii buckthorn aphid	*Rhamnus cathartica* (common buckthorn)	several herbaceous plant species, including potato	15
Brachycaudus helichrysi leafcurling plum aphid	*Prunus domestica* (common plum) *Prunus spinosa* (blackthorn)	mainly Compositae; also *Veronica* and *Myosotis* and others	16
Cavariella aegopodii carrot aphid	*Salix* (willow)	numerous Umbelliferae, e.g. carrot, celery, caraway	15
Myzus persicae green peach aphid	*Prunus persicae* (peach) other *Prunus* spp. (esp. *P. serotina*, wild black cherry) *Lycium halimifolium* (boxthorn)	over 400 plant species	113
Rhopalosiphum padi bird cherry aphid	*Prunus padus* (european bird cherry)	cereals and grasses	15

*Data from Fritzsche *et al.* 1972.

Soil-inhabiting nematodes, as well as the soil-borne viruses which they transmit, may simultaneously infect crop and wild plants. Because of the relative immobility of nematode vectors only those alternative hosts that grow themselves in or with their roots in the arable fields are of direct concern, for instance weeds, sometimes including trees in hedgerows. *Arabis* mosaic virus and its vector *Xiphinema diversicaudatum* may come from a *Prunus spinosa* tree into a strawberry crop (Fig. 3, Harrison and Winslow, 1961).

Similarly, *Olpidium brassicae*, the vector of lettuce big-vein virus, is indigenous in various wild plants. The same holds for the virus, which seems almost restricted to Compositae. Such weeds may harbor both vector and virus (Campbell, 1965), but the virus might even persist in the fungus on hosts of the fungus not susceptible to virus.

1.6.6 Sources of Crop Resistance to Viruses

Wild plants and vegetation may also contain elements beneficial to crops. Because of long-time association between wild plants and local disease-inducing agents, such as viruses, genotypic resistance (including tolerance) may have developed through natural selection. Such natural resistance is often found especially in gene centers of cultivated plants and their wild progenitors (Holmes, 1954; Leppik, 1970). Such wild material is now increasingly used in breeding programs as a source of resistance to viruses and other pathogens.

The first striking success was the crossing of cultivated sugarcane (*Saccharum officinarum*) with its wild relative *S. spontaneum*, and the development in 1921 of the famous POJ 2878. The new cultivar was resistant to the devastating sereh disease, most probably caused by virus. In 1930, the cultivar covered almost all the Javanese area of sugarcane (200,000 ha) (Jeswiet, 1936) and saved the sugar industry from ruin. Other examples of resistance to viruses derived from wild relatives are of more recent dates (for reviews see Holmes, 1954, and especially Russell, 1978).

However viruses, that are endemic and of minor importance, if at all, in the region where the new cultivars are bred and finally exploited, may not be present in the center of origin of the wild species. Then the species may be highly susceptible and sensitive to the virus and the resulting cultivar may provide the endemic virus with a chance to come to the fore. In California, turnip mosaic virus showed up in new cultivars of lettuce (*Lactuca sativa*) resistant to downy mildew (*Bremia lactucae*). This was ascribed to inadvertent introduction of mosaic susceptibility by crossing with *Bremia*-resistant *L. serriola*, where the resistance was even genetically linked to susceptibility to turnip mosaic (Zink and Duffus, 1969).

I have already referred to the other side of the coin, the attendant risks of virus dissemination with such wild germplasm (Section 1.6.3). Much of the wild resistance may be tolerance to infection. Although not suffering from infection, the material may harbor viruses and other pathogens. Leppik (1968, 1970) concluded that gene centers of plants not only serve as main sources of resistance to diseases but also of new pathogens or virulent races (or strains) that may start serious epidemics. Attention, indeed, should not only be given to new viruses but also to new strains. Moreira *et al.* (1978) recently discovered a resistance-breaking strain of potato virus X in wild germplasm. The strain might easily escape detection if *Gomphrena globosa* be used as indicator host, since it does not there cause local lesions.

Breeders are not always aware of the risks, as obvious from Knott and Dvořák's (1976) recent review on "alien germplasm as a source of resistance to disease". They do not even mention the risks and do not refer to Leppik's papers. Such lack of awareness makes the risk graver. For further details see Hewitt and Chiarappa (1977).

1.7 RELATIVE CONTRIBUTION OF WILD PLANTS TO VIRUS DISEASES

For crop management one needs to decide whether, when and to what extent wild plants (and weeds) should be removed to avoid economic damage by viruses in crop. So quantitative data on relative importance would be useful. However, the ecology of viruses and virus diseases is multifactorial (Fig. 1). So far, there is little information on epidemic buildup of virus diseases in crops (Bos, 1979).

As with virus ecology as a whole, the relative importance of wild plants varies widely. It depends on (1) the virus and its pathogenicity (= aggressiveness or infectivity + virulence) and way of spread, (2) crop vulnerability (= susceptibility + sensitivity), (3) vector behavior, efficiency and abundance, (4) other sources of infection than wild species, (5) the wild plants themselves as reservoirs of infection, their susceptibility, sensitivity, number, time of availability and distance from the sensitive crop, and (6) growing conditions. Aspects 1, 4 and 5 require further explanation.

Soil-borne and aphid-borne viruses differ completely in their ecology because of entirely different vector ecology. Aphid-borne viruses, even when transmitted in non-persistent manner, may still have large differences in rates of spread. This has been reported for pepper veinbanding virus (potato virus Y) and cucumber mosaic virus in peppers in Florida, both spread by the same vector, *Myzus persicae*. The two viruses reached maximum concentration in their hosts two weeks after inoculation, but cucumber mosaic virus then dropped rapidly in titer and remained low from three weeks after inoculation (Simons, 1958).

The relative contribution of wild plants as sources of infection also depends on *relative absence of other sources of infection*. Other sources include plants developing from partly infected propagation material, or other infected crops nearby. Viruses that are transmitted by seed or tuber, and of which no absolute freedom of propagation material can be guaranteed, may not need other sources of infection, as for bean common mosaic virus in *Phaseolus* bean, soybean mosaic virus in soybean, and potato viruses in potato. However, the cleaner the propagation material, the more critical is the absence of external sources of infection. Wild plants that are the sole alternate overwintering host of essential virus vectors may be *the* factor determining whether a virus is spread during the next season. In cold winters, *Myzus persicae* and *Aphis fabae* depend on their woody primary hosts.

Efficiency of wild hosts also greatly depends on their infected *number* and thus on their abundance and rate of infection, which depends on virus pathogenicity, and on *susceptibility and chances of survival* of the wild host and so on tolerance to infection, the host's lifespan and possible virus transmission through seed. Whether buildup of infection in the wild host causes an epidemic depends also on the presence of other susceptible species and on their frequency. Heavy infection is often brought about by influx of inoculum from nearby susceptible crops like tobacco with cucumber mosaic virus (Häni, 1971; Section 1.6.4).

Numbers of infected wild plants are usually high with soil-borne nematode-transmitted viruses. These viruses infect a wide range of weeds within the crop. Virus perennation is then often through weed seeds and, according to crop susceptibility during the crop rotation, the weed population may be recharged with virus through nematode-transmission from the crop. This situation is similar to that of the aphid-borne cucumber mosaic virus (Häni, 1971) in Switzerland in tobacco fields with the virus mainly overwintering in seeds of *Stellaria media* (Fig. 5).

Numbers of infected wild plants may be low if alien viruses are introduced from wild germplasm. Their number may be so low that they escape notice and do not directly lead to epidemic outbreak of economically damaging disease. Through germplasm collections, breeding programs, field trials, multiplication and distribution of commercial propagation material or by escape, introduction may eventually lead to epidemic outbreak of a new disease. The chance should not be neglected.

An important quantitative factor is *distance* between wild source of infection and crop. It has to be short if non-persistent insect transmission is required. The viruses are not usually carried much farther than a few hundred meters. For example, Wellman (1937) found that 85-95% of the celery plants became diseased with southern mosaic (cucumber mosaic virus) if 1 to 9 m from the weed sources (mainly *Commelina nudiflora*, *Phyllacca rigida* and *Physalis* spp.), but only 4% of the plants if 36 m away. This clearly illustrates the potential of weeds around fields, but even more of weeds within the fields, especially when viruses hibernate within such weeds or are seed-borne in them, as is cucumber mosaic virus in *Stellaria media* and some other weeds in lettuce fields in England (Tomlinson and Carter, 1970) and in tobacco fields in Switzerland (Häni, 1971). Proximity is determinant for soil transmission. With nematode-borne viruses, the immediate presence of various weed hosts makes up for the low mobility of the vector.

Volunteer plants and groundkeepers are of considerable importance, especially volunteer potatoes. Under wheat after a potato crop, Doncaster and Gregory (1948) found an almost full crop of volunteer potatoes and occasionally some remained even six years after the crop, despite cultivation. When another crop of potatoes were grown, the volunteers became a prime source of infection. They can cause much concern in seed potato production.

1.8 IMPLICATIONS FOR CONTROL OF VIRUS DISEASES

The crucial question now is how the available information may assist protection of crops from infection by viruses. Crops can be protected from damage by viruses by (1) avoiding or removing sources of infection, (2) preventing or reducing virus spread, and (3) improving crop resistance.

Wild plants are now recognized as important direct sources of crop viruses and of vectors of viruses. Their removal eliminates sources of infection, reduces virus spread in seeds (if the virus is seed-transmitted) and prevents vectors from breeding on them. As soon as weeds were found to harbor viruses of crops, successful attempts were made to remove them as sources of infection.

A few early examples were the removal of several weed hosts of cucumber mosaic virus. In Wisconsin, Doolittle and Walker (1926) advised removal of the weeds throughout the season if they were within 45 to 70 m (50 to 75 yards) of the cucumber crop. In Florida, where *Commelina nudiflora* had been found to be the main source of infection, severe southern celery mosaic could be nearly eliminated by removing any plant of this weed within 25 m of seedbeds (Wellman, 1937). Thung (1934) found that tobacco leafcurl in Indonesia could be successfully controlled by removing *Ageratum conyzoides, Synedrella nodiflora* and *Vernonia cinerea* from ruderal strips up to 50 m from the tobacco fields (Fig. 6).

Stylet-borne potato virus Y in peppers was controlled in Florida (pepper veinbanding mosaic) by isolating crops from weed sources (Simons, 1957). In cold weather, the aphids carried the virus only 50 to 70 m but, in warm weather, plants up to about 270 or 330 m (800 to 1000 ft) away could still be infected.

More recently in Australia, necrotic yellows could be prevented from moving into lettuce fields from borders where common sowthistle (*Sonchus oleraceus*) abounded if the sowthistle was destroyed (Stubbs *et al.*, 1963).

In the Yakima Valley of the northwest United States, Wallis and Turner (1969) destroyed weeds in drainage ditches by burning a 22-square mile area near sugar-beet fields during the winter. Consequently populations of *M. persicae* in sugar-beet fields were reduced by 51 to 91% and the number of yellows-diseased (mainly beet western yellows-infected) beet plants by 77 to 84%. The resulting increase in sugar-beet yield was calculated to be about 3.8 ton/ha (0.38 kg/m^2).

These few examples illustrate simple local applications. Simons *et al.* (1959) found that weed elimination in an area in Florida before planting peppers was more effective than insecticides on the crop itself, provided all growers in the region cooperated in weed control. An approach involving whole regions is even more important with persistently transmitted viruses like sugar-beet yellows viruses (mainly beet western yellows virus) and potato leafroll virus in a 500-mile

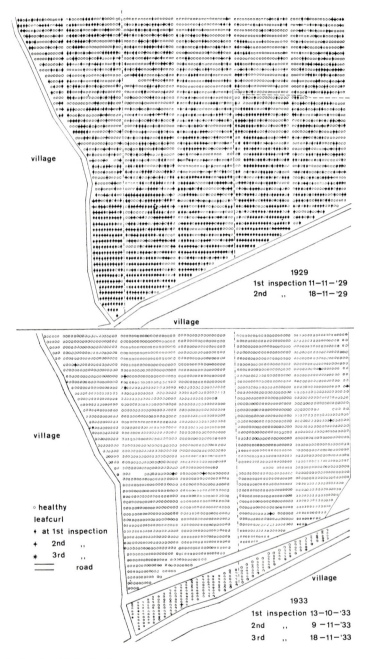

FIG. 6. Leaf-curl infection in tobacco field near Klaten, Java, in 1929 without weed control (upper picture), and in 1933 after removal of *Ageratum conyzoides, Synedrella nodiflora* and *Vernonia cinerea* from ruderal strips around the fields (lower picture). After Thung (1934).

area in the Columbia Basin of Washington State (Powell and Wallis, 1976). Here weeds on ditch banks were not only burned but peach trees in orchards were sprayed as well (Powell and Wallis, 1976). Another example of regional approach is that of ecologically altering the breeding areas of the leafhopper of beet curly-top vector in the western United States. Weed areas that had resulted from intermittent farming, land abandonment, overgrazing and burning were resown with perennial grasses and shrubs and restored to natural vegetation that was unfavorable to reproduction of the leafhopper (Piemeisel and Chamberlin, 1936; Piemeisel, 1954).

Large-scale direct leafhopper control with insecticides has been applied by the California State Department of Agriculture on wild hosts of the curly-top virus and its vector in canyons and foothills of the San Joaquin Valley to control beet curly top. The program has been in operation since 1943 at a cost of up to $350,000 a year. Since then in the area covered there have been few disastrous epidemics in any susceptible crop (Bennett, 1971).

Some of the measures were meant to reduce vector populations in order to reduce virus dissemination. Removal of weeds may also adversely affect crop health.

Wolcott (1928) observed that weed destruction increased insect-transmitted mosaic in sugarcane because of forced movement of vectors from wilting weeds. Carter (1939) noted that handweeding of *Emilia sonchifolia*, a host of tomato spotted wilt virus and its vector *Thrips tabaci*, in pineapple fields exposes nearby pineapple plants to infection, especially if the weeds are left in the field.

With nematode-transmitted viruses, incidence of crop infection may also be higher in weed-free crops, since the vectors have only crop plants to feed on. Removal or destruction of infected weed plants forces viruliferous nematodes to migrate to crop plants. This explains why Cooper and Harrison (1973) found tobacco rattle virus infection in potato (spraing disease) in plots kept free of weeds for 1.5 years to be 3.4 times that in weed-infected plots, though *Trichodorus* numbers did not differ appreciably between the two kinds of plot.

Another example is corn stunt (now ascribed to a spiroplasma) in Mississippi State. Pitre and Boyd (1970) found that more maize plants became infected when the plots were weed-free than when infested with *Brachiara platyphylla*, a preferred food plant of the leafhopper vector *Graminella nigrifrons*, even if there were less of the vectors. These examples illustrate how weed plants may be beneficial in distracting vectors from crop plants and from infecting them with viruses. Weeds then act as 'trap plants' or dilute the vector population on crop plants.

Another positive effect of weeds may be their influence on crop density. Incidence of virus diseases often decreases with increasing plant density. This may explain why weeds between the rows of beet steckling beds reduced infection with beet yellows virus (Hull, 1952).

A problem in avoiding or eradicating wild hosts of viruses is the lack of symptoms, so that elimination of abnormal wild plants is far from sufficient. Wild hosts may taxonomically be remote from the crop to be protected and information on natural wild host range of crop viruses is still limited. It is often insufficient to remove particular species of wild plant. The virus to be controlled may also be hidden in other species, and the wild vegetation in general may even have a reservoir of "new" viruses in store.

For the majority of the crop viruses known, wild plants have been shown to play a major role in the ecology of the crop diseases. Despite some positive effects of weeds on plant health, crops should in general be kept free from all sorts of weed, including volunteer plants and groundkeepers. Weed control should start before crop emergence or before planting to avoid forcing vector populations from weed hosts to crop, so spreading the virus (Carter, 1973). Special attention should also be paid to possible overwintering wild hosts of viruses along the hedgerows, and near the fields. The distance from the crop that need be searched depends on virus, vector, virus-vector relationships and conditions. Certain overwintering hosts may have to be removed as far as several hundred meters from the crop in waste land or along road sides.

The international spread and introduction of alien viruses with germplasm derived from wild plants, and distributed through botanic gardens or special gene banks needs emphasis. Quarantine may reduce the chances of such virus spread in vegetatively propagated material but most virus infections escape visual inspection, especially in dormant material. Virus detection still requires expensive facilities and much expertise. Since a continuously increasing number of viruses turns out to be seed-borne in certain hosts, although often in small proportions, virus spread in breeding material cannot be overcome by using seed. Seed testing is still cumbersome and there are latent infections by seed-borne viruses that react on special test plants only. Absolute safeguards can never be provided. Duffus's (1971) review of the role of weeds in the incidence of virus diseases optimistically concluded that "although wild plants may have played a role in the origin of virus diseases of crop plants, and serve as significant factors in their perpetuation —through modern weed control procedures and their utilization as breeding material—they may ultimately serve in the control of these destructive diseases." Present intensive international transfer of wild breeding material and the difficulties still encountered in virus detection still entail risks of further spread of viruses that have been of mere local significance or were hidden in local vegetation.

1.9 NATURAL COMPLEXITY

Wild plants, especially those that grow within the crop, are usually harmful to crops for various reasons. They compete with crop plants for space, water, food and light. They harbor viruses, other pathogens and a multitude of pests, which may severely reduce yield. Farmers are justified in eradicating weeds in and around their crops and in doing so thoroughly.

Since they may harbor non-persistent aphid-borne viruses, certain wild plants around the fields and in nearby waste land, in hedgerows and along roadsides may also have to be eradicated. Conflicts with the interests of nature conservation are likely, since wild plants are enjoying increasing public goodwill.

The correct balance is not only a diplomatic question of keeping in good terms with human society. It is a philosophical question as well. For instance, radical removal of weeds from crops can increase damage by certain viruses (Section 1.8). Nature is immensely complex. Within its dynamic complexity, the effect of several, if not most, measures may be unpredictable. There may be side effects or delayed effects from interference.

Myxomatosis virus was intentionally introduced into Australia. It drastically reduced the rabbit population there, which allowed sowthistle to become more prevalent. So the natural reservoir of lettuce necrotic yellows virus increased, and severe outbreaks of the disease occurred in lettuce (Stubbs, cited by Matthews, 1970). In California selective herbicides have increased weeds of the Compositae, especially sowthistle. As a consequence beet yellow stunt and sowthistle yellow vein viruses have increased on lettuce in the Salinas Valley (Duffus, 1971).

Natural diversity, although at low levels of yield for a crop, as in gene centers, tends to keep life in some sort of balance. With increasing uniformity, modern crops become more vulnerable and epidemics may more rapidly develop. Man is compelled to learn from nature and to exploit natural self-regulating mechanisms. We now know that wild plants may indeed be beneficial. Besides virus vectors, they harbor their predators and superparasites. Weeds are major components of agro-ecosystems and definitely contribute to the richness of the fauna within those systems. Altieri et al. (1977) have recently reviewed the great potential of weeds as biological components of some pest management systems. But the subject remains controversial and van Emden and Williams (1974) suggest that any beneficial contribution of weeds in pest problems never completely compensates for their competitive effect. Likewise, the opinion (or dogma?) that diversity creates stability remains controversial in ecological theory (van Emden and Williams, 1974).

Nature is delicately balanced and can readily be disturbed. Man is the main disturber but he has increasing power to manipulate biological entities (Wilson, 1969). Pathogens that may threaten crops from their wild hosts seem also to have a potential for biological weed control, and imported pathogens are most likely to be effective. Recent reviews (Wilson, 1969; Zettler and Freeman, 1972; Templeton and Smith, 1977) hardly mention viruses. After deliberate infestation with viruses, some of the infected individual weeds may survive and develop tolerance. Introduced viruses are unlikely to possess specificity to a particular weed. The virus may then ultimately become established in non-target weeds, and eventually damage crops. So deliberate introduction of viruses for weed control seems unacceptable. Some research has been done on weed control by viruses for control of blue-green algal blooms in atrophied bodies of fresh water by cyanophages that resemble viruses of bacteria and do not infect higher plants. Results under controlled conditions are promising (Desjardins et al., 1978).

The ambivalence of human interference with nature is also obvious from man's increasing exploitation of wild germplasm for genes of resistance to viruses, other pathogens, and pests with the attendant risks of transfer and introduction of alien viruses and harmful organisms.

So nature haunts human culture. Many viruses may still be hidden in wild plants within crops, nearby or even on other continents. Man has to keep on the alert. No matter how artificial crops may seen, they remain highly complicated ecosystems dependent on the natural environment at large. Within the agro-ecosystem, wild plants will continue to be major factors that can never be completely evaded except in sterile systems *in vitro*.

Cautious management of agro-ecosystems requires detailed knowledge of the involvement and interactions of viruses, micro-organisms, nematodes, mites, insects, birds, rodents and other animals and wild plants. Wild plants will remain a major factor. *Weed research* should not merely study wild plants in their competitive effect on crop plants. It must bring in mycology, bacteriology, virology, entomology and nematology, and fully participate in crop protection research (as is increasingly appreciated, e.g. van der Zweep, 1976). Results can benefit crop production and give a better view of the complexity of life.

1.10 REFERENCES

Ainsworth, G. C., and Ogilvie, L. (1939). Lettuce mosaic. *Ann. Appl. Biol.* 26: 279-297.

Altieri, M. A., Schoonhoven, A. van, and Doll, J. (1977). The ecological role of weeds in insect pest management systems: A review illustrated by bean (*Phaseolus vulgaris*) cropping systems. *PANS* 23: 195-205.

Ashby, J. W., Bos, L., and Huijberts, N. (1979). Yellows of lettuce and some other vegetable crops in the Netherlands caused by beet western yellows virus. *Neth. J. Plant Path.*: 85: 99-111.

Bennett, C. W. (1952). Origin and distribution of new or little known virus diseases. *Pl. Dis. Reptr.*, Suppl. 211: 43-46.

Bennett, C. W. (1967). Epidemiology of leafhopper-transmitted viruses. *Ann. Rev. Phytopath.* 5: 87-108.

Bennett, C. W. (1971). The curly top disease of sugarbeet and other plants. *Am. Phytopath. Soc. Monogr.* No. 7. 81 pp.

Blattný, C., and Schmelzer, K. (1976). Virologische Beobachtungen an Alpinum-Pflanzen im Botanischen Garten der Karls-Universität, Prag. *Preslia, Praha* 48: 76-80.

Bode, O. (1977). Fragen der Quarantäne bei Kartoffeln. *Potato Res.* 20: 349.

Bos, L. (1977). Seed-borne viruses. *In* "Plant Health and Quarantine in International Transfer of Genetic Resources" (W. B. Hewitt and L. Chiarappa, eds.), pp. 39-69. CRC Press, Inc., Cleveland, Ohio.

Bos, L. (1978). Role of wild plants in the ecology of virus diseases. *Abstr. Papers 3rd Int. Congr. Plant Path.*, München 16-23 Aug., 1978. p. 22.

Bos, L. (1980). Ecology and control of plant viruses and virus diseases, a model and general principles. In manuscript.

Brčák, J., and Polák, Z. (1966). Importance of wild hosts of plant viruses. *Meded. Rijksfac. Landbouww. Gent* 31: 967-971.

Bruckart, W. L., and Lorbeer, J. W. (1976). Cucumber mosaic virus in weed hosts near commercial fields of lettuce and celery. *Phytopathology* 66: 253-259.

Campbell, R. N. (1965). Weeds as reservoir hosts of lettuce big vein virus. *Can. J. Bot.* 43: 1141-1149.

Carter, W. (1939). Populations of *Thrips tabaci* with special reference to virus transmission. *J. Anim. Ecol.* 8: 261-276.

Carter, W. (1973). Insects in relation to plant disease. 2nd edition. Wiley & Sons, New York, London, Sydney, Toronto. 759 pp.

Chessin, M., and Lesemann, D. (1972). Distribution of cactus viruses in wild plants. *Phytopathology* 62: 97-99.

Christie, S. R., and Crawford, W. E. (1978). Plant virus range of *Nicotiana benthamiana*. *Pl. Dis. Reptr.* 62: 20-22.

Cooper, J. I. (1976). The possible epidemiological significance of pollen and seed transmission in the cherry leafroll virus/*Betula* spp. complex. *Mitt. Biol. BundesAnst. Berlin-Dahlem* 170: 17-22.

Cooper, J. I., and Harrison, B. D. (1973). The role of weed hosts and the distribution and activity of vector nematodes in the ecology of tobacco rattle virus. *Ann. Appl. Biol.* 73: 53-66.

Desjardins, P. R., Barkley, M. B., Swiecki, S. A., and West, S. N. (1978). Viral control of blue-green algae. *Calif. Water Resources Center Univ. Calif.* Contr. No. 169. 29 pp.

Diener, T. O., and Raymer, W. B. (1971). Potato spindle tuber 'virus'. *CMI/AAB Descriptions of Plant Viruses* No. 66. 4 pp.

Doncaster, J. P., and Gregory, P. H. (1948). The spread of virus diseases in the potato crop. *Agric. Res. Council Rep.*, Ser. 7. 189 pp.

Doolittle, S. P., and Gilbert, W. W. (1919). Seed transmission of cucurbit mosaic by the wild cucumber. *Phytopathology* 9: 326-327.

Doolittle, S. P., and Walker, M. N. (1925). Further studies on the overwintering and dissemination of cucurbit mosaic. *J. Agric. Res.* 31: 1-58.

Doolittle, S. P., and Walker, M. N. (1926). Control of cucumber mosaic by eradication of wild host plants. *Dep. Bull. USDA* No. 1461. 14 pp.

Doolittle, S. P., and Wellman, F. L. (1934). *Commelina nudiflora*, a monocotyledonous host of a celery mosaic in Florida. *Phytopathology* 24: 48-61.

Dorst, H. J. M. van, and Hoof, H. A. van (1965). *Arabis*-mozaïekvirus bij komkommer in Nederland (with Engl. summ: *Arabis* mosaic virus of cucumber in the Netherlands). *Neth. J. Plant Path.* 71: 176-179.

Drijfhout, E., and Bos, L. (1977). The identification of two new strains of bean common mosaic virus. *Neth. J. Plant Path.* 83: 13-25.

Duffus, J. E. (1971). Role of weeds in the incidence of virus diseases. *Ann. Rev. Phytopath.* 9: 319-340.

Duffus, J. E. (1973). The yellowing virus diseases of beet. *Adv. Virus Res.* 18: 347-386.

Duffus, J. E. (1977). Aphids, viruses, and the yellow plague. *In* "Aphids as Virus Vectors" (K. F. Harris and K. Maramorosch, eds.), pp. 361-383. Academic Press, New York, San Francisco, London.

Dunn, J. A. (1965). Studies on the aphid, *Cavariella aegopodii* Scop. I. on willow and carrot. *Ann Appl. Biol.* 56: 429-438.

Dunn, J. A., and Kirkley, J. (1966). Studies on the aphid, *Cavariella aegopodii* Scop. II. on secondary hosts other than carrot. *Ann. Appl. Biol.* 58: 213-217.

Emden, H. F., van, and Williams, G. I. (1974). Insect stability and diversity in agro-ecosystems. *Ann. Rev. Entomol.* 19: 455-475.

Everett, I. R., and Lamey, H. A. (1969). Hoja blanca. *In* "Viruses, Vectors, and Vegetation" (K. Maramorosch, ed.), pp. 361-377. Wiley-Interscience, New York, London, Sydney, Toronto.

Fritzsche, R., Karl, E., Lehmann, W., and Proeseler, G. (1972). Tierische Vektoren pflanzenpathogener Viren. Fischer, Jena. 541 pp.

Häni, A. (1971). Zur Epidemiologie des Gurkenmosaikvirus im Tessin. *Phytopath. Z.* 72: 115-144.

Häni, A., Pelet, F., and Kern, H. (1970). Zur Bedeutung von *Stellaria media* (L.) Vill. in der Epidemiologie des Gurkenmosaikvirus. *Phytopath. Z.* 68: 81-83.

Harris, K. F., and Maramorosch, K. (eds.) (1977). Aphids as virus vectors. Academic Press, New York, San Francisco, London. 559 pp.

Harrison, B. D., and Winslow, R. D. (1961). Laboratory and field studies on the relation of arabis mosaic virus to its nematode vector *Xiphinema diversicaudatum* (Micoletzky). *Ann. Appl. Biol.* 49: 621-633.

Heathcote, G. D. (1970). Weeds, herbicides and plant virus diseases. *Proc. 10th Br. Weed Control Conf.* pp. 934-941.

Heathcote, G. D. (1973). Control of viruses spread by invertebrates to plants. *In* "Viruses and Invertebrates" (A. J. Gibbs, ed.) pp. 587-609. North Holland (Amsterdam, London) and American Elsevier (New York).

Hein, A. (1953). Die Bedeutung der Unkräuter für die Epidemiologie pflanzlicher Virosen. *Dt. Landw.*, Berlin 4: 521-525.

Heinze, K. (1939). Zur Biologie und Systematik der Virusübertragenden Blattläuse. *Mitt. Biol. Reichsanst. Land- u. Forstw. Berlin* 59: 35-48.

Hewitt, W. B., and Chiarappa, L. (eds.) (1977). Plant health and quarantine in international transfer of genetic resources. CRC Press Inc., Cleveland, Ohio. 346 pp.

Holmes, F. O. (1954). Inheritance of resistance to viral diseases in plants. *Adv. Virus Res.* 2: 1-30.

Howell, W. E., and Mink, G. I. (1977). The role of weed hosts, volunteer carrots, and over-lapping growing seasons in the epidemiology of carrot thin leaf and carrot motley dwarf viruses in central Washington. *Pl. Dis. Reptr.* 61: 217-222.

Hull, R. (1952). Control of virus yellows in sugarbeet seed crops. *J. R. Agric. Soc. Engl.* 113: 86-102.

Jeswiet, J. (1936). Results of crossing in sugar-cane and other *Saccharum* species for the sugar-cane culture in Java. *Proc. Intern. Congr. Plant Breeders*, 22-27 June, 1936, II: 68-73.

Jones, R. A. C., and Fribourg, C. E. (1977). Beetle, contact and potato true seed transmission of Andean potato latent virus. *Ann. Appl. Biol.* 86: 123-128.

Jones, R. A. C., and Fribourg, C. E. (1978). Symptoms induced by Andean potato latent virus in wild and cultivated potatoes. *Potato Res.* 21: 121-127.

Kahn, R., and Monroe, R. L. (1970). Virus infection in plant introductions collected as vegetative propagations: I Wild vs. cultivated *Solanum* species. *FAO Plant Prot. Bull.* 18: 97-101.

Kahn, R., and Sowell, G. (1970). Virus infection in plant introductions collected as vegetative propagations: II Wild vs. cultivated "*Arachis*" sp. *FAO Plant Prot. Bull.* 18: 142-144.

Kaiser, W. J., and Mossahebi, G. H. (1974). Natural infection of mungbean by bean common mosaic virus. *Phytopathology* 64: 1209-1214.

Koenig, R., and Bode, O. (1977). In Westeuropa bisher nicht vorkommende Viren aus südamerikanischen Kartoffeln und ihr hochempfindlicher Nachweis mit dem serologischen Latextest. *Mitt. Biol. Bundesanst. Braunschw.* 178: 102 (Abstr.).

Knott, D. R., and Dvořák, J. (1976). Alien germplasm as a source of resistance to disease. *Ann. Rev. Phytopath.* 14: 211-235.

Koot, IJ. van, Slogteren, D. H. M. van, Cremer, M. C., and Camfferman, J. (1954). Virusverschijnselen in Freesia's. (with Engl. summ: Virus diseases in freesias). *Tijdschr. PlZiekt.* 65: 157-192.

Kovachevski, I. (1972). Host plants of bean yellow mosaic in Bulgaria (Bulg. with Engl. summ.). *Plant Sci.* (Sofia) 9: 123-140.

Kröll, J. (1975). Untersuchungen zum Wirtspflanzenkreis des Scharkavirus (SAV). *Zbl. Bakt. Ab. II.* 130: 219-225.

Leppik, E. E. (1968). Introduced seed-borne pathogens endanger crop breeding and plant introduction. *FAO Plant Prot. Bull.* 16: 57-63.

Leppik, E. E. (1970). Gene centers of plants as sources of disease resistance. *Ann. Rev. Phytopathol.* 8: 323-344.

Lister, R. M., and Murant, A. F. (1967). Seed transmission of nematode-borne viruses. *Ann. Appl. Biol.* 59: 49-62.

MacClement, W. D., and Richards, M. G. (1956). Virus in wild plants. *Can. J. Bot.* 34: 793-799.

McWhorter, F. P., Boyle, L., and Dana, B. F. (1947). Production of yellow bean mosaic in beans by virus from mottled gladiolus. *Science* 105: 177-178.

Maramorosch, K. (1972). Origin and classification of viruses and mycoplasmas. In: *A Symposium on Ecosystematics*. R. T. Allen and F. C. James, eds., U. Arkansas: 195-217.

Matthews, R. E. F. (1970). Plant virology. Academic Press, New York, London. 778 pp.

Meiners, J. P., Gillaspie, A. G., Lawson, R. H., and Smith, F. F. (1978). Identification and partial characterization of a strain of bean common mosaic virus from *Rhynchosia minima*. *Phytopathology* 68: 283-287.

Moreira, A., Jones, R. A. C., and Fribourg, C. E. (1978). A resistance-breaking strain of potato virus X that does not cause local lesions in *Gomphrena globosa*. *Abstr. Papers 3rd Int. Congr. Plant Path.*, München, 16-23 Aug., 1978. p. 56.

Murant, A. F. (1970). The importance of wild plants in the ecology of nematode-transmitted plant viruses. *Outl. Agric.* 6: 114-121.

Murant, A. F., and Lister, R. M. (1967). Seed transmission in the ecology of nematode-borne viruses. *Ann. Appl. Biol.* 59: 63-76.

Nienhaus, F., Mack, C., and Schinzer, U. (1974). The isolation of viruses from fern plants. *Z. Pflanzenkrankh.* 81: 533-537.

Noordam, D. (1956). Waardplanten en toetsplanten van het ratelvirus van tabak (with Engl. summ.: Hosts and test plants of tobacco rattle virus). *Tijdschr. PlZiekt.* 62: 219-225.

Noordam, D., Bijl, M., Overbeek, S. C., and Quiniones, S. S. (1965). Virussen uit *Campanula rapunculoides* en *Stellaria media* en hun relatie tot komkommermozaïekvirus en tomaat—"aspermy"—virus. *Neth. J. Plant Path.* 71: 61 (Abstr.).

Oosten, H. J. van (1970). Further information about the herbaceous host range of sharka (plum pox) virus. *Annls. Phytopath.* no. hors. Série 195-201.

Orsenigo, J. R., and Zitter, T. A. (1971). Vegetable virus problems in South Florida as related to weed science. *Fl. State Hort. Soc. Proc.* 84: 168-171.

Phatak, H. C. (1974). Seed-borne plant viruses—identification and diagnosis in seed health testing. *Seed Sci. Technol.* 2: 3-155.

Piemeisel, R. L. (1954). Replacement and control; changes in vegetation in relation to control of pests and diseases. *Bot. Rev.* 20: 1-32.

Piemeisel, R. L., and Chamberlin, J. C. (1936). Land improvement measures in relation to a possible control of the beet leafhopper and curly top. *U.S. Dept. Agr. Circ.* 416: 1-24.

Pierce, W. H., and Hungerford, C. W. (1929). A note on the longevity of the bean mosaic virus. *Phytopathology* 19: 605-606.

Pitcher, R. S., and Jha, A. (1961). Distribution and infectivity with arabis mosaic virus of a dagger nematode. *Plant Path.* 10: 67-71.

Pitre, H. N., and Boyd, F. J. (1970). A study of the role of weeds in corn fields in the epidemiology of corn stunt disease. *J. Econ. Entom.* 63: 195-197.

Powell, D. M., and Wallis, R. L. (1976). Spraying peach trees and burning weed hosts in managing green peach aphid to reduce the incidence of virus in potatoes and sugar-beets. *Proc. Tall Timbers Conf. Ecol. Anim. Control Habitat Manege* 6: 87-98.

Proctor, V. W. (1968). Long-distance dispersal of seeds by retention in digestive tract of birds. *Science* 168: 321-322.

Provvidenti, R., and Braverman, S. W. (1976). Seed transmission of bean common mosaic virus in phasemy bean. *Phytopathology* 66: 1274-1275.

Provvidenti, R., and Cobb, E. D. (1975). Seed transmission of bean common mosaic virus in tepary bean. *Pl. Dis. Reptr.* 59: 966-969.

Provvidenti, R., and Granett, A. L. (1974). Sweet violet, a natural host of bean yellow mosaic virus. *Pl. Dis. Reptr.* 58: 155-156.

Provvidenti, R., and Schroeder, W. T. (1972). Natural occurrence of bean yellow mosaic virus in *Proboscidea jussieui. Pl. Dis. Reptr.* 56: 548-550.

Provvidenti, R., and Uyemoto, J. K. (1973). Chlorotic leaf spotting of yellow summer squash caused by the severe strain of bean yellow mosaic virus. *Pl. Dis. Reptr.* 57: 280-282.

Quantz, L. (1962). Untersuchungen über das gewöhnliche Bohnenmosaikvirus und das Sojamosaikvirus. *Phytopath. Z.* 43: 79-101.

Quiot, J. B., Douine, L., Marchoux, G., and Devergne, J. C. (1976). Ecologie du virus de la mosaique du concombre dans le sud-est de la France. I. Recherches des plantes spontanées hotes du C. M. V., caractérisation des populations virales. *Poljopr, zuanstv. Smotra (Agric. Consp. Sci.)* 39(49): 533-540.

Russell, G. E. (1978). Plant breeding for pest and disease resistance. Butterworths, London, Boston, Sydney, Wellington, Durban, Toronto. 485 pp.

Salatzar, L. F., and Harrison, B. D. (1978). Host range, purification and properties of potato virus T. *Ann. Appl. Biol.* 89: 223-235.

Schmelzer, K. (1957). Untersuchungen über den Wirtspflanzenkreis des Tabakmauche-Virus. *Phytopath. Z.* 30: 281-314.

Schmelzer, K., Schmidt, H. E., and Beczner, L. (1973). Spontane Wirtspflanzen des Luzerne-mosaik-Virus. *Biol. Zentralbl.* 92: 211-227.

Schmelzer, K., Gippert, R., Weissenfels, M., and Beczner, L. (1975). Spontane Wirtspflanzen des Ackerbohnenwelke-Virus (broad bean wilt virus). I. Mitteilung. *Zbl. Bakt. Abt. II* 130: 696-703.

Schmidt, H. E. (1977). Krautige Wildpflanzen. *In* "Pflanzliche Virologie, Band 4, Die Virosen an Zierpflanzen, Gehölzen und Wildpflanzen in Europa" (M. Klinkowski *et al.*, eds.), pp. 406-528. Akademie Verlag, Berlin.

Schwarz, R. (1959). Epidemiologische Untersuchungen über einige Viren der Unkraut- und Ruderalflora Berlins. *Phytopath. Z.* 35: 238-270.

Severin, H. H. P. (1934). Weed host range and overwintering of curly-top virus. *Hilgardia* 8: 263-277.

Silberschmidt, K. (1954). Potato viruses in the Americas. *Phytopathology* 44: 415-420.

Simons, J. N. (1957). Effects of insecticides and physical barriers on field spread of pepper veinbanding mosaic virus. *Phytopathology* 47: 139-145.

Simons, J. N. (1958). Titers of three nonpersistent aphid-borne viruses affecting pepper in south Florida. *Phytopathology* 48: 265-268.

Simons, J. N. (1959). Potato virus Y appears in additional areas of pepper and tomato production in south Florida. *Pl. Dis. Reptr.* 43: 710-711.

Simons, J. N., Conover, R. A., and Walter, J. M. (1956). Correlation of occurrence of potato virus Y with areas of potato production in Florida. *Pl. Dis. Reptr.* 40: 531-533.

Simons, J. N., Orsenigo, J. R., Stoll, R. E., and Thayer, P. L. (1959). Potato virus Y in peppers and tomatoes. *Mimeogr. Rept. Everglades Exp. Sta. Univ. Fla.*

Storey, H. H. (1925). Streak disease of sugar-cane. *Union of S. Africa Dept. Agric. Sci. Bull.* 39: 30 pp.

Storey, H. H., and McClean, A. P. D. (1930). The transmission of streak disease between maize, sugarcane and wild grasses. *Ann. Appl. Biol.* 17: 691-719.

Stubbs, L. L., Guy, J. A., and Stubbs, K. J. (1963). Control of lettuce necrotic yellows virus disease by destruction of common sowthistle (*Sonchus oleraceus*). *Austral. J. Exp. Agr. Anim. Husb.* 3: 215-218.

Swenson, K. G. (1968). Role of aphids in the ecology of plant viruses. *Ann. Rev. Phytopathol.* 6: 351-374.

Templeton, G. E., and Smith, R. J. (1977). Managing weeds with pathogens. *In* "Plant Disease, an Advanced Treatise. I. How Disease Is Managed" (J. G. Horsfall and E. B. Cowling, eds.), pp. 167-176. Academic Press, New York, San Francisco, London.

Thung, T. H. (1932). De krul- en kroepoekziekten van tabak en de oorzaken van hare verbreiding (with Engl. summ.: The curl- and crinkle-diseases of tobacco and the causes of their dissemination). *Meded. Proefst. Vorstenl. Tabak, Klaten, Java*, 72: 1-54.

Thung, T. H. (1934). De bestrijding der krul- en kroepoekziekten (with Engl. summ.: The control of the curl- and crinkle-diseases of tobacco). *Meded. Proefst. Vorstenl. Tabak, Klaten, Java*, 78: 18 pp.

Tomlinson, J. A., and Carter, A. L. (1970). Studies on the seed transmission of cucumber mosaic virus in chickweed (*Stellaria media*) in relation to the ecology of the virus. *Ann. Appl. Biol.* 66: 381-386.

Tomlinson, J. A., and Walker, V. M. (1973). Further studies on seed transmission in the ecology of some aphid-transmitted viruses. *Ann. Appl. Biol.* 73: 293-298.

Walkey, D. G. A., and Cooper, V. C. (1971). Virus diseases of celery. *Rept. Natn. Veg. Res. Stn* Wellesbourne, Warwick for 1970: 111-113.

Wallis, R. L. (1967). Some host plants of the green peach aphid and beet western yellows virus in the Pacific Northwest. *J. Econ. Entom.* 60: 904-907.

Wallis, R. L., and Turner, J. E. (1969). Burning weeds in drainage ditches to suppress population of green peach aphids and incidence of beet western yellows disease in sugarbeets. *J. Econ. Entom.* 62: 307-309.

Watson, M. A., and Serjeant, E. P. (1963). The effect of motley dwarf virus on yield of carrots and its transmission in the field by *Cavariella aegopodiae* Scop. *Ann. Appl. Biol.* 53: 71-93.

Webster, N. (1976). Webster's Third New International Dictionary of the English Language (unabridged). Merriam Co., Springfield, Mass., U.S.A. 1976: 2662 pp.

Wellman, F. L. (1937). Control of celery mosaic by eradicating wild hosts. *Tech. Bull. U.S. Dept. Agric.* No. 548. 16 pp.

Wilson, C. L. (1969). Use of plant pathogens in weed control. *Ann. Rev. Phytopath.* 7: 411-433.

Wolcott, G. N. (1928). Increase of insect transmitted plant disease and insect damage through weed destruction in tropical agriculture. *Ecology* 9: 461-466.

Zettler, F. W., and Freeman, T. E. (1972). Plant pathogens as biocontrols of aquatic weeds. *Ann. Rev. Phytopath.* 10: 455-470.

Zink, F. W., and Duffus, J. E. (1969). Relationship of turnip mosaic virus susceptibility and downy mildew (*Bremia lactucae*) resistance in lettuce. *J. Amer. Soc. Hort. Sci.* 94: 403-407.

Zweep, W. van der (1976). Weeds, weed control, pests and plant diseases. *IAC–Intern. Course Plant Prot.*, Wageningen. 13 pp. manuscript.

Chapter 2

ENZYME-LINKED IMMUNOSORBENT ASSAY (ELISA): PRINCIPLES AND APPLICATIONS FOR DIAGNOSIS OF PLANT VIRUSES

M. Bar-Joseph and S. M. Garnsey

Virus Laboratory, Agricultural Research Organization, Volcani Center, Bet Dagan, Israel, and Horticultural Research Laboratory, AR, SEA, U.S. Department of Agriculture, Orlando, Florida

2.1 INTRODUCTION

The effective control of plant virus diseases is based mainly on: (i) Use of virus-free propagative material. (ii) Control of mechanical and vector transmission. (iii) Breeding and selection for resistance or tolerance.

Accurate and rapid identification of viral pathogens is essential to most control programs. Diagnostic methods differ for various host-virus combinations, but they are based primarily on specific symptoms in selected indicator plants, or on direct assay via microscopy, biochemical analysis, or serology.

Serological techniques (Matthews, 1970; Purcifull and Batchelor, 1977; and van Regenmortel, 1978) are frequently favored for plant virus diagnosis when antisera are available, because these techniques are specific, rapid, and can be standardized. However, conventional serological techniques have not been used for practical identification of many important plant viruses, or have been used only in special circumstances, because of limitations such as low virus concentration, unsuitable particle morphology, the presence of deleterious substances in plant extracts, lack of sufficient quantities of antisera, or the need for specialized equipment or techniques.

The recent adaptation of enzyme-linked immunosorbent assay (ELISA) to plant virus detection (Voller *et al.*, 1976; Clark and Adams, 1977) has created new interest in serological diagnosis of plant viruses. The ELISA procedure is relatively simple, rapid, very sensitive, and requires small amounts of antiserum. It has been rapidly adopted for many diagnostic applications in plant virology, especially where other serological techniques have been difficult to use.

This chapter describes some basic concepts of ELISA, the current methodology for plant viruses, and provides some examples of application.

2.2 ANTIBODY LABELING

Enhancing the sensitivity of immunological reactions by coupling various markers to antibodies is a well-established concept. Fluorescent dyes and radio-isotopes have been used for this purpose for many years, and radioimmunoassay is a well-established diagnostic method in clinical laboratories. There are few published reports on radioimmunoassay for plant viruses (Ghabriel *et al.*, 1980). The technique has not been applied routinely for plant virus diagnosis, probably because it has been difficult to label undissociated plant viruses *in vitro* with radioactive markers, the need for costly reagents with short shelf life and expensive detection equipment, and restrictions and hazards associated with isotope use. Immunofluorescence has been used for locating viral antigens in plant cells and protoplasts, and a few reports describe its application as a diagnostic tool (Tsuchizaki *et al.*, 1978). Immunofluorescent assays can be rather tedious and time consuming on a large scale, and results are hard to quantify precisely.

Enzyme-labeled antibodies were first used for light microscopic detection of antigens in tissues (Nakane and Pierce, 1966; Wicker and Avrameas, 1969). Engvall and Perlmann (1971) and van Weeman and Schuurs (1971) demonstrated that enzyme-labeling could yield quantitative assays with a sensitivity comparable to radioimmunoassay. Voller *et al.* (1974) introduced the microplate method of ELISA which has subsequently been used for diagnosing a wide variety of antigens (Voller, 1978; Voller and Bidwell, 1977; Voller *et al.*, 1976, 1977, 1978a, 1978b).

Several enzymes have been used to label antibodies, and these have been coupled by several procedures (Voller *et al.*, 1977; Wisdom, 1976; and Avrameas *et al.*, 1978; Nakane, 1979). Important criteria for selecting enzymes include high activity in a coupled state, stability, availability, simple assay, nonhazardous substrates, and cost.

For plant virus detection, antibodies have usually been labeled with alkaline phosphatase by a one-step glutaraldehyde conjugation (Clark and Adams, 1977). Most of the material in the following sections pertains to the alkaline phosphatase system.

2.3 METHODOLOGY

2.3.1 The Double Antibody Sandwich Method

The common ELISA procedure for plant virus detection has been the double antibody sandwich method (Clark and Adams, 1977; Voller *et al.*, 1977). The basic steps are illustrated in Fig. 1 and summarized here:

1) Virus-specific antibodies are adsorbed to a solid surface such as the wells of a polystyrene Microtiter® plate. 2) A test sample, suspected to contain the viral antigen, is incubated with the antibody-sensitized solid phase. Viral antigens recognized by the antibodies are serologically bound. 3) Enzyme-labeled specific antibodies are added after the test sample is removed. The labeled antibodies bind to the viral antigen already bound to the coating antibodies (step 1) and form the double antibody sandwich. 4) Enzyme substrate is added. 5) The reaction is stopped and read. The color change in the substrate is proportional to the amount of enzyme present, which in turn is proportional to antigen concentration. Some important features of each step are described below.

2.3.1.1 Preparation of Specific Antibodies. Plant virus antibodies are found in the γ-globulin fraction of whole serum. Usually fractionated, and partially purified γ-globulins are used for ELISA, although purification is not always essential and ammonium sulfate-precipitated γ-globulins have been used successfully without further purification (Rochow and Carmichael, 1979; Voller *et al.*, 1977; Saunders, 1979). Commonly, γ-globulins are precipitated by ammonium sulfate and partially purified by chromatography on DEAE-cellulose columns

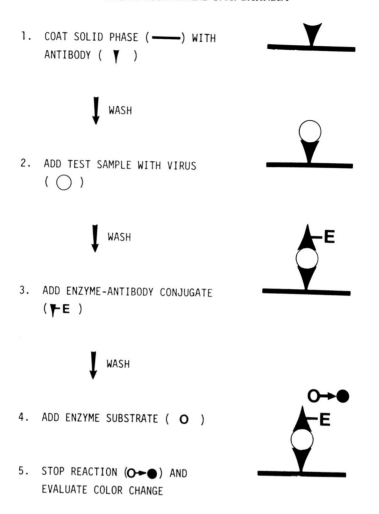

1. COAT SOLID PHASE (———) WITH
   ANTIBODY (**Y**)

   WASH

2. ADD TEST SAMPLE WITH VIRUS
   (◯)

   WASH

3. ADD ENZYME-ANTIBODY CONJUGATE
   (**Y-E**)

   WASH

4. ADD ENZYME SUBSTRATE (**O**)

5. STOP REACTION (**O→●**) AND
   EVALUATE COLOR CHANGE

FIG. 1. Diagram of double antibody sandwich method of enzyme-linked immunosorbent assay. (After Voller *et al.*, 1976.)

(Clark and Adams, 1977). The authors have also purified γ-globulins by adsorption on Protein A-Sepharose CL-4B (Pharmacia, Uppsala[1]), followed by elution with 0.1 N glycine –HCl, pH 3.0.

High-titered antisera free of antibodies to host plant antigens are desirable, but sera of modest titer have been used successfully (Clark and Adams, 1977; Maat and De Bokx, 1978b; Lister, 1978). Antibodies to host components can be

[1] Mention of a chemical or proprietary product does not constitute a recommendation for use by the U.S. Department of Agriculture to the exclusion of other products that may also be suitable.

removed by addition of host protein prior to fractionation. Alternatively, host-reacting antibodies can be adsorbed from antibody conjugates just prior to use (Lister, 1978). Specificity is less important for coating γ-globulin than for γ-globulin used to prepare conjugates, since nonspecific antigens bound by the coating antibodies are not reactive unless they also bind the conjugate.

The immunization period needed to obtain antibodies for ELISA tests was examined by Koenig (1978). Conjugates of antibodies to three plant viruses pre-pared from early bleedings (after immunization periods of 1 to 2 weeks) did not react in ELISA tests with their homologous antigens, although the antisera had high titers in immunodiffusion tests. Antibodies to Andean potato latent virus (APLV) obtained 1 month after immunization were considerably less efficient for ELISA than antibodies collected after 2 months. Rochow and Carmichael (1979) successfully used antisera from early bleedings to detect barley yellow dwarf, but did not make comparisons with other bleedings.

2.3.1.2   The Solid Phase and Its Coating by Antibodies.   An essential step in ELISA is sensitizing a solid phase (surface) with antibodies. Plastics are common-ly used because they adsorb proteins via an essentially irreversible hydrophobic interaction (Engvall and Ruoslahti, 1979). Nonionic detergents prevent this in-teraction, but do not reverse it, and are added in later steps to prevent nonspecif-ic binding. The solid phase commonly used for plant virus detection has been polystyrene microtiter plates; however, polystyrene beads (Precision Plastic Ball Co., Chicago, Ill., 60641)[1], tubes, and stirring rods, as well as other types of plastics have also been used for ELISA (Engvall and Ruoslahti, 1979; Saunders 1979). Recently, polystyrene sample cells have been produced which become cuvettes in an automated system (Gilford Instrument Labs., Inc., Oberlin, Ohio, 44074)[1]. Other solid phases used as carriers in radioimmunoassay can also be used for ELISA (Voller *et al.*, 1977).

The cells in Microtiter® plates require only a 0.2-ml sample, whereas breads or sticks can be dipped in large volumes of test sample which contain less antigen per unit volume. Reliability, cost, and convenience influence choice of solid phase for various applications. Additional developments in this area can be ex-pected.

Passive sensitization of the solid phase is achieved by exposing it to a solution of partially purified γ-globulin in sodium carbonate buffer, pH 9.6, and allowing antibodies to adsorb to the solid phase. The optimal conditions (e.g. concn of γ-globulin, time and temperature) are determined experimentally. The schedule reported by Clark and Adams (1977) has been widely used with minor modifica-tions. Recently, McLaughlin, Barnett and Burrows (1980) have done consider-able work to optimize ELISA reactions. They found that 30 min at 30°C or 1 h at 5°C was more efficient for coating than longer incubation periods. Short incu-bation periods were also found advantageous by Saunders (1979), and Pesce *et al.* (1978) showed that coating periods over 4-6 h are deleterious. The optimal concentration of γ-globulin in the coating solutions is usually 1-10 $\mu g/ml$. Concentrations exceeding 10 $\mu g/ml$ usually reduce the strength of the virus specific reactions, and increase intensity of nonspecific reactions (Clark and Adams, 1977). Lower concentrations of antibodies were better for differentiating closely related antigens, whereas higher coating concentrations were favored for detecting more distantly related antigens (Koenig, 1978).

Microtiter plates and polystyrene beads have been successfully reused as many as six times by treating the plates and beads for 30-60 min with 0.2 M glycine –HCl, pH 2.2, between uses (Bar-Joseph et al., 1979c, 1980a). This treatment dissociates the antigen-antibody bonds, but does not remove the coating antibody from the solid phase. Antigen-antibody complexes of different viruses differ in their dissociability by acidic or basic solutions. Also, some unspecified properties of the solid phase affect plate reuse.

2.3.1.3 Washing the Plate. Thorough washing between steps is essential to prevent carryover of reactants that are not part of the solid phase-double antibody-antigen complex. Usually, the solid phase is washed at least three times with phosphate-buffered saline containing 0.05% Tween-20 (PBST) and it is often left for several minutes in the wash solution (Clark and Adams, 1977). Tapwater has been used for washing in certain systems with good results. For safety, the wash solution should not contain NaN_3.

2.3.1.4 Preparing the Test Sample. Virus concentration, virus stability, and host sap components (presence of oxidative enzymes, tannins, etc.) can affect sample preparation. For certain virus-host combinations, e.g. pepper infected by potato virus Y (PVY) and cucumber mosaic virus (CMV), diagnosis was accomplished by immersing leaf discs in PBS-buffer without prior homogenization (Marco and Cohen, 1979). Tobacco mosaic virus (TMV) in several hosts was detected by gently crushing leaf pieces with a glassbar in the microplate wells, followed by addition of 0.2 ml of extraction buffer (Bar-Joseph, 1980). Generally, sample tissues must be ground in the extraction medium. Tough and difficult-to-grind plant material can be processed with a dispersion homogenizer such as a Polytron PT 10/35[1] (Brinkman Instruments, N.Y.) (Converse, 1978), or a T-18 Ultra Turrax[1] (Janke and Kunkel KG, Staufen, West Germany) (Bar-Joseph et al., 1979a), which is rapidly cleaned between samples in a wash solution.

Clark and Adams (1977) reported that sap constituents of several herbaceous hosts had little effect on test results. Readings were no different for purified plum pox virus (PPV), diluted directly in PBST buffer or diluted in a 1:100 (w/v) extract of healthy *Nicotiana clevelandii*. In contrast, extracts of blackcurrant buds prevented the detection of Arabis mosaic virus (ArMV) at a dilution of 1:10 but not at a dilution of 1.100. Extracts of *Prunus* leaves collected in mid-summer and extracts of leaves of some other perennials caused nonspecific reactions. The addition of 1 to 2% (w/v) polyvinyl pyrrolidone (PVP) mol. wt 25,000 or 44,000 to the extraction buffer (PBST) reduced or eliminated these nonspecific reactions (Clark and Adams, 1977).

While most research workers have used the PBST-PVP extraction medium (Clark and Adams, 1977), the effects of buffers and additives on new virus-host combinations should be examined to optimize ELISA results.

McLaughlin *et al.* (1979a) tested the effects of combinations of PBST, 2-mercapthoethanol (2-ME), diethyldithiocarbamate (DIECA), and PVP on purified virus and extracts of clover yellow vein virus (CYVV) and white clover mosaic virus (WCMV)-infected white clover and red clover leaves. Most buffer systems were satisfactory for purified virus, but addition of 2-ME gave the best results. With sap extracts, DIECA gave the best results for both host-virus combinations. The PVP decreased nonspecific background reactions in both virus systems, but also decreased the specific reaction in the CYVV system. Addition of 0.25 M potassium phosphate or 0.1 M ethylenediaminetetraacetate (EDTA) to PBST gave much better results for papaya ringspot virus than PBST alone (D. Gonsalves, 1980). Best results were obtained with 0.25 M potassium phosphate, pH 7.5.

The length and condition of incubation may also affect results. With a short incubation period (3 h at 33°C), water extracts of citrus tristeza-infected lime plants gave similar reactions to extracts in PBST-PVP, but for longer incubation periods (14 h at 6°C), extracts in PBST-PVP gave somewhat higher readings (Sharafi and Bar-Joseph, 1980).

Addition of chloroform to the extraction medium increased ELISA values with the RPV isolate of barley yellow dwarf virus (Lister and Rochow, 1979). Flegg and Clark (1979) overcame interference with ELISA tests for apple chlorotic leaf spot virus by using dilute extracts and by adding the enzyme conjugate simultaneously with the test sample.

Clark and Adams (1977) reported that long incubation periods (18 h) at 6°C with PPV-infected leaf extracts gave better results than incubation periods of 1, 2 or 4 h at 37°C. Incubating test samples overnight at low temperature has been convenient for routine testings.

2.3.1.5   Preparation and Use of the Enzyme-Labeled Antibody (Conjugate). The one-step glutaraldehyde method yields satisfactory conjugates of antibody and alkaline phosphatase and horseradish peroxidase, although the ratio of enzyme molecules to antibody varies somewhat (Avrameas, 1969; Avrameas *et al.*, 1978). The two-step glutaraldehyde method produces conjugates with an equal ratio of enzyme and antibody molecules and may yield more sensitive assays, especially with peroxidase (Avrameas *et al.*, 1978; Clark, 1980). An adverse effect of glutaraldehyde conjugation on binding ability of antibodies has been reported (Ghabrial and Shepherd, 1980).

Additional strategies for conjugation have been described (Avrameas *et al.*, 1978; Nakane, 1979; Voller *et al.*, 1977; Saunders, 1979) and may be useful for some applications. An approach to ultrasensitive assay using fluorogenic methyl umbelliferin substrates with a conjugate of FAB′ fragments and β– D– galactosidase (Kato *et al.*, 1976; Ishikawa and Kato, 1978) is noteworthy, although not yet used for plant viruses.

Concentration of conjugate, the length of incubation period, and temperature affect the results of ELISA. Twofold dilutions of conjugate in the range of 1:400 to 1:2000 usually halved ELISA values over a wide range of antigen concentrations, and this effect could be partly compensated by increasing the incubation period (Clark and Adams, 1977). Conjugate incubation for 12 h at 5°C gave better results for CYVV than shorter incubation periods or a similar period at 30°C (McLaughlin *et al.*, 1980). Conjugate can be reused up to four times where infection rates are low ($<5\%$) and still retain more than 60% of its original specific activity (Bar-Joseph *et al.*, 1979b). Since nonspecific antibodies are adsorbed, reused conjugate may become more specific.

2.3.1.6 Substrate and Evaluation of Results. The final step is addition of the enzyme substrate. The substrate should be freshly prepared at the desired concentration and free of color. The substrate for alkaline phosphatase, p-nitrophenyl phosphate, is available in convenient tablet form (Sigma, St. Louis, Mo.)[1] and generally prepared at 0.6 to 1.0 mg/ml in 10% diethanolamine, pH 9.8.

Alkaline phosphatase activity is stopped by addition of NaOH when sufficient color change has occurred, normally after 30 to 60 min. When plates cannot be read for several hours they may be stored at 4°C. Plates can be evaluated visually and photographed for a permanent record. A light box with 5500 K fluorescent tubes is convenient for this purpose. Visual scoring is sensitive to an OD_{405} of approximately 0.15 and this is adequate for many routine applications. A more precise quantitative evaluation is made spectrophotometrically at an appropriate wavelength (405 nm for the alkaline phosphatase system).

Because healthy extracts or buffer solutions yield a low but detectable reading and presence of host-specific antibodies can cause an appreciable background reading, a threshold value for a positive reaction must be defined. Often an arbitrary value is selected, such as a reading twice the mean background, or an absorbance value 0.1 greater than the background. More accurately, the threshold is established by statistical analysis for the desired confidence level. Appropriate controls and standards are essential.

2.3.1.7 Test Conditions and Equipment. Although approximate guidelines are available, a certain amount of experimentation and practice is necessary to establish optimum conditions for each new application of ELISA. Since application of ELISA to plant virus detection is still in its infancy, much room exists for modification and innovation in procedures and equipment. As indicated by Saunders (1979), some flexibility in approach is highly desirable.

The ELISA procedure can be done with rather simple equipment and rudimentary facilities, especially when a prepared conjugate is available and results are evaluated visually. Corrections for variable conditions can be made with appropriate reference standards. Where high precision is required and large numbers of samples must be processed, automated processing and reading equipment is desirable (Ruitenberg and Brosi, 1978; Saunders et al., 1979). A considerable amount of automation is now available for microtiter plate systems, or is in development (Gilford Instrument Labs., Inc., Oberlin, Ohio; Dynatech Labs., Inc., Alexandria, Virginia, Flow Labs Inc., McLean, Virginia; R. Casper, 1979).

2.3.2 Sensitivity and Quantitative Assays

The approximate concentrations of several viruses that can be detected by ELISA are shown in Table I. Although variable limits are reported for different viruses, it is clear that ELISA provides a sensitive assay. Sensitivity should also increase with further experience and research.

The approximate sensitivity of ELISA and several other common plant virus detection procedures are shown in Table II. Although concise and complete comparisons are lacking for all these procedures with the same virus-host systems, it is clear that ELISA compares well in sensitivity. The sensitivity of ELISA is especially remarkable, because it does not require elaborate equipment and measurements can be made on crude extracts without clarification or concentration. The sensitivity of ELISA permits the batch testing of samples, a procedure of particular value where the incidence of virus infection is low, such as in seed certification tests.

TABLE I.   Some Concentrations of Different Plant Viruses Detectable by Enzyme-Linked Immunosorbent Assay

Virus	Group	ng/ml	Dilution[a]	Reference
		Concentration		
Apple chlorotic leaf spot	Closterovirus	1-5	10^{-4}	Flegg & Clark, 1979
Apple mosaic	Ilar virus		10^{-4}	Clark and Adams, 1977
Barley yellow dwarf	Luteovirus	30		Lister & Rochow, 1979
Cauliflower mosaic	Caulimovirus	25		Ghabrial & Shepherd, 1980
Citrus tristeza	Closterovirus		10^{-3}	Bar-Joseph et al., 1979a
Clover yellow mosaic	Potexvirus		10^{-6}	McLaughlin & Barnett, 1978
Cucumber mosaic	Cucumovirus	0.01		Gera et al., 1978
Lettuce mosaic	Potyvirus	9		Jafarpour et al., 1979
Maize chlorotic dwarf	-	115		Reeves, et al., 1978
Plum pox	Potyvirus	< 1	10^{-5}	Clark & Adams, 1977
Tobacco mosaic	Tobamovirus	50		Bar-Joseph & Salomon, 1980
Tobacco rattle	Tobravirus	30		Gugerli, 1978
Tobacco ringspot	Nepovirus	15		Lister, 1978

[a] Dilution of extracts of infected plants.

TABLE II. Approximate Sensitivity of Some Plant Virus
Detection Methods[a]

Method	Detectable concn μg/ml
ELISA	< .01
Immunodiffusion	1.00
Microprecipitin	.50
Latex flocculation	.01
Electron microscopy (EM)	.10
Serologically specific (immune) EM	< .01
Infectivity assay	.05

[a]From several sources, including Clark and Adams, 1977; Matthews, 1970; and Derrick, 1973.

The ELISA values can be related quantitatively to virus concentration, as shown for dilutions of several purified viruses and extracts from virus-infected hosts (Clark and Adams, 1977; Converse, 1978; Thresh *et al.*, 1977; Koenig, 1978; Bar-Joseph *et al.*, 1979a). The relationship is linear only for certain ranges of virus concentration, and may be influenced by specificity and concentration of coating and conjugate antibodies (Fig. 2).

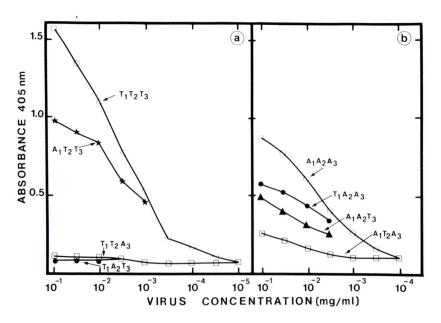

FIG. 2. Reaction of different concentrations of viral antigens tested with homologous and heterologous combinations of coating and enzyme-linked (conjugated) antibodies. T and A refer to common and A-strains of TMV, 1, 2, and 3 refer to coating antibody, antigen, and conjugate, respectively. Coating done with γ-globulin concentration of 3/μg/ml. Conjugate used at 1/400 dilution of stock. (From Bar-Joseph and Salomon, 1980.)

2.3.3 Differentiating Closely Related Antigens

Koenig (1978) found that ELISA was very sensitive not only for detecting low concentrations of virus, but also for differentiating closely related virus strains. Strains of several isometric and elongated plant viruses, whose heterologous and homologous titers differed fourfold or more in precipitin tests, often failed to cross react in ELISA. However, some heterologous antigens were detected which differed considerably. Similar specificity was also reported for other viruses by McLaughlin et al., 1978; Rochow and Carmichael, 1979; and Barbara et al., 1978b. In Fig. 2a, the results of homologous and heterologous ELISA tests with the common TMV and an avocado isolate (TMV-A) are depicted (Bar-Joseph and Salomon, 1980). In homologous reactions (T_1, T_2 and T_3 - where T refers to TMV and 1, 2 and 3 refer to coating, antigen and conjugate components of the double sandwich, respectively), 1 $\mu g/ml$ of TMV gave an OD_{405} of 0.53, whereas in $T_1 A_2 T_3$ (A refers to TMV-A), a concentration 100 times higher could not be detected. In the heterologous antibody reactions, replacement of homologous coating-antibodies with the heterologous TMV-A antibodies ($A_1 T_2 T_3$) reduced ELISA values by 14 to 37% depending on antigen concentration. Replacement of the homologous conjugated antibodies by the heterologous conjugate ($T_1 T_2 A_3$) almost completely prevented the ELISA reaction even at higher antigen concentrations.

The differentiation was less specific when TMV-A antibodies were used (Fig. 2b), indicating that a common antigen in TMV-A particles probably elicited more antibody than that in TMV particles. The reasons for the greater selectivity of the conjugate antibodies are not fully understood. The conjugation process may cause spatial impairments or conformational changes in the combining sites and decrease the avidity of the antibodies (Koenig, 1978). Conjugating bovine serum albumin to TMV and TMV-A antibodies used for coating impaired their activity in homologous reactions, but did not render then completely specific in the heterologous systems (Bar-Joseph and Salomon, 1980). Specificity among isolates of several viruses (Koenig, 1978; Rochow and Carmichael, 1979) in ELISA complicates use of the procedure for assaying field-collected samples because different globulins or mixtures of globulins (Koenig et al., 1979a; Barbara et al., 1978b) must be used. Barbara et al. (1978b) detected strains of necrotic ringspot virus (NRSV) in hop and plum by ELISA. Apple mosaic virus (ApMV) and NRSV serotypes cross reacted weakly or not at all in ELISA. These serotypes were used concurrently to detect and differentiate NRSV and ApMV isolates and an apparent intermediate isolate found in hop. However, biologically different strains may not show strong serological specificity. Antiserum to a single strain of citrus tristeza virus reacted well in ELISA tests with a diverse range of biological isolates.

Detienne et al. (1979) were able to detect eight different fruit tree strains of apple chlorotic leafspot virus (CLSV) by ELISA, and it appears, as for citrus tristeza virus (CTV), that ELISA does not differentiate strains of CLSV.

2.4 APPLICATIONS

One of the immediate applications of ELISA in plant virology has been for diagnosis. Some of these applications are reviewed in this section and illustrate the advantages of ELISA. One can note, especially, its use for survey and for detecting viruses where low concentrations are expected.

2.4.1 Use of ELISA for Diagnosing Virus Diseases of Major Crops

2.4.1.1 Citrus. For more than four decades, tristeza (caused by an aphid-transmitted, phloem-limited closterovirus), has been an important citrus virus disease, especially in areas with extensive plantings on tristeza-sensitive sour orange (*Citrus aurantium*) rootstock. More than 300,000 trees are indexed annually for CTV in eradication and suppression programs in California and Israel. Where CTV is widespread, indexing is important to research on rootstock and varietal resistance, cross protection, vector relationships, and therapy procedures. Graft inoculation of Mexican lime (*C. aurantiifolia*) has been the main CTV-detection procedure for many years, despite requirements for extensive amounts of slow-growing plant material, good glasshouse and screenhouse facilities, skilled personnel to inoculate and read indicator plants, and difficulty in detecting mild isolates in warm weather. Following the successful development of CTV antisera (Gonsalves *et al.*, 1978), the SDS-immunodiffusion procedure (Garnsey *et al.*, 1979), and the ELISA technique (Bar-Joseph *et al.*, 1979a) were successfully tested for field detection of CTV in Israel and Florida. A broad range of CTV isolates, including very mild, ordinary, and severe seedling yellows strains, was detected with antiserum to a single isolate.

Over 82,000 samples have been indexed by ELISA during 1978-79 in Israel, and ELISA has nearly replaced the lime test for routine indexing. A survey of 2,811 groves, with 5 trees tested per grove, revealed 4 new centers of infection, mainly symptomless trees. A second survey of 50 trees in each of 176 groves revealed a single new center of infection (Bar-Joseph *et al.*, 1980a).

The virus could be assayed from many phloem-containing tissues during warm and cold weather, but best results were normally obtained with fruit pedicel bark (Fig. 3). Two Shamouti orange (*C. sinensis*) groves were tested for CTV infection by sampling picked fruit. The predicted infection rates (1% and 11%) obtained from fruit samples were in reasonable agreement with the observed rates of 1% (15/1400) and 16% (324/2053), respectively (Bar-Joseph *et al.*, 1978). Monitoring easily obtainable packinghouse samples reduces the need for expensive field surveys.

Experimental tests with ELISA are in progress in the California tristeza eradication project (D. Cordas, 1980), and in Spain (Cambra *et al.*, 1979).

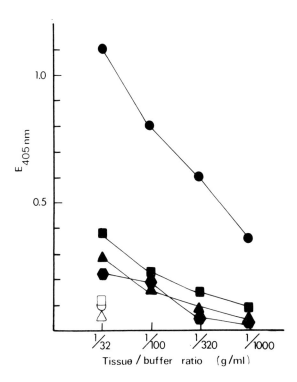

FIG. 3. Reaction of extracts from different tissues of Shamouti orange trees in enzyme-linked immunosorbent assay. Pedicel bark = ●, young stem bark = ■, leaf = ▲, and leaf mid-vein = ◆. Solid symbols = citrus tristeza virus-infected, open symbols = healthy. All extracts prepared in phosphate-buffered saline (PBS) containing 0.05% Tween 20 and 2% polyvinyl-pyrrolidone, MW = 40,000. (From Bar-Joseph *et al.*, 1979a.)

Casper (1980) demonstrated detection of citrus leaf rugose virus by ELISA, and Clark *et al.* (1978) used ELISA to detect *Spiroplasma citri*, the causal agent of citrus little-leaf (stubborn) disease in culture and in plant extracts. The organism was readily detected in seed coats collected from infected trees in January and April. However, positive reactions were obtained with extracts of leaf mid-veins only from young citrus leaves collected in July. Similar results were also reported by Saillard *et al.* (1978).

2.4.1.2 Deciduous Fruits. Several viruses are harbored in propagative material of deciduous fruit crops, and some of these were among the first plant viruses successfully detected by ELISA.

Plum pox virus (PPV). Plum pox (Sharka), a potyvirus transmitted nonpersistently by aphids, is a major threat to fruit-growing in several countries. Eradication of the disease depends upon early detection of all infected trees before secondary spread occurs (Adams, 1978a, b). Although PPV causes conspicuous fruit symptoms on sensitive varieties, detection in nurseries and orchards based on leaf symptoms is difficult. In a comprehensive assessment of ELISA for PPV, Clark *et al.* (1976b) examined over 2,000 samples of *Prunus* material from field and glasshouse. The PPV was detected in leaves, dormant buds, roots, and fruits. The ELISA test was as sensitive as immune electron microscopy (Kerlan *et al.*, 1978) and more sensitive than several other immunological methods (Clark and Adams, 1977).

The use of ELISA has greatly increased the speed and reliability of PPV testing. In Britain, several thousands of samples have been tested by this method and new PPV outbreaks were confirmed within a few days of their discovery. It is applicable to a wide variety of hosts (Adams, 1978b) and for detection of mild isolates (Kerlan *et al.*, 1978).

Apple chlorotic leafspot virus (CLSV). Apple chlorotic leafspot virus, a closterovirus for which no vector is yet known, is widespread in both commercial and ornamental woody hosts. Difficulties were encountered in CLSV detection by the standard ELISA procedure, probably due to virus instability (Flegg and Clark, 1979). A modified procedure where test sample and conjugate were incubated simultaneously was approximately 1,000 times more sensitive than tube precipitin tests for detecting partially purified CLSV preparations. The virus was detected in apple leaves, petals and calyx tissue of freshly picked and cold-stored fruits.

Apple mosaic virus (ApMV). ApMV an ILAR virus has been detected by ELISA (Clark *et al.*, 1976a; Hardcastle and Gotlieb, 1977). It was found in apple leaves collected in spring and early summer but not after a prolonged hot period. Even so, ELISA is advantageous over previous diagnostic methods, including infectivity assays or double-diffusion tests which enabled detection only from petals.

Arabis mosaic virus (ArMV) and hop mosaic virus (HMV). Arabis mosaic virus (ArMV) has a wide host range including fruit crops and other woody species. Previous diagnosis was based on mechanical inoculation to *Chenopodium* spp. and serological identification in extracts from the indicator by double-diffusion tests. Using ELISA, Clark *et al.* (1976a) identified ArMV directly in dormant blackcurrant buds, and Thresh *et al.* (1977) identified ArMV in spring-collected, developing buds of hop plants. Bulk sampling for ArMV in hop saved time and effort. Detecting HMV by grafting on sensitive indicators requires skill and glasshouse space, and weeks or months elapse between inoculation and symptom expression. Using ELISA, Thresh *et al.* (1977) were able to identify HMV in HMV-sensitive and HMV-tolerant (symptomless) varieties.

Tomato ringspot virus (TomRSV). Tomato ringspot, a nematode-transmitted (NEPO) virus, was detected by ELISA (Converse, 1978) in leaf stem, bud, and root samples of red raspberry (*Rubus idaeus*). The virus was detected in extracts of raspberry leaves diluted up to 4,000-fold, whereas variable results were obtained by agar double-diffusion tests with extracts diluted only 5-fold. Assay of bulk collections of 50 to 200 samples appeared feasible.

Peach rosette mosaic virus (PRMV). Ramsdell *et al.* (1979) compared ELISA and infectivity assays on *Chenopodium quinoa* for detection of PRMV in vines of *Vitis labrusca* 'Concord'. While as little as 10 μg of purified PMRV could be detected, more consistent results from extracts were achieved by infectivity assay.

Potatoes. Considerable losses are caused by several potato virus diseases, and seed potato certification programs are important for control. Several serological techniques have been used in potato certification programs for a number of years (Shephard, 1972).

Richter *et al.* (1977) applied ELISA to potato virus S, and Maat and de Bokx (1978a) used it for potato virus A (PVA) and potato virus Y (PVY). Young, succulent potato sprouts were good sources for PVA indexing, but glasshouse-grown leaves gave inconsistent results. Using a serum to the tobacco veinal necrosis strain of PVY (PVY[n]), PVY[n] was detected in extracts from glasshouse plants, but the common strain was not. Presence of PVY common strain in doubly infected tissues interfered with detection of PVY[n]. Several sera or polyvalent antisera will be needed to detect all expected PVY strains. Maat and de Bokx (1978a) and de Bokx and Maat (1979) investigated factors that could affect ELISA results for PVY[n] infection in tubers. They found no differences in virus titer between the rose and the heel end of tubers shortly after lifting. Brief storage of tubers increased ELISA values. Several potato viruses, including PVX, PVY, PVS, PVT, APLV, TMV, and Andean potato mottle virus, are now diagnosed by ELISA at the International Potato Center, Lima, Peru. Multiple strain and mixed virus infections could be detected with high sensitivity by using a mixture of antisera (Salazar, 1978, 1979).

Casper (1977b), Mehrad *et al.* (1978), Maat and de Bokx (1978b), Clarke *et al.* (1980), and Rowhani and Stace-Smith (1979) used ELISA to detect potato leafroll virus (PLRV), a luteovirus transmitted persistently by many aphid species, and a major disease problem in many potato-growing countries. Low virus titer in infected plants caused difficulties in obtaining sufficient virus to immunize test animals, and when a potent serum was achieved, the titer of antigen in infected tissues was usually too low for detection by common serological assays. The PLRV was detected in potato and *Physalis floridana* extracts by ELISA. Higher PLRV concentrations were found in potato roots than in leaf stem and tubers by Casper (1977b). Maat and de Bokx (1978b), Rowhani and Stace-Smith (1979), and Clarke *et al.* (1980), all easily detected PLRV in potato foliage, but encountered problems in detecting PLRV in tuber extracts, possibly due to interference from particulate matter such as starch grains in the extracts (Rowhani and Stace-Smith, 1979).

2.4.1.4 Legumes. Several seed-transmissible viruses occur in soybean (Shepherd, 1972). Tobacco ringspot virus (TRSV) and soybean mosaic virus (SMV), are seed-transmitted at a high rate from infected plants, but large scale tests for seed certification programs have been difficult because sensitive methods applicable for screening numerous large bulk samples were not available. Lister (1978) reported that TRSV and SMV in extracts of individual soybean seeds diluted 1/2500 and 1/160 (w/v), respectively, could be easily detected by ELISA. In a comparative sero-diagnosis study for pea seedborne mosaic (PSbMV), a potyvirus under eradication in North America, Hamilton and Nichols (1978) were able to detect PSbMV in composite leaf samples containing 5-10% infected tissue, whereas samples containing less than 25% infected tissue could not be detected confidently by SDS immunodiffusion. Serologically specific electron microscopy (SSEM) was more sensitive than ELISA for detecting seed infection, but presence of some host antibodies probably reduced sensitivity limits of ELISA.

Several forage legume viruses, including alfalfa mosaic virus (AMV), clover yellow mosaic virus (CYMV), CYVV, WCMV, and peanut stunt mosaic, were surveyed by McLaughlin and Barnett (1978). Screening tests for these viruses on indicator plants required several plants and considerable time and greenhouse space. Comparable tests were completed in a matter of hours by ELISA. The indexing potential of ELISA was extended by mailing sensitized microplates to research cooperators (McLaughlin and Barnett, 1978; McLaughlin *et al.*, 1979b). This procedure extended the use of ELISA in a cooperative regional improvement program.

2.4.1.5 Cereals.

Barley yellow dwarf virus (BYDV). The diagnosis of BYDV, a phloem-restricted luteovirus that occurs worldwide in several grains and grasses, has been extremely difficult because vector transmission was necessary for correct identification (Rochow, 1979a). Recently, Lister and Rochow (1979) detected BYDV in barley leaves by ELISA, and the reaction was strain specific without cross reactivity between MAV and RPV isolates. Comparative aphid transmission trials, and ELISA tests of field-collected BYDV samples demonstrated that in most cases, especially in mixed infections, ELISA was much simpler and more sensitive than differentiation by aphid transmissibility (Rochow, 1979a).

Maize dwarf mosaic (MDMV) and maize chlorotic dwarf mosaic (MCDV) Viruses. Studenroth (1979) applied ELISA for identification of maize dwarf mosaic virus (MDMV), a nonpersistent potyvirus in indigenous plant species. Comparative studies with two isolates, MDMV-A and MDMV-B, showed positive identification of both strains in crude sap diluted 10^{-3}, with negligble cross reactivity. Reeves *et al.* (1978) used ELISA to assay corn and Johnson grass plants for MDMV and for MCDV.

2.4.1.6 Bulbs and Flowers. The flower industry involves active international trade of propagative material, and is dependent on virus-free basic stocks. Maat *et al.* (1978) used ELISA to identify Nerine latent virus (NLV), an aphid-transmitted carlavirus in two cultivars of *Nerine bowdenii*, in *N. flexnosa*, and in naturally infected *Hippeastrum hybridum*. Koenig *et al.* (1979b) compared several diagnostic methods for another carlavirus, chrysanthemum B (CVB). The ELISA test enabled reliable detection of CVB in mature plants throughout the year, with a much greater sensitivity than the microprecipitin test. ELISA was useful for screening CVB in several chrysanthemum varieties and for selection of virus-free plantlets from meristem culture. Stein *et al.* (1979) detected cucumber mosaic virus (CMV) and bean yellow mosaic virus (BYMV) in gladiolus plants by ELISA, and found it more sensitive than sap inoculation of indicator plants for diagnosis of both viruses, especially for CMV from corms and cormlets.

2.4.1.7. Sugar Beets. Beet curly top virus (BCTV) was detected in beet plants and viruliferous leafhoppers by ELISA (Mumford and Thornley, 1980). Sensitivity of ELISA for BCTV was estimated at 1 μg/ml, and ELISA was much more sensitive than latex flocculation assay or bioassay for detecting BCTV in clarified extracts.

2.4.2 The Use of ELISA for Preventing Disease Transmission

2.4.2.1 The Application of ELISA in Epidemiological Studies. The sensitivity and specificity of ELISA suggested its use for virus detection in virus vector studies. Gera *et al.* (1978, 1979) reported detection of CMV in single apterous *Aphis gossypii* by ELISA. Aphids probing plants infected with a nontransmissible strain of CMV failed to acquire CMV and short (1.5 min) feeding on healthy plants rendered viruliferous aphids nonreactive in ELISA tests. These results suggest use of ELISA for CMV epidemiological studies (Gera *et al.*, 1978); however, application to winged aphids that acquire and carry the virus under natural conditions was not demonstrated. Cambra, *et al.* (1979) detected CTV by ELISA in aphids collected from CTV-infected trees. Clarke *et al.* (1980) detected PLRV in viruliferous aphids, and Mumford and Thornley (1980) detected BCTV in single leafhoppers (*Circulifer tenellus*). Mumford *et al.* (1980) have suggested monitoring field populations of *C. tenellus* for BCTV by ELISA to avoid crop losses.

The potential of ELISA is obvious for diagnosing pathogens which multiply within their vectors and exist at high concentrations for prolonged periods. Banttari (1979) detected oat blue dwarf virus in individual aster leafhoppers (*Macrosteles fascifrons*). Bar-Joseph and Townsend (1980) tested extracts (0.6 ml/insect) of individual *Euscelidius variegatus* leafhoppers injected with ca. 1.5 $\times$ 10^3 CFU of *S. citri*, and obtained ELISA values of 0.78 ± 0.05 (OD_{405}) with insects 6 and 10 days following injection compared with only 0.19 ± 0.02 in insects after 0 and 2 days. Bové *et al.* (1979) found 7 leafhopper species out of 34 tested to harbor *S. citri* naturally in Morocco. Two of these species, *Psammotettix striatus* and *Laodelphax striatellus*, were found in great numbers and were often infected.

The ELISA procedure is readily adapted to large-scale surveys. In addition to the surveys for CTV and PPV already cited, ELISA has been used by Converse (1978) to map a red raspberry field for TomRSV, and by Thresh *et al.* (1977) to survey spread of HMV into virus-free hop plantings. Frozen samples collected in New York for 20 consecutive years were tested by ELISA to show the fluctuations of the relative amount of different BYDV isolates (Rochow, 1979b). The ELISA method was used by Studenroth (1979) to locate the species carrying MDMV among a large number of weeds.

2.4.2.2 Seed Transmission. Several important virus diseases are seed-transmitted (Shepherd, 1972), and accurate diagnosis and estimation of infection rate in seed lots are of great importance for limiting spread of such diseases by seed certification programs. Simple, rapid, and sensitive procedures are needed, since many lots must be tested and low rates of infection must be detected in bulk samples. One approach has been to use serologically specific electron microscopy (SSEM) (Brlansky and Derrick, 1979).

The ELISA test has been used in testing for TRSV and SMV in soybean seed (Lister, 1978). Less than 1% seed infection with TRSV and about 2.4% with SMV can be detected by testing seed batches. Germination of seed for 1 week in the dark enhanced SMV concentrations and enabled detection of lower infection rates. Bossennec and Maury (1977 and Bowers and Goodman (1979) dissected soybean seeds and tested for SMV in mature seeds. The virus was detected in testae, cotyledons and embryos of a variety known to transmit SMV through seeds. Seed transmission, with a few possible exceptions, is dependent on embryo infection, and testing seed parts free of testae might be the most appropriate procedure for such viruses. Lister *et al.* (1978) detected three barley stripe mosaic virus (BSMV) isolates in extracts of seed and seed parts at 10^{-3} – 10^{-5} dilutions, and estimates of infection rate were close to those based on seedling symptoms. Casper (1977a) detected prune dwarf virus (PDV) in individual and batch samples of *Prunus avium* seed. Jafarpour *et al.* (1979) used the ELISA test to detect bean common mosaic virus (BCMV) in pinto bean seed and lettuce mosaic virus (LMV) in lettuce seed. Batch samples of one BCMV-infected embryo in 2,000 healthy embryos could be detected by ELISA, and LMV was detected in homogenized batch mixtures containing one infected lettuce seed among 1,400 healthy seeds. Ghabrial and Shepherd (1980) encountered problems in detecting low levels of LMV by ELISA and reported improved results with a radioimmunosorbent assay (RISA). Their RISA procedure is similar in principle to ELISA and can be used in conjunction with it.

2.4.3 Breeding Programs

Marco and Cohen (1979) suggested the use of ELISA in breeding pepper resistant to CMV and PVY. ELISA values agreed well with the visual evaluation of resistance to these viruses, and ELISA provided an objective and more detailed evaluation of resistance that enabled discrimination between levels of resistance not differentiated visually.

2.5 CONCLUDING REMARKS

In a recent prefatory chapter, D. L. Bailey (1966) cited the following sentences of Wigglesworth (1955): "Pure science provides new tools that can be used by the applied biologist. Where and when these new tools are going to be produced is utterly unpredictable and is not amenable to planning."

The validity of these sentences is illustrated by the short history of the ELISA technique. Following a basic immunological study which showed that it is possible to chemically link antibody and enzyme while retaining both immunological and enzymic activity (Avrameas, 1969), diagnostic assays have been developed and applied to the detection of a wide variety of antigens including plant viruses.

While the value of serology for plant virus detection has been recognized for many years, serology has not been fully utilized. Useful existing serological techniques have not always been adopted where they could have been used to advantage. However, in other cases, serological techniques with sufficient sensitivity and simplicity have not been available. For the latter, ELISA has created numerous new opportunities. Many of the first reports on use of ELISA for plant virus detection have been directed to such difficult applications.

It seems unlikely that ELISA will be only a passing fad. It has been widely and enthusiastically applied to situations where no alternatives so promising exist. Because ELISA has been only recently introduced to plant virus detection, further improvements in materials and techniques can be anticipated. Applications will undoubtedly expand, especially as a tool in plant virology research.

The ELISA technique can be readily learned and applied, even with limited experience in plant virus serology. Development of prepared and standardized plates, reagents, and conjugates could further extend use of ELISA to persons not equipped to prepare and standardize their own materials, and to those who need to make only limited tests for a particular virus.

The serological specificity of ELISA may allow more precise differentiation of virus serotypes and strains. However, this same specificity may hamper use of ELISA for virus surveys where multiple strains are expected. Further research on the nature of specificity and the production and use of multivalent antisera is needed. Combination of ELISA with other detection procedures (Koenig *et al.*, 1979a; Ghabrial and Shepherd, 1980) will also be desirable in certain situations.

2.6 ACKNOWLEDGMENTS

The authors thank O. W. Barnett, L. Bos, M. Clark, J. A. de Bokx, J. Dunez, S. A. Ghabrial, D. Gonsalves, R. M. Goodman, R. I. Hamilton, H. T. Hsu, D. L. Mumford, H. A. Neely, L. F. Salazar, and J. C. Studenroth for providing information, unpublished data, and unpublished manuscripts.

Support for the studies on citrus tristeza virus came, in part, from the United States-Israel Binational Science Foundation Grant No. 1716.

2.7 REFERENCES

Adams, A. N. (1978a). The detection of plum pox virus in *Prunus* species by enzyme-linked immunosorbent assay (ELISA). *Ann. Appl. Biol.* 90: 215-221.

Adams, A. N. (1978b). The incidence of plum pox virus in England and its control in orchards. *In* "Plant Disease Epidemiology" (P. R. Scott and A. Bainbridge, eds.), pp. 213-219. Blackwell Sci. Pubs. Ltd., Oxford.

Avrameas, S. (1969). Coupling of enzymes to proteins with glutaraldehyde. Use of the conjugates for the detection of antigens and antibodies. *Immunochemistry* 6: 43-52.

Avrameas, S., Ternynck, T., and Guesdon, J. L. (1978). Coupling of enzymes to antibodies and antigens. P. 7-23. *In* "Quantitative Enzyme Immunoassay" (E. Engvall and A. J. Pesce, eds.), Vol. 8, Suppl. 7, *Scand. J. Immunol.* Blackwell Sci. Pubs., Ltd., Oxford.

Bailey, D. L. (1966). Whither pathology. *Ann. Rev. Phytopathol.* 4: 1-8.

Banttari, E. E. (1979). Detection of oat blue dwarf virus in plants and in aster leaf hopper using enzyme-linked immunosorbent assay. *In* "Abstracts IX Int. Cong. Plant Protection," Wash. D.C. No. 215.

Barbara, D. J., Clark, M. F., Thresh, J. M., and Casper, R. (1978b). Rapid detection and serotyping of prunus necrotic ringspot virus in perennial crops by enzyme-linked immunosorbent assay. *Ann. Appl. Biol.* 90: 395-399.

Bar-Joseph, M., and Salomon, R. (1980). Heterologous reactivity of tobacco mosaic virus strains in enzyme-linked immunosorbent assays. *J. Gen. Virol.* 47: 509-512.

Bar-Joseph, M., Sacks, J. M., and Garnsey, S. M. (1978). Detection and estimation of citrus tristeza virus infection rates based on ELISA assays of packing house fruit samples. *Phytoparasitica* 6: 145-149.

Bar-Joseph, M., Garnsey, S. M., Gonsalves, D., Moscovitz, M., Purcifull, D. E., Clark, M. F., and Loebenstein, G. (1979a). The use of enzyme-linked immunosorbent assay for the detection of citrus tristeza virus. *Phytopathology* 69: 190-194.

Bar-Joseph, M., Sharafi, Y., and Moscovitz, M. (1979b). Re-using the non-sandwiched antibody-enzyme conjugates of two plant viruses tested by enzyme-linked immunosorbent assay (ELISA). *Pl. Dis. Rep.* 63: 204-206.

Bar-Joseph, M., Moscovitz, M., and Sharafi, Y. (1979c). Re-use of coated enzyme-linked immunoassay plates. *Phytopathology* 69: 424-426.

Bar-Joseph, M. (1980). Unpublished.

Bar-Joseph, M., Oren, J. and Sacks, J.M. (1980a). Unpublished.

Bar-Joseph, M., Garnsey, S. M., Gonsalves, D., and Purcifull, D. E. (1980b). Detection of citrus tristeza virus, I. Enzyme-linked immunosorbent assay (ELISA) and SDS-immunodiffusion methods. *In* "Proc. 8th Conf. Int. Organ. Citrus Virol." (E. C. Calavan *et al.*, eds.). IOCV, Riverside, Calif. (In press.)

Bokx, J. E. de, and Maat, D. Z. (1979). Detection of potato virus in tubers with the enzyme-linked immunosorbent assay (ELISA). *Med. Fac. Landbown, Rijksuniv. Gent.* 44: (In press.)

Bossennce, J. M., and Maury, Y. (1977). Le probleme de la mise en évidence du virus de la mosaique du soja dans les semences de soja: proposition d'une méthode rapide de detection. *Ann. Phytopath.* 9: 223-226.

Bové, J. M., Moutous, G., Saillard, C., Fos, A., Bonfils, J., Vignault, J. C., Nhami, A., Abassi, M., Kabbage, K., Hafidi, B., Mouches, C., and Viennot-Bourgin, G. (1979). Mise en evidence de spiroplasma citri l'agent causal de la maladie du "Stubborn" des agrumes dans 7 cicadelles au Maroc. *C. R. Acad. Sci.*, Ser. D, 288: 335-338.

Bowers, G. R., and Goodman, R. M. (1979). Soybean mosaic virus: Infection of soybean seed parts and seed transmission. *Phytopathology* 69: 569-572.

Brlansky, R. H., and Derrick, K. S. (1979). Detection of seedborne plant viruses using serologically specific electron microscopy. *Phytopathology* 69: 96-100.

Cambra, M., Moreno, P., and Navarro, L. (1979). Deteccion rapida del virus de la tristeza de los citricos (CTV), mediante tecnicas immunoenzimaticas (ELISA-sandwich). *Anal. INIA, Ser. Proteccion Vegetal.* 12: 115-125.

Casper, R. (1977a). Testung von *Prunus avium*—Samen auf prune dwarf virus mit dem ELISA-verfahren. *Phytopath. Z.* 90: 91-94.

Casper, R. (1977b). Detection of potato leafroll virus in potato and in *Physalis floridana* by enzyme-linked immunosorbent assay (ELISA). *Phytopath. Z.* 90: 364-368.

Casper, R. (1977c). Andwendung eines neuen serologischen verfahren (ELISA) zum nachweis pflanzenpathogener Viren. 41 Deut. Pflanzenschutz-Tagung in Munster. October 1977. *Mitt. Biol. Bundesanst. Land und Forstw. Berlin-Dahlem* 178: 265.

Casper, R. (1978). New developments in plant virus serology. *Abstracts 3rd Int. Cong. Plant Path.*, Munchen, p. 16.

Casper, R. (1979). Geräte für die rationale massentestung mit dem enzyme-linked immunosorbent assay (ELISA). Gesunde *Pflanzen* 31: 291-293.

Clark, M. F., and Adams, A. N. (1977). Characteristics of the microplate method of enzyme-linked immunosorbent assay for the detection of plant viruses. *J. Gen. Virol.* 34: 475-483.

Clark, M. F., Adams, A. N., and Barbara, D. J. (1976a). The detection of plant viruses by enzyme-linked immunosorbent assay (ELISA). *Acta Hort.* 67: 43-49.

Clark, M. F., Adams, A. N., Thresh, J. M., and Casper, R. (1976b). The detection of plum pox and other viruses in woody plants by enzyme-linked immunosorbent assay (ELISA). *Acta Hort.* 67: 51-57.

Clark, M. F., Flegg, C. L., Bar-Joseph, M., and Rottem, S. (1978). The detection of *Spiroplasma citri* by enzyme-linked immunosorbent assay (ELISA). *Phytopath. Z.* 92: 332-337.

Clark, M. F. (1980). Personal communication.

Clarke, R. G., Converse, R. H., and Kojima, M. (1980). Enzyme-linked immunosorbent assay to detect potato leafroll virus in potato tubers and viruliferous aphids. *Pl. Dis.* 64: 43-45.

Converse, R. H. (1978). Detection of tomato ringspot virus in red raspberry by enzyme-linked immunosorbent assay (ELISA). *Pl. Dis. Rep.* 62: 189-192.

Cordas, D. (1980). Personal communication.

Derrick, K. S. (1973). Quantitative assay for plant viruses using serologically specific electron microscopy. *Virology* 56: 652-653.

Detienne, G., Delbos, R., and Dunez, J. (1979). Use and versatility of the immunoenzymatic ELISA procedure in the detection of different strains of apple chlorotic leafspot virus. *XI Int. Symp. Fruit Tree Virus Diseases.* Budapest, July 1979.

Engvall, E., and Perlmann, P. (1971). Enzyme-linked immunosorbent assay (ELISA). Quantitative Assay of Immunoglobulin G. *Immunochemistry* 8: 871-874.

Engvall, E., and Ruoslahti, E. (1979). Principles of ELISA and recent applications to the study of molecular interactions. *In* "Laboratory and Research Methods in Virology and Medicine, Vol. 3. Immunoassays in the Clinical Laboratory" (R. M. Nakamara, W. R. Dito, and E. S. Tucker III, eds.), pp. 89-97. Alan R. Liss, New York.

Flegg, C. L., and Clark, M. F. (1979). The detection of apple chlorotic leafspot virus by a modified procedure of enzyme-linked immunosorbent assay (ELISA). *Ann. Appl. Biol.* 91: 61-65.

Garnsey, S. M., Gonsalves, D., and Purcifull, D. E. (1979). Rapid diagnosis of citrus tristeza virus by sodium dodecyl sulfate-immunodiffusion procedures. *Phytopathology* 69: 88-95.

Gera, A., Loebenstein, G., and Raccah, B. (1978). Detection of cucumber mosaic virus in viruliferous aphids by enzyme-linked immunosorbent assay (ELISA). *Virology* 86: 542-545.

Gera, A., Loebenstein, G., and Raccah, B. (1979). Protein coats of two strains of cucumber mosaic virus affect transmission by *Aphis gossypii*. *Phytopathology* 69: 396-399.

Ghabrial, S. A., and Shepherd, R. J. (1980). A sensitive radioimmunosorbent assay for detection of plant viruses. *J. Gen. Virol.* (In press.)

Gonsalves, D. (1980). Unpublished.

Gonsalves, D., Purcifull, D. E., and Garnsey, S. M. (1978). Purification and serology of citrus tristeza virus. *Phytopathology*, 68: 553-559.

Gugerli, P. (1978). The detection of two potato viruses by enzyme-linked immunosorbent assay (ELISA). *Phytopath. Z.* 92: 51-56.

Hamilton, R. I., and Nichols, C. (1978). Serological methods for detection of pea seedborne mosaic virus in leaves and seeds of *Pisum sativum*. *Phytopathology* 68: 539-543.

Hardcastle, T., and Gotlieb, A. R. (1977). Detection of the yellow birch strain of apple mosaic virus (ApMV) using an enzyme-linked immunosorbent assay (ELISA). *Proc. Ann. Phytopath. Soc.* 4: 188. (Abstr.)

Ishikawa, E., and Kato, K. (1978). Ultrasensitive enzyme immunoassay. *In* "Quantitative Enzyme Immunoassay" (E. Engvall and A. J. Pesce, eds.), pp. 43-55. Vol. 8, Suppl. 7,

Scand. J. Immunol. Blackwell Sci. Pubs., Ltd., Oxford.

Jafarpour, B., Shepherd, R. J., and Grogan, R. G. (1979). Serologic detection of bean common mosaic and lettuce mosaic viruses in seed. *Phytopathology* 69: 1125-1129.

Kato, K., Fukui, H., Hamaguchi, Y., and Ishikawa, E. (1976). Enzyme-linked immunoassay: Conjugation of the FAB′ fragment of rabbit IgG with β-d-galactosidase from E. coli and and its use for immunoassay. *J. Immunol.* 116: 1554-1560.

Kerlan, C., Maison, P., Lansac, M., Dunez, J., Delbos, R., Massonie, G., and Marenaud, C. (1978). Obtention de ouches attenuées du virus de la Sharka à partir d'une population naturelle fortement pathogène. *Ann. Phytopathol.* 10: 303-315.

Koenig, R. (1978). ELISA in the study of homologous and heterologous reactions of plant viruses. *J. Gen. Virol.* 40: 309-318.

Koenig, R., Fribourg, C. E., and Jones, R. A. C. (1979a). Symptomatological, serological and electrophoretic diversity of isolates of Andean potato latent virus from different regions of the Andes. *Phytopathology* 69: 748-752.

Koenig, R.,Dalchow, J., Lesemann, D. E., and Preil, W. (1979b). ELISA as a reliable tool for the detection of chrysanthemum virus B in chrysanthemums throughout the year. *Pl. Dis. Rep.* 63: 301-303.

Lister, R. M. (1978). Applicaiton of enzyme-linked immunosorbent assay for detecting viruses in soybean seed and plants. *Phytopathology* 68: 1393-1400.

Lister, R. M., and Rochow, W. F. (1979). Detection of barley yellow dwarf by enzyme-linked immunosorbent assay. *Phytopathology* 69: 649-654.

Lister, R. M., Wright, S. E., and Kloots, J. M. (1978). Sensitive detection of barley stripe mosaic virus in barley seeds and embryos by ELISA. *Phytopathol. News* 12: 198.

Maat, D. Z., and Bokx, J. A. de (1978a). Enzyme-linked immunosorbent assay (ELISA) for the detection of potato viruses A and Y in potato leaves and sprouts. *Neth. J. Plant Path.* 84(5): 167-173.

Maat, D. Z., and Bokx, J. A. de (1978b). Potato leafroll virus: antiserum preparation and detection in potato leaves and sprouts with the enzyme-linked immunosorbent assay (ELISA). *Neth. J. Plant Path.* 84: 149-156.

Maat, D. Z., Huttinga, H., and Hakkaart, F. A. (1978). Nerine latent virus: some properties and serological detectability in *Nerine bowdenii*. *Neth. J. Plant Path.* 84: 47-59.

Marco, S., and Cohen, S. (1979). Rapid detection of viruses and evaluation of their titer in pepper by enzyme-linked immunosorbent assay (ELISA).*Phytopathology* 69: 1259-1262.

Matthews, R. E. F. (1970). Plant virology. Academic Press, New York, London. 778 pp.

McLaughlin, M. R., and Barnett, O. W. (1978). Enzyme-linked immunosorbent assay (ELISA) for the detection and identification of forage legume viruses. *Proc. 35th Conf. Southern Pasture and Forage Crop Improvement*, Sarasota, Florida.

McLaughlin, M. R., Barnett, O. W., and Gibson, P. B. (1978). Detection and differentiation of bean yellow mosaic virus and clover yellow vein virus by ELISA. *Phytopathol. News* 12: 198.

McLaughlin, M. R., Barnett, O. W., and Gibson, P. B. (1979a). The influence of plant sap and antigen buffer additives on the enzyme-immunoassay of two plant viruses. *In* "Abstracts IX Int. Cong. Plant Protection," Wash., D.C. No. 246.

McLaughlin, M. R., Barnett, O. W., and Gibson, P. B. (1979b). Mailing ELISA plates extends virus indexing potential. *Phytopathology* 69: 1-A7. (Abstr.)

McLaughlin, M. R., Barnett, O. W., and Burrows, T. M. (1980). Unpublished.

Mehrad, M., Lapierre, H., and Maury, Y. (1978). Le virus de l'enroulement de la pomme de terre: purification, detection serologique et dosage dans la plante. *C. R. Acad. Sci.*, Ser. D, 286: 1179-1182.

Mumford, D. L., and Thornley, W. R. (1980). Detection of beet curly top virus in plant tissue and leafhopper vector by enzyme-linked immunosorbent assay. In manuscript.

Mumford, D. L., Thornley, W. R., and Agosta, G. G. (1980). Unpublished.

Nakane, P. K. (1979). Preparation and standardization of enzyme-labeled conjugates. *In* "Laboratory and Research Methods in Virology and Medicine, Vol. 3. Immunoassays in the Clinical Laboratory" (R. M. Nakamara, W. R. Dito, and E. S. Tucker III, eds.), pp. 81-87. Alan R. Liss, N.Y.

Nakane, P. K., and Pierce, G. E., Jr. (1966). Enzyme-labeled antibodies: preparation and application for the localization of antigens. *J. Histochem. and Cytochem.* 14: 929-931.

Pesce, A. J., Ford, D. J., and Gaizutis, M. A. (1978). Qualitative and quantitative aspects of immunoassays. *In* "Quantitative Enzyme Immunoassay" (E. Engvall and A. J. Pesce, eds.), pp. 1-6. Vol. 8, Suppl. 7, *Scand. J. Immunol.* Blackwell Sci. Pubs., Ltd., Oxford.

Purcifull, D. E., and Batchelor, D. L. (1977). Immunodiffusion tests with sodium dodecyl sulfate (SDS)-treated plant viruses and plant viral inclusions. *Fla. Agr. Exp. Sta. Bull.* No. 788 (Tech.). 39 pp.

Ramsdell, D. C., Andrews, R. W., Gillett, J. M., and Morris, C. E. (1979). A comparison between enzyme-linked immunosorbent assay (ELISA) and *Chenopodium quinoa* for detection of peach rosette mosaic virus in 'Concord' grapevines. *Pl. Dis. Rep.* 63: 74-78.

Reeves, J. T., Jackson, A. O., Paschke, J. D., and Lister, R. M. (1978). Use of enzyme-linked immunosorbent assay (ELISA) for serodiagnosis of two maize viruses. *Pl. Dis. Rep.* 62: 667-671.

Richter, J., Augustin, W., and Kleinhempel, H. (1977). Nachweis des kartoffel-S-virus mit hilfe des ELISA-testes. *Arch. Phytopathol.* 13: 289-292.

Rochow, W. F. (1979a). Comparative diagnosis of barley yellow dwarf by serological and aphid transmission tests. *Pl. Dis. Rep.* 63: 426-430.

Rochow, W. F. (1979b). Field variants of barley yellow dwarf virus, detection and fluctuation during twenty years. *Phytopathology* 69: 655-660.

Rochow, W. F., and Carmichael, L. E. (1979). Specificity among barley yellow dwarf viruses in enzyme immunosorbent assays. *Virology* 95: 415-420.

Rowhani, A., and Stace-Smith, R. (1979). Purification and characterization of potato leafroll virus. *Virology* 98: 45-54.

Ruitenberg, E. J., and Brosi, B. J. M. (1978). Automation in enzyme immunoassay. *In* "Quantitative Enzyme Immunoassay" (E. Engvall and A. J. Pesce, eds.), pp. 63-72. Vol. 8, Suppl. 7, *Scand. J. Immunol.* Blackwell Sci. Pubs., Ltd., Oxford.

Saillard, C., Dunez, J., Garcia-Jurado, O., Nhami, A., and Bové, J. M. (1978). Détection de *Spiroplasma citri* dans les agrumes et les prevenches par la technique immunoenzymatique–ELISA. *C. R. Acad. Sci.* Ser. D, 286: 1245-1248.

Salazar, L. F. (1978). Simultaneous diagnosis of potato viruses by enzyme-linked immunosorbent assay (ELISA). *Abstracts 3rd Int. Cong. Plant Path.*, Munchen, p. 17.

Salazar, L. F. (1979). Applicacion de la tecnica serologica conjugados enzymaticos (ELISA) para diagnosticar virus de la papa. *Fitopatologia* 14: 1-9.

Saunders, G. C. (1979). The art of solid-phase enzyme immunoassay including selected protocols. *In* "Laboratory and Research Methods in Virology and Medicine, Vol. 3. Immunoassays in the Clinical Laboratory (R. M. Nakamara, W. R. Dito, and E. S. Tucker, III, eds.), pp. 99-118. Alan R. Liss, N.Y.

Saunders, G. C., Campbell, S., Sanders, W. M., and Martinez, A. (1979). Automation and semiautomation of enzyme immunoassay instrumentation. *In* "Laboratory and Research Methods in Virology and Medicine, Vol. 3. Immunoassays in the Clinical Laboratory (R. M. Nakamara, W. R. Dito, and E. S. Tucker III, eds.), pp. 111-138. Alan R. Liss, N.Y.

Schuurs, A. H. W. M., and van Weemen, B. K. (1977). Enzyme immunoassay. *Clin. Chim. Acta* 81: 1-40.

Sharafi, Y., and Bar-Joseph, M. (1980). Unpublished.

Shephard, J. F. (1972). Gel-diffusion methods for the serological detection of potato virus X, S and M. *Montana State Univ. Bull.* No. 662, Bozeman. 72 pp.

Shepherd, R. J. (1972). Transmission of viruses through seeds and pollen. *In* "Principles and Techniques in Plant Virology" (C. I. Kado and H. O. Agrawal, eds.), pp. 267-292. Van Nostrand-Reinhold Co., New York.

Stein, A., Loebenstein, G., and Koenig, R. (1979). Detection of cucumber mosaic virus and bean yellow mosaic virus in gladiolus by enzyme-linked immunosorbent assay (ELISA). *Pl. Dis. Rep.* 63: 185-188.

Studenroth, J. C. (1979). Some aspects of the epidemiology of maize dwarf mosaic in New York. Ph. D. Thesis, Cornell University.

Thresh, J. M., Adams, A. N., Barbara, D. J., and Clark, M. F. (1977). The detection of three viruses of hop (*Humulus lupulus*) by enzyme-linked immunosorbent assay (ELISA). *Ann. Appl. Biol.* 87: 57-65.

Tsuchizaki, T. A., Sasaki, A. A., and Saito, Y. (1978). Purification of citrus tristeza virus from diseased citrus fruits and the detection of the virus in citrus tissues by fluorescent antibody techniques. *Phytopathology* 68: 139-142.

van Regenmortel, M. H. V. (1978). Applications of plant virus serology. *Ann. Rev. Phytopathol.* 16: 57-81.

van Weemen, B. K., and Schuurs, A. H. M. W. (1971). Immunoassay using antigen-enzyme conjugates. *FEBS Letters* 15: 232-236.

Voller, A. (1978). The enzyme-linked immunosorbent assay (ELISA). *La Ricerca* 7: 289-297.

Voller, A., and Bidwell, A. D. (1977). Enzyme immunoassays and their potential in diagnostic virology. *In* "Comparative Diagnosis of Viral Diseases, Vol. II. Human and Related Viruses" (E. Kurstak, ed.), Part B, pp. 449-457. Academic Press, New York, San Francisco, London.

Voller, A., Bidwell, D., Huldt, G., and Engvall, E. (1974). A micro plate method of enzyme-linked immunosorbent assay and its application to malaria. *Bull. World Health Organ.* 51: 209-211.

Voller, A., Bartlett, A., Bidwell, D. E., Clark, M. F., and Adams, A. N. (1976). The detection of viruses by enzyme-linked immunosorbent assay (ELISA). *J. Gen. Virol.* 33: 165-167.

Voller, A., Bidwell, D. E., and Bartlett, A. (1977). The enzyme-linked immunosorbent assay (ELISA). Flowline Pubs., Guernsey, Europe. 48 pp.

Voller, A., Bartlett, A., and Bidwell, D. E. (1978a). Enzyme immunoassays with special reference to ELISA techniques. *J. Clin. Pathol.* 31 (6): 507-520.

Voller, A., Bartlett, A., and Bidwell, D. E. (1978b). The use of the enzyme-linked immunosorbent assay in the serology of viral and parasitic diseases. *In* "Quantitative Enzyme Immunoassay" (E. Engvall and A. J. Pesce, eds.), pp. 125-129. Vol. 8, Suppl. 7, *Scand. J. Immunol.* Blackwell Sci. Pubs., Ltd., Oxford.

Wicker, R., and Avrameas, S. (1969). Localization of virus antigens by enzyme-labeled antibodies. *J. Gen. Virol.* 4: 465-471.

Wigglesworth, V. B. (1955). The contribution of pure science to applied biology. *Ann. App. Biol.* 42: 34-44.

Wisdom, G. B. (1976). Enzyme-immunoassay. *Clin. Chem.* 22: 1243-1255.

Chapter 3

MYCOPLASMALIKE ORGANISMS AND PLANT DISEASES IN EUROPE

Pătru G. Ploaie

Laboratory of Phytopathology
Plant Protection Institute
Academy of Agricultural and Silvicultural Sciences
Bucharest, Romania

Copyright © 1981 by Academic Press, Inc.
ISBN 0-12-470240-6

3.1 INTRODUCTION

The detection in 1967 of two classes of plant pathogens—viroids and mycoplasmalike organisms (MLO)—has started new chapters in the history of plant
pathology. The new directions of investigation have shown that numerous plant
diseases, assumed to be of viral etiology, are actually associated with viroids or
MLO. For example, several diseases like spindle tuber, exocortis disease of citrus,
chrysanthemum stunt, pale fruit of cucumber or cadang-cadang of coconut palm
are now known to be caused by free low molecular weight RNAs (Diener and
Raymer, 1967; Semancik and Weathers, 1972; Selydiyko and Reifman, 1978).

The presumed viral etiologies of most of the so-called yellows and witches'
broom-type diseases have been increasingly questioned by findings of mycoplasma-
like bodies in sieve elements of yellows infected plants and by the therapeutic effect of tetracycline antibiotics used in Japan (Doi *et al.*, 1967; Ishiie *et al.*, 1967;
Nasu *et al.*, 1967), U.S.A. (Granados *et al.*, 1968; Maramorosch *et al.*, 1968),
and Europe (Giannotti *et al.*, 1968a,b; Ploaie, 1968a,b; Ploaie *et al.*, 1968).

During the past 12 years, in more than 100 plant diseases (from different continents) with symptoms of chlorosis, little leaf, dwarfing, green petal, proliferation, phyllody or decline, organisms resembling mycoplasmas have been detected
by electron microscopy in the phloem of diseased plants or in the insect vectors.
Many papers which suggest that MLOs might be the causal agents of plant diseases have already been reviewed (Ploaie, 1968c; 1971a,d, 1972, 1973a, 1977;
Casper, 1969; Giannotti *et al.*, 1969b; Shikata *et al.*, 1969; Maramorosch, 1974,
1976, 1979; Maramorosch *et al.*, 1968, 1970, 1975; Bos, 1970; Horvath, 1970;
Quoirin *et al.*, 1970; Whitcomb and Davis, 1970; Davis and Whitcomb, 1971;
Hull, 1971, 1972; Kleinhempel *et al.*, 1971; Belli *et al.*, 1972; Hampton, 1972;
Vago and Giannotti, 1972; Bjφrn, 1973; Whitcomb, 1973; Caudwell, 1978).

These reports have stimulated a massive effort to isolate and cultivate the causal agents on artificial media. Consequently, isolates similar to the species *Acholeplasma laidlawii* were obtained in Europe from white clover plants with symptoms of phyllody (Giannotti *et al.*, 1972; Kleinhempel *et al.*, 1972), but these
isolates were not pathogenic to plants.

Electron microscopy investigation revealed helical motile filaments in the
juice extracted from corn plants infected with the corn stunt agent, which led to
the introduction of the term "spiroplasma" in the literature (Davis *et al.*, 1972;
Davis and Worley, 1973). Similar helical mycoplasmas isolated from stubborn-
diseased citrus materials collected in Morocco were well characterized, and a new
genus and species—*Spiroplasma citri*— was proposed (Saglio *et al.*, 1973). This
first isolate, known as *Morocco R8-A2* and deposited in ATCC as 27556, was
followed by a second isolate, *California strain* (C.189) ATCC 27563 (Fudl-Allah
and Calavan, 1973). An identical spiroplasma was isolated from citrus with little-
leaf disease from Israel and grown in a cell-free medium (Markham *et al.*, 1974),
while a new isolate tentatively named *Spiroplasma kunkelii* (ATCC 29594),

slightly different from *S. citri*, was obtained from the ornamental cactus *Opuntia tuna monstrosa* (Kondo *et al*., 1976) and from lettuce with symptoms of aster yellows, the latter one being denominated *S. kunkelii* var. *callistephii* ATCC 29747 (Maramorosch and Kondo, 1978).

During the past decade, spiroplasmas have also been isolated from Bermuda grass (Chen *et al*., 1977), honeybees and spring flowering trees (Davis *et al*., 1976; Davis *et al*., 1977; Clark, 1977; Davis, 1978a,b), rice plants with yellow-dwarf symptoms and green leaf bugs (*Trigonotylus ruficornis*) such as isolates LB-10 and LB-12, the latter considered to be different from *S. citri* (Su *et al*., 1978).

The isolation of mycoplasmas and spiroplasmas from non-surface sterilized floral elements of plants which had no MLOs in sieve elements suggests that certain non-plant-pathogenic spiroplasmas and mycoplasmas are harbored externally on flowers (Davis, 1978a). These new data may change the opinion that, in the case of plants, mycoplasmas and spiroplasmas are phloem restricted.

Structures resembling rod shaped viruses were also electromicroscopically detected on clover dwarf, stolbur, aster yellows and clover phyllody MLOs (Ploaie 1968b, 1971b; Allen, 1972; Vago and Giannotti, 1971; Gourret *et al*., 1973). These virus-like particles were similar to those isolated from *Acholeplasma laidlawii* (Gourlay *et al*., 1971; Liss and Maniloff, 1971. Cultured *S. citri* is infected by a rod-shaped virus and by bacteriophages of a classical type B morphology, a condition unknown among mycoplasmas (Cole *et al*., 1973; Cole, 1978).

Electron microscopic investigations have also shown that rickettsia-like organisms (RLOs) are associated with several plant diseases, as reviewed by Whitcomb (1973), Maramorosch (1974) and Maramorosch *et al*. (1975).

From 1967 until today our knowledge about this new group of plant pathogenic agents has broadened. The number of papers has grown from about 10 in 1968 to 852 in 1974, and 1,813 in 1977. Several books (Ploaie, 1973a; Grunewaldt -Stöcker and Nienhaus, 1977; Whitcomb and Tully, 1979) and two lists of bibliographies on the role of MLOs in plant disease are available now (Müller *et al*., 1975; Elmendorf, 1977).

3.2 THE ROLE OF MLOS IN PLANT DISEASE

3.2.1 Early Researches on Yellows Diseases in Europe

The aster yellows-type diseases have certainly been present in Europe for a long time. The modification of flower morphologies was first studied only by botanists or naturalists. Long ago, such terms as virescence or phyllody, liable to be interpreted as symptoms caused by MLOs, have been reported in botanical literature. Thus, Ulisse Aldrovandi, a professor in Bologna, Italy was the first to describe *Aquilegia* virescence, in his "Monstruorum Historia" published in 1642. White clover phyllody was observed in Great Britain as early as the 17th century by Merrett (1666), quoted by Carr and Large (1963). Later, Augustin Pyrame de Candole (1778-1841), famous botanist and professor at Montpellier and Geneva, as well as the author of a vast work about the planet's flora, described white clover plants with virescence and phyllody as a new variety named *Trifolium repens* L. f. *viviparum*. He observed that virescent flowers could be easily rooted to give rise to new plants in contact with humid soil, a phenomenon specific only to alsike or Ladino clover infected with clover phyllody agent, as was demonstrated experimentally in U.S.A. (Kreitlow, 1963) and in Italy by Grancini (Bos and Grancini, 1965).

In the herbarium of the Popular Museum of Prague, Blattný (1958) found several specimens of *Anagallis arvensis* L., collected by Opitz in 1817, Knaf in 1837 and Čelakowský in 1870, with reduced and virescent flowers. These abnormal flowers were described as variations (Opitz: var. *corollis minoribus*; Knaf: var. *parviflora*) or monstrosities (Čelakowský: *viridiflora*). Experimental transmission of stolbur and related diseases of aster yellows-type to *Anagallis arvensis* L. proved the infectious nature of such modifications (Blattný, 1960).

Plants of *Alliaria officinalis* L. with different abnormalities had been collected by Suringar (1882) at Leyden, the Netherlands. The infectious nature of these modifications (virescence, abnormal growth, monstrosities, phyllody) was first suspected by Hugo de Vries (1896), on the basis of an epidemic of virescence discovered in 27 species of plants in his Amsterdam garden. The paper of De Vries remained unknown to plant pathologists until 1957 when Bos called attention to it.

Delacroix (1908), in his book of plant pathology gave a detailed description of several monstrosities of flower organs of *Trifolium* and *Pelargonium*, which take on the shape and functions of leaves. He called these modifications "virescence" or "chloranthie," and the phenomenon "regressive metamorphosis," which botanists often use to support the argument that flowers evolved from leaves. The virescence or phyllody of white clover was also described as a teratological phenomenon (Penzig, 1921). In Romania and in other European countries several botanists have also considered these morphological deviations to be a teratological phenomenon, as reviewed by Ploaie (1972), ignoring the possible role of pathogens. More than 400 species of plants from the spontaneous flora in Europe have such malformations (abnormal phytomorpha, terata, monstrosities); a detailed morphological (botanical) analysis of these phenomena, accompanied by a list of species, was made by Dihoru and Dihoru (1973).

In 1902, Smith described aster yellows in U.S.A. This disease became the prototype of the yellows-type diseases after Kunkel (1926) reported that he had experimentally transmitted aster yellows virus with *Cicadula divisa* Uhl., first to 170 species in 150 genera belonging to 38 families of plants, and later to 300 different species belonging to 48 families of plants (Kunkel, 1953). Kunkel's hypothesis (1926) that aster yellows disease is caused by a virus transmitted by leafhoppers has increased the interest of European plant pathologists in the study of several symptoms in various crops, e.g., premature wilting of potato, pepper and egg plant, phyllody of clover and tobacco, and stolbur of tomato or witches' broom of potato.

The stolbur disease was first recognized in Georgia, USSR, and subsequently in the Crimea (Pilipenko, 1959) and in Bavaria in 1927-1928 (Böning, 1958). Symptoms of virescence and proliferation of tobacco, assumed to be of an infectious nature, were recorded in Romania (Ghimpu, 1931). A similar symptom called "female sterility" of tobacco came to the attention of pathologists in the USSR (Kostoff, 1933) in the same year as another disease, "fruit woodiness of tomato" (Ryshkov *et al.*, 1933), both transmissible by grafting and found to be identical with the stolbur disease.

Extensive investigations have been carried out on stolbur disease in the Soviet Union and *Hyalesthes obsoletus* Sign., has been established as its vector (Sukhov and Vovk, 1946; 1949). Further detailed studies have shown that stolbur is a devastating disease for potato, tomato, tobacco and other cultivated species of plants in southeastern Europe, e.g., Czechoslovakia (Valenta *et al.*, 1961), Romania (Săvulescu and Pop, 1956; Ploaie, 1960; Săvulescu and Ploaie, 1964; 1967a), Bulgaria (Kovachevsky *et al.*, 1964), Yugoslavia (Panjan, 1950), Italy (Ciccarone, 1951), France (Cousin and Grison, 1966a), Spain (Orad and San Roman, 1954), and in other countries.

Aster yellows, observed and studied in central and southern Europe, is very similar by its symptoms to the American disease (Heinze and Kunze, 1955; Albouy *et al.*, 1967; Ploaie, 1969a,b; Ploaie, 1971c).

In the same yellows-type group was also included virescence and phyllody of clover, identical with green petal of strawberry, known to occur widely in England (Frazier and Posnette, 1957), the Netherlands and Italy (Bos and Grancini, 1965), France (Albouy *et al.*, 1967), Czechoslovakia (Musil and Valenta, 1958), Romania (Săvulescu and Ploaie, 1961; Ploaie, 1966, 1969a, 1970), Belgium (Tahon, 1963), Switzerland (Bovey, 1957), Germany (Kreczal, 1960) and Scandinavia (Tapio, 1970).

The yellows-type disease group was enriched by adding to it potato witches' broom, which can easily be distinguished from stolbur through a greater tendency to flower proliferation, virescence and phyllody in tomato (Valenta, 1956), aerial stolons, and profuse development of axillary buds at the top of potato plants to give an erect growth (Todd, 1958). Witches' broom of potato was first reported in Russia by Jaczewski (1926) and more extensively studied in Czechoslovakia and Scotland, but the position of this entity among other yellows-type diseases in Europe is not clear. Symptoms of witches' broom were distinguished in plants with clover phyllody as a new entity—clover witches' broom disease (Frazier and Posnette, 1957). Also, the witches' broom of wooden plants, e.g., apple proliferation (Rui, 1950), witches' broom of *Vaccinium* sp. (Bos, 1960), and apple rubbery wood (Beakbane *et al.*, 1971) are considered as distinct, different diseases.

In 1958, a new disease named Crimean yellows, which induced flower proliferation of dodder, was described (Valenta, 1958). Later, two diseases, clover dwarf and parastolbur, differing from stolbur and from each other, yet closely resembling, by their symptoms in *Vinca rosea*, the European aster yellows described by Heinze and Kunze (1955) and Ploaie (1969a,b), were studied in detail in Czechoslovakia (Valenta and Musil, 1963).

In the literature, this group of diseases has been reported in several combinations: southern or typical stolbur, northern stolbur, synonymous with potato witches' broom, pseudo-classic stolbur, different from parastolbur, and metastolbur or virescence of clover (Valenta *et al.*, 1961). New diseases such as stolbur C, M, SM and P were described (Marchoux *et al.*, 1969), but these strains were not compared with the typical stolbur over a large host range. Several diseases with symptoms of chlorosis or decline, such as grape yellows (flavescence dorée) (Caudwell, 1964), pear decline (Bălăşcuţă *et al.*, 1978), apoplexy of apricots (Morvan *et al.*, 1973) and plum decline (Ionică *et al.*, 1978), must be included in this group.

The relationships of the various European aster yellows type diseases to their host plants and vectors have been the objects of rather extensive comparative studies (Valenta *et al.*, 1961; Valenta and Musil, 1963; Albouy *et al.*, 1967; Cousin *et al.*, 1968; Săvulescu and Ploaie, 1967a; Ploaie, 1969a, 1971c). These diseases have several characteristics, such as symptomatology (virescence, phyllody, proliferation, witches' broom, chlorosis) and pathogen transmission (grafting, dodder, leafhoppers), in common with similar diseases described in North America, Japan, Africa and Australia.

Although they are considered to be "yellows type diseases," known in Europe by the name of "European yellows type diseases" (Valenta *et al.*, 1961), this term is not proper, because chlorosis is not a recurrent symptom (Kunkel, 1931). The most frequent symptoms are virescence, phyllody, reduction of leaves, proliferation or dwarfing (Ploaie, 1969a, 1971c, 1973a). The name of "witches' broom diseases" proposed by Bos (1957), and that of "proliferating diseases," introduced by me (Ploaie, 1973a), seem to be more appropriate.

The characterization of causal agents by techniques such as enzyme-linked immunosorbent assay (ELISA), DNA homology studies, or electrofocusing-electrophoretic properties of proteins in polyacrylamide gels will improve our knowledge of the number of such agents in Europe and other parts of the world.

3.2.2 Proofs for MLOs as Plants Pathogens

In spite of the earlier assumption that aster yellows type diseases were induced by viruses, the morphological, antigenic, biophysical and biochemical properties of the so-called yellows type viruses and, consequently, the degree of their inter relatedness have remained practically unknown for more than 60 years.

Difficulties associated with aster yellows "virus" filtration through Berkefeld V and N filters and its purification, and sedimentation of the agents after 1 min at 500 rpm led Black (1943) to the conclusion that the infectious agent had particles larger than those of known plant viruses. Lee and Chiykowski (1963), attempting purification of the pathogen from the leafhopper *Macrosteles facifrons* (Stål) after centrifugation of insect extract at 15,000 rpm and Steere (1967), using agar gel filtration, arrived at the same conclusion. These findings, together with the *in vitro* instability of sedimented agent (Black, 1955), were the first proofs that aster yellows agent had properties unlike those of viruses. Nevertheless, virologists in Europe and North America were intent on seeing viruses in plant tissues and not any other microorganisms, overlooking the possibility of an etiology different from the presumed viral one.

In 1955, Vovk and Nikiforova and later Protsenko (1958b) showed that in the homogenates of different plants with stolbur or aster yellows symptoms, particles with a diameter of 50 nm did exist. Because the authors did not work with purified preparations or with experimentally infected plants, their results were questioned (Maramorosch, 1960; Valenta *et al.*, 1961).

Numerous attempts have been made by us since 1964 to purify aster yellows, stolbur and clover phyllody agents isolated in Romania. Large heterogenous particles 50-110 nm, with a high degree of instability *in vitro* have been purified from plants of *Nicotiana rustica* L., *N. tabacum* L., *Lycopersicum esculentum* Mill., *Vinca rosea* L., *Cuscuta campestris* Yunck., and *Sinapis alba* L. experimentally infected with stolbur and clover phyllody. A high concentration of particles was found in the interphase after precipitation with chloroform and butanol. The particles appeared slightly compressed from the top downwards or collapsed, followed by aggregation (Ploaie, 1973a). The images obtained by us were similar to those obtained with mycoplasmas from animals (Lemcke, 1972).

In 1965, electron micrographs of ultrathin sections through *Vinca rosea* L. infected with aster yellows revealed typical particles resembling MLOs, although the resolution of the electron microscope used by us (table EM Tesla 242A) was poor. The particles of clover phyllody purified from *Sinapis alba* L. and interpreted by me as viruses, were 90 nm in diameter and after negative staining showed an outer membrane (Ploaie, 1967). No such particles were found in healthy plants.

The idea that the causal agent of aster yellows type diseases is a virus was too strongly rooted in the minds of scientists. Consequently, all materials that showed particles other than viruses were considered inconclusive.

After carefully analyzing Doi's micrographs of thin sections through several species of plants infected with yellows agents (Doi *et al.*, 1967), Dr. Koshimizu, a microbiologist from the Institute of Veterinary Sciences in Tokyo, was the first to recognize their resemblance to *Mycoplasma gallisepticum* and to incriminate mycoplasmas in the etiology of plant diseases (Maramorosch, 1976; Leeuw, 1977). Koshimizu suggested that his colleagues try tetracycline treatment to induce a cure or remission of disease symptoms. The presence of particles resembling mycoplasmas and remission of symptoms (Doi *et al.*, 1967; Ishiie *et al.*, 1967) suggested a possible new etiology.

The strong evidence in favor of this new hypothesis, obtained between 1967 and 1968 by a group of virologists, including myself, under the guidance of Dr. Karl Maramorosch at the Boyce Thompson Institute in Yonkers, N.Y., questioned Kunkel's viral hypothesis advanced at this institute forty years earlier. Although our effort to purify the corn stunt agent failed, particles resembling mycoplasmas were found in thin sections of corn plants infected with the corn stunt agent, as well as in *Nicotiana rustica* L. and *Callistephus chinensis* Nees. experimentally infected with the aster yellows agent (Granados *et al.*, 1968; Maramorosch *et al.*, 1968). We also found mycoplasma-like organisms in several samples of periwinkle plants experimentally infected with the aster yellows agent in Romania; I took this sample with me to Boyce Thompson Institute (November, 1967) to continue the investigations started at the Institute of Biology of the Romanian Academy. For the first time MLOs were detected in plants with European aster yellows, stolbur, parastolbur, clover dwarf, Crimean yellows and *Vinca* proliferation (aster yellows). These results were reported by Maramorosch *et al.* (1968) at a meeting of the Division of Microbiology of the New York Academy of Sciences on March 22, 1968 (Maramorosch *et al.*, 1968) and, subsequently, by Ploaie *et al.* (Ploaie, 1968a,b; Ploaie *et al.*, 1968; Ploaie and Maramorosch, 1969). Our results, together with those of Giannotti *et al.*, (1968a,b) and Maillet *et al.* (1968) on clover phyllody, presented to the French Academy at the end of April and in May 1968, supported the hypothesis of Doi *et al.* (1967). Electron microscopical investigations in Japan, USA and Europe played an important role in substantiating this new field of research.

The following evidence supports the theory that MLOs are the causal agents:

a. Detection of MLOs in the phloem tissue of different species of plants showing the same disease symptoms, in distinct climatic zones of Europe and other continents. Examples are numerous—one being clover phyllody association with MLOs as demonstrated in Romania, East Germany, France, Italy, Czechoslovakia, the Netherlands and Scandinavia. The same symptom has also been associated with MLOs in Canada and Australia. A similar situation applies to stolbur, a European disease related to big-bud of tomato plants and also associated with MLOs in South America and Australia. Such findings strongly suggest MLO etiologies for these different diseases.

b. The presence of MLOs in the insect vectors that transmit the presumptive causal agents of the diseases. The presence of same type of agent in both vectors and plants and its absence in healthy ones is rather convincing evidence, which cannot be neglected in assessing etiology.

c. Detection of MLOs in sieve elements of different species of dodder used for experimental transmission of the disease agents.

d. The presence of MLO in wild plants, e.g., *Convolvulus arvensis* L. or *Trifolium repens* L., known to serve as natural host plants (reservoirs) for stolbur or clover phyllody disease agents.

e. Temporary suppression of disease symptoms following treatment of plants with tetracycline antibiotics.

f. Suppression of the vectors' capacities to transmit the disease agents by feeding or injection with tetracycline antibiotics.

g. Recovery of diseased plants after thermotherapy.

h. Reproduction of the disease using MLOs isolated in axenic culture as inoculum of insect vectors.

i. Reisolation of MLOs from such inoculated and diseased plants.

3.2.3 Proliferating Diseases Associated with MLO

Several European proliferating diseases, known for a long time, have been characterized by their symptomatology, as well as by the transmissions and host-plant ranges of their causal agents. From 1967 until now, all these diseases (Table I) have been found to be associated with MLOs (Ploaie, 1973a).

Structures resembling MLOs have also been observed by electron microscopy in numerous cultivated plants or weeds in Europe showing symptoms of dwarfing, little leaf, witches' broom or proliferation (Ploaie et al., 1977; Kleinhempel and Müller, 1978; Ploaie, 1978). The association of these symptoms to one of the better known diseases remains obscure. In Romania, the following symptoms

TABLE I.   Proliferating Diseases Associated with MLOs in Europe

Disease	Host Plants	Reference*
Stolbur	*Vinca rosea*	Ploaie, 1968a
Parastolbur	*Vinca rosea*	Ploaie, 1969a; Ploaie and Maramorosch, 1969
Clover dwarf	*Vinca rosea*	Ploaie and Maramorosch, 1969
Aster yellows	*Vinca rosea*	Ploaie, 1968a,b
Crimean yellows	*Vinca rosea*	Ploaie and Maramorosch, 1969
Witches' broom of potato	*Solanum tuberosum*	Harrison and Roberts, 1969
Clover phyllody	*Trifolium repens*	Giannotti et al., 1968a
Green petal of strawberry	*Fragaria* sp.	Cousin et al., 1970
Mal azul	*Lycopersicum esculentum*	Borges and David-Ferreira, 1968
Onion proliferation	*Allium cepa*	Petre and Ploaie, 1971
Rice yellows	*Oryza sativa*	Pellegrini et al., 1969
Wheat chlorosis	*Vinca rosea*	Ploaie et al., 1975
Oats pseudorosette	*Avena sativa*	Fedotina, 1977
Grape yellows	*Vitis vinifera*	Caudwell et al., 1971
Apple proliferation	*Malus sylvestris*	Giannotti et al., 1968b
Plum yellows	*Prunus domestica*	Ionică et al., 1978
Apoplexy of apricot	*Vinca rosea*	Morvan et al., 1973
Pear decline	*Pyrus communis*	Blattný and Váňá, 1974
Apple rubbery wood	*Malus domestica*	Beakbane et al., 1971
Apple chat fruit	*Malus domestica*	Beakbane et al., 1971
Blueberry witches' broom	*Vaccinium myrtillus*	Kegler et al., 1973
Rubus stunt	*Rubus* sp.	Murant and Roberts, 1971
Black currant reversion	*Ribes* sp.	Silvere, 1970

*First detection of MLO by electron microscopy.

have constantly been associated with MLOs: little leaf *Tropaeolum majus* L., proliferation of leek, green petal and proliferation of *Dianthus barbatus* L. and *D. caryophyllus* L., proliferation of *Phlox drummondii* Hook., proliferation of *Gaillardia pulchella* Fouh., chlorosis and degeneration of grape, virescence of *Cichorium inthybus* L., proliferation of *Aquilegia* sp., virescence of *Gladiolus* spp., green petal of strawberry, malformation of *Ammi visnaga* L., and proliferation of apple (Ploaie *et al.*, 1977; Ploaie, 1978).

In East Germany, MLOs have been detected in the following species of plants: *Allium cepa* L., *Anagalis arvensis* L., *Aquilegia vulgaris* L., *Brassica oleracea* L., *Descurainia sophia, Gaillardia pulchella* Fouh., *Hydrangea macrophylla* L., *Lactuca sativa* L., *Limonium sinuatum* L., *Malus domestica* L., *Nigella damascena* L., *Papaver somniferum* L., *Phlox drummondii* Hook., *Phlox paniculata* L., *Primula* sp., *Raphanus sativus* L., *Rubus idaeus* L., *Rudbeckia purpurea* L., *Stelaria media* L., *Thlaspi arvense* L., *Vaccinium myrtillus* L., *Pseudolysimachium spicatum* L. (Kleinhempel and Müller, 1978).

MLOs were found in gladiolus and hyacinth plants with symptoms of grassy top or "Lissers" in the Netherlands (van Slogteren, 1974), *Alisma plantago* L., *Leersia oryzoides* L. and *Panicum* Crus-Galli affected by giallume in Italy (Amici *et al.*, 1973), citrus plants with symptoms of stubborn, and tobacco with floral regression in Spain (Borges and Martins, 1972; Gimenez, 1975). In France, particles resembling MLOs were also found in lavenders with yellows symptoms (Cousin *et al.*, 1971), in *Opuntia subulata "minor"* Engelm. (Delay and Darmananden, 1973) and *Daucus carota* L., with symptoms of proliferation (Giannotti *et al.*, 1974a).

In the Soviet Union, MLOs were detected in *Mentha piperita* L. and *Erysimum plantain* L. with symptoms of witches' broom (Surgucheva and Protsenko, 1971; Protsenko and Surgucheva, 1973), grape affected by *"Marbour"* (Milkus, 1974) and *Orobanche aegyptiaca* L. which had parasitized stolbur infected tomato plants (Fedotina, 1973).

Numerous other plants were also found associated with MLOs, e.g., *Dactylorhiza majalis* L. (Marwitz and Petzold, 1972), *Linum austriacum* L. (Westphal and Heitz, 1971), *Opuntia tuna "monstrosa"* (Lesemann and Casper, 1970), species of *Helenium, Cirsium* and *Epilobium* (Begtrup, 1975; Begtrup and Thomsen, 1975). Also suspected was *Humulus* sp. with symptoms of necrotic mosaic (Chod *et al.*, 1974), although micrographs were not convincing.

3.3 SYMPTOMATOLOGY

3.3.1 External Symptoms

The disorders and abnormalities caused by MLOs in different species of plants under natural and experimental conditions are of a systemic nature, several symptoms being characteristic only of diseases caused by these prokaryotic plant pathogens. The severity of symptoms depends on the virulence of the MLO strains and the species of infected plants.

The abnormal growth and development caused by MLOs are a consequence of the interference of the pathogenic agent with normal plant morphogenesis. MLOs develop in the sieve tubes and consume the substances translocated from the leaves to different parts of the plant, especially sucrose and indole-3-acetic acid (IAA). A high accumulation of MLOs in phloem cells can physically block normal translocation in sieve tubes.

The decreases in sucrose and indole acetic acid (IAA) concentrations have a great influence on replication of DNA and protein synthesis, resulting in hormonal imbalances and modifications of biosynthetic pathways. First affected by these changes are the flowers (Fig. 1), which develop virescence and sterility. The petals, if formed, become green through the development of chloroplasts in these organs that are normally colored by non-green pigments. This process is called virescence, and flowers with green petals fail to produce seeds. For crops cultivated especially for seeds, such as cereals or clover, MLO infections could be disastrous. A second symptom is transformation of virescent petals and other floral organs such as sepals, bracts and ovaries into leaves, a phenomenon commonly known as phyllody or antholysis (Fig. 1b; Fig. 2b and c). The ovary may proliferate into leaves and clusters of phylloid flowers may appear in one flower (Fig. 1d; Fig. 3b), while the entire inflorescence shows strong proliferation (Fig. 1d; Fig. 3b). When affected by MLOs, the normal inflorescence of sessile flowers of numerous species of plants bears green flowers with remarkably elongated flower stalks measuring 2-7 cm (Fig. 1d).

A strong malformation of flowers with a tumor-like growth of inflorescences was detected by us in *Ammi visnaga* L. (Fig. 4). An enormous mass of material is accumulated in the malformed inflorescence which drops down.

Reduction of leaves is another symptom observed in many species of plants (Fig. 3c and d). On the stems of several species of plants affected by MLOs, rosette-like leaf clusters are formed. Many of these symptoms can easily be produced in periwinkle plants experimentally infected with any of several MLO strains (Fig. 2).

The aforementioned symptoms are often accompanied by other degenerative processes such as stunting, wilting and leaf yellowing, with the last one being more specific to MLO infected grapevines, peach trees and cereals.

In Europe, the symptomatologies of suspected or known MLO-associated diseases have been described in detail and illustrated (Valenta *et al.*, 1961; Valenta and Musil, 1963; Bos and Grancini, 1965; Ploaie, 1966, 1969a, 1971c, 1973a; Albouy *et al.*, 1967; Ploaie *et al.*, 1975, 1977). Several diseases were named according to the typical symptom produced in host plants. Thus, changes in growth (dwarfing and stunting) are characteristic of clover dwarf disease in Europe, corn stunt in America. Modifications of plant color are known for such diseases as

FIG. 1. Different types of symptoms experimentally induced by MLO at floral level: a) big bud and hypertrophy of calix of stolbur infected tomato plant; b) virescence and phyllody caused by clover phyllody mycoplasma at white clover; c) symptoms of green petal of strawberry; d) proliferation and phyllody of onion; e) malformation of *Sinapis alba* L. flowers caused by clover phyllody mycoplasma. (Original.)

FIG. 2. Serial transformations of *Vinca rosea* plant after inoculation with onion proliferation mycoplasma: a) normal plant; b) diseased plant, 3 months after inoculation by grafting; c) transformation of normal (left) flower to leaves; c) reduction and chlorosis of the leaves. (Original.)

FIG. 3. Proliferation of flowers and reduction of leaves caused by MLO: normal inflorescence a) and strongly proliferated flower b) after infection of *Gaillardia pulchella*; c) reduction of the leaves of nasturtium; d) reduction of the leaves of tomato plant after inoculation with aster yellows mycoplasma. (Original).

rice yellow dwarf in southern Asia, purple top of potato, and blue disease (mal azul) of tomato (Ploaie, 1973a).

Experiments with European strains of MLOs involving a large host range, have shown that the symptoms caused by a single MLO strain may differ widely from one host plant species to another. According to our experience, several types of symptoms can be distinguished: a) premature wilting and death (as in stolbur-infected potato, pepper and eggplants); b) flower changes (e.g., virescence, phyllody, proliferation, antholysis and tumor-like growths caused by clover dwarf, parastolbur, aster yellows and wheat chlorosis), and c) chlorosis and die back (as in pear decline and apricot apoplexy). Moreover, the absence of symptoms, although MLOs are present in phloem cells, has been reported for *Nicotiana glauca* Grah. (Valenta *et al.*, 1961; Ulrychová and Limberk, 1967). Therefore, for the identification of MLO strains, the appropriate indicator host plants should be used (Valenta *et al.*, 1961; Valenta and Musil, 1963; Ploaie, 1973a).

3.3.2 Histological Alterations

Abnormal internal modifications of tissues in infected plants, consisting mainly of irregularities in the phloem, were first observed by Michailova (1935) in tomato plants with symptoms of stolbur.

Fluorescence microscopy was later used to demonstrate the presence of internal lesions in different species of plants affected by yellows diseases. These studies have revealed that in diseased plants infected with stolbur, clover phyllody, aster yellows, grapevine flavescence dorée, apple proliferation, apricot and pear decline and others, pronounced anatomical changes occur in the phloic tissue, especially in secondary phloem, e.g. hypertrophy, hyperplasia, and proliferation or necrosis of cells (Rasa and Esau, 1961; Carle, 1965; Cousin and Grison, 1966b; Kaminska, 1971; Bălășcuță *et al.*, 1979). Observations made under fluorescent light have also shown an abnormal accumulation of substances in the form of chromophilous spots stainable with methylene blue or ammoniacal gentian violet at the level of libriform bundles of plants (Cousin and Grison, 1966b). Hiruki and Shukla (1973b) have demonstrated that the phloem cells of witches' broom affected plants, showing fluorescent material, correspond to those phloem elements filled with mycoplasma cells as revealed by electron microscopy of ultrathin sections of the same plant. Simultaneously with light microscopical detection of abundant fluorescence in periwinkle plants infected with mycoplasmalike bodies of the sandal spike disease, numerous mycoplasmas were also seen in the ultrathin sections from the same phloem area by electron microscopy, which is suggestive of a positive correlation between the two phenomena (Hiruki and Djikstra, 1973a). A similar fluorescence was observed in bark samples of apple and pear trees with proliferation or pear decline symptoms caused by the presence of mycoplasmas (Seemüller, 1976).

FIG. 4. Tumorlike malformations of the inflorescence of *Ammi visnaga* caused by MLO, with total a) or partial b) aspermy. (Original.)

Another pathological feature of infected tissue is an abnormal accumulation of callose, due to infection with mycoplasmas, which contributes largely to increased fluorescence. Accumulation of callose on phloem walls, degeneration of middle lamellae, collapsed sieve elements, (abnormal in size and shape), and increased production of secondary phloem are typical mycoplasma-induced modifications of the sieve elements and phloem parenchyma (Cousin et al., 1968; Hiruki and Djikstra, 1973; Bălășcuță et al., 1978). These degenerative effects, especially cell necrosis, excessive callose formation and mycoplasma accumulation in sieve elements, are visible under a fluorescence microscope after staining with aniline blue. However, for rapid diagnosis, a direct fluorescence method is recommended (Seemüller, 1976; Leeuw, 1977) in which the fluorchrome 4', 6-diamidino-2-phenylindol (DAPI) is used for its specific reaction with nucleic acids from mycoplasmas.

Autoradiographic studies based on H^3-thymidine incorporation have shown that multiplication of the clover phyllody mycoplasma in the vector *Euscelis lineolatus Brullé* occurs particularly in the cells of the midgut and in the salivary gland. Hypertrophy of midgut cells caused by multiplication of mycoplasmas, and a correlation between the incorporation of tritiated thymidine and the presence of a Feulgen-positive cytoplasm were established by cytological studies (Maillet and Gouranton, 1971).

Cell alterations at the ultrastructural level in white clover affected by clover dwarf, e.g., an increased content of ribosomes, degeneration of mitocondria and chloroplasts, and vacuolization of cytoplasm, were detected by electron microscopy (Lombardo et al., 1970). In periwinkle plants experimentally infected with clover phyllody or clover dwarf mycoplasmas, lesion-like necrosis, obliteration of cells and formations of large numbers of protein crystals occurred progressively (Gourret, 1970; Ploaie, 1976).

Modifications of the shape of the nucleus of young sieve elements of *Phlox drummondii* Hook., showing symptoms of proliferation, were also observed (Ploaie, 1973a). In *Opuntia subulata* f.h. *"minor,"* infected with mycoplasma, the configuration of the cells is modified and takes the form of spindles or stars, the process of necrosis starting with the degradation of the plasmalemma (Delay and Dermanaden, 1973). A strong accumulation of photosynthetic starch and reduction of thylakoids in chloroplasts of stolbur-infected tomato plants followed by decreased chlorophyll content is also characteristic of mycoplasma infection (Plavšić et al., 1978).

3.3.3 Metabolic Changes

Abnormal morphogenesis and anatomical aberrations in plants experimentally infected with different isolates of mycoplasmalike organisms must be considered as a consequence of a strong interference between the pathogenic agent and the

normal biosynthetic pathways in higher plants. However, although these diseases
have been known for a long time, literature on the biochemical and physiological
status of diseased plants is rather poor. Some metabolic disturbances were found
in tomato plants infected with potato witches' broom, e.g., an increase of alkaline metals (Na and K), alkaline earth (Ca and Sr) and boron, and a decrease in
the content of aluminum, zinc, silicon, phosphorus and nitrogen fractions
(Ulrychová and Limberk, 1964). The content of ammoniacal nitrogen was 7
times higher in the stem of diseased plants than in healthy plants. In the diseased
tomato flowers the total phosphorus content was decreased by 38%, that of ammoniacal nitrogen by 54% and total nitrogen by 22% (Ulrychová and Limberk,
1964). The content of amino acids in the flowers of diseased plants was similar
to that in the stems of healthy plants with low concentration of glutamine
(Ulrychová and Limberk, 1964).

A general decrease in the nitrogen fraction was also detected in *Nicotiana
glauca* Grah., a symptomless carrier of potato witches' broom; but in this species,
as in infected tomato plants, a 67% rise of the low molecular phosphorous fraction associated with phosphorylated sugars and nucleotides, including the ATP/
ADP system, is the most important metabolic disturbance (Ulrychová and
Limberk, 1967). It should be noted that, according to Rodeia *et al.* (1973), in
tomato plants with symptoms of "mal azul" (blue sickness) in Portugal the phosphorus content in leaflets was usually lower than in healthy plants. But in discolored purplish blue infected plants, anthocyanin increased by more than 1,000
percent. A phosphorus deficiency supposedly gave rise to the increased anthocyanin content (Ulrychová and Sosnova, 1970).

A rapid translocation of indole-3-acetic acid, malic acid, succinic acid and also
of glucose, manose, and sulfate or rubidium ions was observed in sieve elements
of tobacco plants infected with aster yellows agent (Granados, 1965).

Several years ago we gave special attention to *Vinca rosea* L., one of the best
host plants for aster yellows agent, and a source of more than 50 alkaloids, several of which are used successfully in the treatment of different types of cancer.
Our investigations showed that in the roots of periwinkle plants experimentally
infected with a strain of aster yellows, the total content of alkaloids was three
times that of healthy plants. This means that MLOs can induce an increase in
alkaloid biosynthesis. However, in infected plants, synthesis of the most effica-
cious antitumor alkaloid, vinkaleukoblastine (VLB), no longer occurred (Ghiorghiu *et al.*, 1966). Our chromatographic analysis showed that the biosynthetic
pathway of VLB was totally blocked in infected plants. The increase of the total
content of alkaloids in periwinkle plants after infection with mycoplasma could
be used in the future for production of yohimbine and other alkaloids.

A disturbance of the hormonal balance also seems to take place in infected
plants. Small fruit caused by chat fruit disease of apple in England was traced to
the absence of *Malus* auxin 2 in the pulp of fruit (Luckwill, 1957).

In white clover plants, infected by clover phyllody mycoplasma, an increase
in cytokinin level (different forms of zeatine and N^6 $-(\Delta^2$ isopentenyl)-adenine)

in phylloid flowers was detected, but its concentration in the leaves was lower than in healthy plants (Gaborjányi and Sziráki, 1978). Electron microscopy of stolbur infected tomato plants that were treated with kinetin revealed a progressive degradation of the mycoplasma, a decrease of starch accumulation, as well as development of thylakoids and a rise in the content of photosynthetic pigments (Plavšić et al., 1978).

The mechanism of metabolic changes caused by mycoplasmas or spiroplasmas is still obscure, although some authors incriminate the production of toxins in this process (Daniels and Meddins, 1974).

3.3.4 Crop Losses

The plant diseases associated with mycoplasmalike organisms are the most devastating diseases of cultivated plants all over the world.

Every year in Europe, almost all cultivated plants are affected by tremendous crop losses produced by stolbur (in tomato, egg plants, potato, tobacco and lavander), apple proliferation, grape yellows, citrus stubborn, pear decline, plum yellows, apoplexy of apricots, wheat and barley chlorosis, onion proliferation, clover phyllody, green petal of strawberry, gladiolus yellows, dwarf growth of hyacinths, witches' broom disease of bilberry and raspberry, apple rubbery wood, rice yellows, mal azul of tomato, and many others (Ploaie, 1973a; Grunewald-Stöcker and Nienhaus, 1977; Leeuw, 1977; Ploaie et al., 1977).

Although few papers give information on crop losses and no methods are available for the quantitative assessment of these diseases in the field, several data will be discussed in brief. Between 1945-1946 alone, more than 50,000 trees were killed by pear decline in Italy (Refatti, 1967) and the disease is still in progress in Europe (Bâlâşcuţă et al., 1979). The sudden die-back of apricots, known as apoplexy, is the most serious disease of this species in Central Europe. In Hungary, 30-40% of the 5-6 year-old orchards were destroyed by apoplexy (Rozsnyay, 1963). A similar situation has been noticed in France and Romania.

The Central European states with fairly warm annual summers lose millions of pounds of potatoes due to stolbur. Thus, in 1956, Czechoslovakia alone recorded a 20-30% loss of potato yield or an estimated annual loss of 100,000 tons of potato, a loss equivalent to the production of about 10,000 hectars (Spaldon, 1958). In 1959, that same country lost more than 42,000 trucks of potato (84 million crowns) to stolbur (Bojňanský and Kosljarová, 1964), while seed development was 41% lower in apple trees infected with apple proliferation than in healthy trees (germination of seeds from diseased trees was reduced by 23-91%) (Seidl and Komárková, 1976). Between 1960 and 1965, cultures of *Vinca rosea* L. and *Ammi visnaga* L. introduced in Romania for pharmaceutical purposes were entirely destroyed by stolbur and aster yellows agents (Ploaie, 1969a).

Starting in 1958, concerns have been expressed at the N.A.A.S. Grassland Officer Conference over the increasing incidence of phyllody in white clover seed crops in East Anglia. From 1959 to 1962, phyllody was reported to occur in about 40 per cent of the crops (Carr and Large, 1963).

These data can be correlated with similar reports from other continents: 30-50% losses of celery crop in Arroyo Grande Valley in California from 1951 to 1953 (Freitag *et al*., 1962) or 100,000 pear trees killed in 1960 in the same area (Nichols and Shalla, 1960) or the over 4 million coconut palms destroyed since 1962 in Jamaica by lethal yellowing disease with an estimated loss of 500 coconut palms per day (Waters *et al*., 1978). If to these losses we added those of citrus, grape and cereals, the question of the harmfulness of these diseases acquires still wider international scope and breadth.

3.4 TRANSMISSION

The biology of mycoplasmas can be appreciated by studying the process of their transmission in the field or, in experimental conditions, in the greenhouse.

Experimental investigations of the host plants of different strains of mycoplasmas can be conducted by known methods: mechanical, grafting, dodder or insect transmission.

3.4.1 Mechanical Transmission

Although numerous attempts at mechanical transmission through the juice of diseased plants were made using European strains of mycoplasmas, the results obtained were negative (Valenta *et al*., 1961; Ploaie, 1969a, 1973a). In our opinion, this could be accounted for by the fact that a) the mycoplasmas which are harbored in phloem elements at high pressure can be damaged by osmotic and pressure changes when extracted from plant tissues and b) the mycoplasma cannot be introduced into sieve elements by simply rubbing the juice from diseased plants on the leaves of host plants (Ploaie, 1973a). However, mechanical transmission of mycoplasmas by microinjecting the hemolymph from infected insects to healthy ones is common practice, owing to the methods developed by Maramorosch (1956) and Maramorosch and Jernberg (1970). The latter technique is also a useful tool for studying transmission of mycoplasmas by, and their multiplication in, vectors.

3.4.2 Transmission by Grafting

Grafting has been widely used in determining the host ranges of several European strains of mycoplasmas, e.g., stolbur (Valenta *et al*., 1961; Cousin and Grison, 1966a; Ploaie, 1969a), clover phyllody (Valenta and Musil, 1963; Ploaie,

1969a), clover dwarf (Valenta and Musil, 1963; Ploaie, 1969a) and aster yellows (Ploaie, 1969a, 1971c).

The methods employed in our laboratory (diseased scion bound to the plant stock, chip budding, and binding cut surfaces of the plant stem together without severing either of them from its own roots) yielded good results, but only within the limits of one plant family. Transmission from *Vinca rosea* L. to *Vinca rosea* L., using more than 15 strains of mycoplasma from our collection, was 100% positive. Similar results were obtained for transmission from tomato to tomato or from tobacco to tobacco. A good union between stock and scion is always very important, although some positive results were obtained even when the scion did not remain alive till the development of symptoms.

The incubation period in *Vinca rosea* L., at temperatures of 20-22°C, varied between 20 and 60 days for many of our strains of mycoplasmas. Sometimes, in the same conditions, the incubation period of several aster yellows strains ranged from 6 months to 1 year or more, even though a good union between very young infected scions and stocks were achieved. These irregularities in transmission are significant and must be borne in mind when inoculations are done with mycoplasmas cultivated on artificial media. Generally, these agents do not have short (20-40 days) incubation periods. Our observations, spanning more than 20 years, have shown that, even under controlled greenhouse conditions, inoculations by grafting to *Vinca rosea* L. yielded positive results only after long incubation periods. Thus, several MLOs associated with onion proliferation, periwinkle proliferation, gladiolus yellows, clover dwarf, and cabbage yellows, grafted on *Vinca rosea* L. between February and August of 1976, did not produce the first symptom, e.g., green flowers or yellowing of veins, till September of 1977.

We have almost come to the conclusion that there are "years of mycoplasmas" and years when symptoms take a long time to appear. This is another time-consuming disadvantage of grafting in the transmission of mycoplasmas, besides the problems of stock and scion compatibility. This situation, observed by us in *Vinca rosea* L., resembles that in woody plants, in which the incubation period of mycoplasmas is also very long.

3.4.3 Transmission by Dodder

The method of transmitting mycoplasmas by dodder was used extensively in earlier studies as reviewed by Fulton (1964) and Hosford (1968), and has become a useful tool in studying the host ranges of mycoplasmas.

Dodder, *Cuscuta campestris* Yunck., parasitizes over 26 genera of 10 plant families (Johnson, 1941). This wide host range makes possible the transmission of mycoplasmas between plant species belonging to different families, making dodder transmission a more efficient approach to the study of mycoplasmas than grafting. Hence, *C. campestris* was successfully used for the transmission of such European strains of mycoplasmas as stolbur (Valenta *et al.*, 1961; Kova-chevsky *et al.*, 1964; Ploaie, 1969a), parastolbur (Valenta and Musil, 1963), Crimean yellows (Valenta, 1958), aster yellows (Protsenko, 1958a), clover phyl-lody (Ploaie, 1966, 1969a), clover dwarf (Valenta and Musil, 1963), onion pro-liferation (Petre and Ploaie, 1971), and die-back of apricot (Ploaie, in prepara-tion). Positive results were obtained in the transmission of stolbur through *C. trifolii* Bab. and Gibs., *C. epilinum* Weihe, (Valenta *et al.*, 1961) and *C. pentagona* Engelm. (Kovachevsky *et al.*, 1964). Another species of dodder (*C. subinclusa* Dur. and Hilg.) was very efficient in the transmission of stolbur (Giannotti *et al.*, 1969a), clover phyllody (Frazier and Posnette, 1957; Mišiga, 1961), clover dwarf and parastolbur (Valenta and Musil, 1963), and die-back of apricot (Morvan *et al.*, 1973). Phyllody of *Vinca rosea* L. was transmitted by *C. monogyna* Auct. and *C. europea* L. (Caudwell, 1965), and clover phyllody by *C. gronovii* Willd. (Frazier and Posnette, 1957).

The above species of dodder were widely used in America, Asia and Australia for the transmission of numerous plant pathogenic mycoplasmas (Ploaie, 1973a). According to our experience, the most susceptible plant to the transmission by dodder was *Vinca rosea* L., with a transmission rate of 90-100% for most of the European strains of mycoplasmas.

The multiplication of stolbur mycoplasma in sieve elements of dodder was first demonstrated by electron microscopy in Europe by Giannotti *et al.* (1969a). Similar results were reported in the USA for aster yellows (Dale and Kim, 1969) and witches' broom of ash (Hibben and Wolanski, 1971).

When different species of dodder are used experimentally as vectors of plant mycoplasmas, they are not usually affected by the mycoplasma passing through them. However, pathological symptoms had occasionally been observed in dod-der grown on plants affected by curly top (Lackey and Bennett, 1949). At the same time, strong symptoms of phyllody were observed when *C. campestris* was allowed to parasitize a big-bud infected tomato plant from the Pacific Northwest (Menzies, personal communication, 1968). Later, strong phyllody and prolifera-tion of *C. campestris* caused by Crimean yellows (a strain labeled stolbur in the USSR) (Valenta, 1958), stolbur (Ploaie, 1969a, 1971c) and clover phyllody (Ploaie, 1966, 1969a) was demonstrated. Malformations of the flowers of *C. europea* L. and *C. monogyna* Auct. were produced by a nonidentified strain of aster yellows (Caudwell, 1965), and modifications of *C. subinclusa* Dur. and Hilg. by apple proliferation agent (Marwitz *et al.*, 1974). Wilting of *C. campestris* oc-curred when this species had been used for the transmission of a strain of aster yellows isolated in Romania (Ploaie, 1969a, 1971c).

An essential condition for observing the appearance of symptoms in dodder is a prolonged contact of dodder with a group of infected plants. Under these con-ditions, the carpel fails to inflate normally and is split apart by the growing ovules. Sometimes the flower axis proliferates and divides, producing a series of phylloid flowers (Fig. 5).

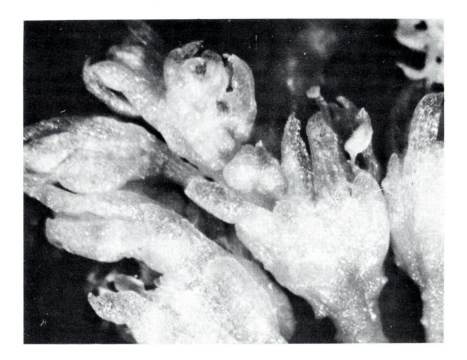

FIG. 5. Proliferation of flowers of dodder (*C. campestris*) induced by stolbur mycoplasma. (Original.)

Transmission and floral modifications involve an association of dodder with sieve elements of diseased plants and multiplication of mycoplasmas in the sieve elements of dodder.

3.4.4   Transmission by Vectors

Plant mycoplasmas and spiroplasmas are propagative arthropod-borne agents which infect and multiply in both vectors and plants. Semiautomated equipment for injection and serial passage techniques, developed by Maramorosch and Jernberg (1970) and successfully used with aster yellows agents and its vector *Macrosteles fascifrons* Stål., opened up a new area of investigations into the relationships between aster yellows agents and their vectors.

It is well known today that most MLOs are transmitted by leafhoppers belonging to the *Homoptera-Auchenorhyncha* group of insects. In the superfamily *Cicadelloidea*, there are 10 families (*Ulopidae, Agaliidae, Iassidae, Macropsidae, Gyponidae, Aphrodidae, Tettigellidae, Coeliddiidae, Cicadellidae* and *Deltocephalidae*) containing a total of 50 vector species (of which 46 belong to *Delto-*

cephalidae) of 27 MLOs on different continents (Ploaie, 1973a). Another super-family, *Fulgoroidea*, has 4 known vector species. In the *Homoptera-Sternorrhyncha* group, 4 species in the family *Psyllidae* are known vectors of MLOs. Species of *Lygus* (*Gimnocerata*) (Neklyudova and Dikii, 1973). *Margarodine* (Coccoidea) (Delattre *et al.*, 1974) and *Phytoptus* (*Eriophyidae*) (Silvere, 1970) were also reported as presumptive MLO transmitters. Since the vectors of each aster yellows type disease has already been discussed in earlier papers (Ploaie, 1973a; Grunewaldt-Stöcker and Nienhaus, 1977) they will not be detailed here.

In Europe, 19 vectors have been studied in greater depth, because of their in-

TABLE II. The Vectors of European Aster Yellows-type Disease

Vector Species	Disease	References
Hyalesthes obsoletus	Stolbur	Sukhov and Vovk, 1946; Razviaskina, 1950; Ploaie, 1960; Valenta *et al.*, 1961; Leclant, 1968.
Hyalesthes mlokosiewiczii	Stolbur	Samundzheva, 1949.
Aphrodes bicinctus	Stolbur,	Valenta *et al.*, 1961
	Clover dwarf	
	Parastolbur	Valenta and Musil, 1963
	Clover yellow	Moreau *et al.*, 1968
Euscelis plebejus	Stolbur	Valenta *et al.*, 1961
	Clover phyllody	Musil, 1965; Ploaie, 1966
	Clover dwarf	Musil, 1965; Lehmann, 1973
	Wheat chlorosis	Ploaie *et al.*, 1977
Euscelis lineolatus	Clover phyllody	Frazier and Posnette, 1957; Maillett, 1970
Euscelidius variegatus	Clover phyllody	Giannotti, 1969
Aphrodes albifrons	Clover phyllody	Frazier and Posnette, 1957
Macrosteles laevis	Aster yellows	Heinze and Kunze, 1955
	Stolbur	
	Parastolbur	Valenta *et al.*, 1961;
	Clover dwarf	Valenta and Musil, 1963
	Gladiolus yellows	Ploaie *et al.*, 1977
Macrosteles viridigriseus	Clover phyllody	Frazier and Posnette, 1957
	Strawberry green petal	
Macrosteles cristatus	Clover dwarf	Lehmann, 1973
Macrosteles quadripunctulatus	Aster yellows	Ploaie, 1968a; 1969a
Scaphoideus littoralis	Grape yellows	Caudwell *et al.*, 1970
Psammotettix alienus	Wheat chlorosis	Ploaie *et al.*, 1975
Macrosteles sexnotatus	Aster yellows	
	"Lissers" of hyacinth	Slogteren, van and Müller, 1972
Phytoptus ribis	Black currant reversion	Silvere, 1970

volvment in the transmission of several dangerous diseases (Table II). Consequently, some biological studies have been done with *Hyalesthes obsoletus,
Euscelis plebejus* and *Aphrodes bicinctus*. *H. obsoletus* is a very efficient vector
of stolbur pathogen (Ploaie, 1969a). Although this species is difficult to rear,
we succeeded in obtaining, under experimental conditions, generations from egg
to egg (Fig. 6a and b) and in demonstrating the capacity of nymphs to transmit
stolbur (Ploaie, 1960). *E. plebejus* (Fig. 6c) is also a highly efficient vector of
clover phyllody, clover dwarf, and parastolbur MLOs and methods for rearing it
in the greenhouse on alternative hosts like *Faba vulgaris* Moench, *Bromus inermis*
Leyes, *Hordeum vulgare* L. and *Trifolium repens* L. have been developed (Musil,
1965; Ploaie, 1966).

In serial transmission by injection of hemolymph to *E. plebejus* the incubation period of different strains of MLO, e.g., clover phyllody, clover dwarf, parastolbur or stolbur, varied from 21 to 45 days (Musil, 1965).

Electron microscopically, MLOs were detected in the salivary glands of *H.
obsoletus* (Ploaie and Ionica, 1968; Ploaie, 1969a) and in different organs of *E.
plebejus, E. lineolatus, Euscelidius variegatus* and *Scaphoideus littoralis* (Giannotti *et al.*, 1968a; Maillet and Gouranton, 1971; Caudwell *et al.*, 1971). According to Maillet and Gouranton (1971) the passage of the mycoplasmas inside or
outside the insect cells is due to endocytotic or exocytotic processes. They consider that MLOs occur extracellulary in the vector intestine and hemolymph, and
plant sieve cells, and intracellulary in plant parenchymatous cells and vector midgut and salivary gland cells.

FIG. 6. Vectors of MLOs in Europe reared in experimental conditions: *Hyalesthes obsoletus*
adult a) and nymph b); c) *Euscelis plebejus* adult. (Original.)

3.4.5   The Circuit of Mycoplasmas in Nature

After numerous observations on and investigations of proliferating diseases in Romania, I concluded that these diseases have natural foci (Ploaie, 1973a). These foci represent areas of close association between mycoplasmas (or spiroplasmas), insect vectors and plants. The plants can be donors or acceptors of mycoplasmas through intermediate vectors. In natural conditions only, the host ranges of mycoplasmas coincide with the food and breeding plants of their insect vectors. In many cases the plants are wild perennial ones, and support both the multiplications of mycoplasmas and breeding of leafhopper vectors. These triadic foci (MLO, vectors and plants) would account for the evolution of disease foci in different geographical and ecological areas, in other words, for either the expansion or limitation of foci. The natural foci of disease form on uncultivated lands covered with spontaneous flora, on lake shores, at the margin of forests, along irrigation canals, and on the boundary between cultivated lands, where numerous gramineous species and other weeds grow.

In natural foci, the stolbur mycoplasma is closely associated with *Convolvulus arvensis* L. or with plants such as *Erigeron canadense* L., *Lepidium draba* L., *Cichorium inthybus* L., or *Daucus carota* L. (Ploaie, 1960; Săvulescu and Ploaie, 1964). Evolution of disease foci depends on the vector *H. obsoletus* which develops on the roots of *Convolvulus arvensis*. The first adult insects appear in June and are already infected (Ploaie, 1960). They transmit stolbur MLO to solanaceous plants or to *C. arvensis* (Ploaie, 1960, 1972; Săvulescu and Ploaie, 1964).

Stolbur occurs in Romania in the plain regions, where temperatures exceed 30°C in July, the hottest month of the year. These regions are well suited for Solanaceae culture and large populations of *H. obsoletus* occur at intervals of several years. The appearance of the insect vector and stolbur has a cyclic character. Stolbur has never occurred in plants growing at heights above 300 m; nevertheless, stolbur symptoms have been noted at this altitude in *C. arvensis*, although *H. obsoletus* was not abundant, (Săvulescu and Ploaie, 1964, 1967a).

The clover phyllody MLO has been recorded in hilly and submountainous regions only, from 300 to 800 m altitudes, with yearly precipitations of 500-800 mm and average temperatures of 16-20°C in July. The distribution of both vectors and MLO is quite different, and their dependence on climatic conditions is obvious from the fact that clover plants infected with phyllody MLO and grown in a Bucharest field, an area propitious to the occurrence of stolbur, lost all phyllody symptoms and appeared healthy; but if transferred to a greenhouse at 16-20°C, some of these plants again showed symptoms of infection (Ploaie, 1966; Săvulescu and Ploaie, 1967a and b).

The clover phyllody mycoplasma is closely associated in natural foci with *Trifolium repens* L., *T. montanum* L., *T. hybridum*, *Taraxacum officinale* L., *Plantago major* L., *Daucus carota* L., and *Cichorium inthybus* L. (Săvulescu and Ploaie, 1961, 1967b; Ploaie, 1966, 1969a, 1972). The vector *E. plebejus*, which has two generations per year, is the most efficient vector and transmits the mycoplasma to strawberry, red clover and wheat or barley (Ploaie, 1966, 1969a; Săvulescu and Ploaie, 1967b; Ploaie *et al.*, 1977).

Other disease foci are comprised of wild gramineous plants and several species of *Macrosteles* and *Psammotettix* leafhoppers that transmit the MLOs to cultivated Gramineae and many other plants (Ploaie *et al.*, 1977).

MLOs may overwinter in vectors as well as host plants. Spring and autumn generations of vectors can be killed by hard winter frosts or cold, and by rainy weather in autumn or spring. Thus, climatic conditions play an important role in the ecology of vectors and, indirectly, in the limitation or expansion of disease foci. It is well known that dry weather in autumn and spring is an essential condition for invasion by stolbur. Several waves of stolbur were recorded in southeastern Europe in the years 1933-1939, 1946-1949, 1953, 1956, and 1959-1961 following dry springs (Krujilin, 1953; Bojňanský and Kosljarová, 1964; Ploaie, 1972). After 1960, no strong stolbur invasion occurred. Agrotechnical control methods used over large areas and facilitated in eastern Europe by a socialist agriculture, have limited the sources of MLOs and their vectors.

Mycoplasmas are often introduced to new areas by the distribution of uncontrolled tubers, bulbs, budwoods, and rootstocks of trees or shrubs. Thus, after the introduction in Romania of gladiolus bulbs from the Netherlands, we have recorded, ever since 1975, invasions of gladiolus yellows disease in this culture early in spring, when such symptoms do not appear in other species of plants. It is well known that in the Netherlands entire gladiolus fields are strongly infected with witches' broom yellows disease (Leeuw, 1977).

Sometimes the autochthonously cultivated plants are more resistant to MLOs or are not visited by vectors. Consequently, natural foci of the diseases are not apparent; they become more evident only after new plant species have been introduced in the respective area. Strong infections were detected by us when *Vinca rosea* from Vietnam, *Ammi visnaga* from Egypt, wheat from Austria, or strawberry from USA were introduced in Romania.

Long distance spread of MLOs can be the work of efficient vectors. For example, *Macrosteles fascifrons*, the vector of aster yellows, comes to the Winnipeg area (Canada) from Kansas, South Dakota or Nebraska, 900 miles away, in a nonstop flight of 20-30 hr (Nichiporick, 1965). This is perhaps the most important instance of expansion of disease foci, into a new area. In a similar way, fruit trees seem to be infected with MLOs. Epidemiological studies on fruit-tree mycoplasma disease in the east of Spain revealed that apple proliferation, European yellows of peach, pear, plum, Japanese plum, apricot and almond, and infected *Convolvulus arvensis* L., or *Cynodon dactylon* L., occur in the same foci, which suggests that all these diseases are caused by a single MLO (Sánches-Capuchino *et al.*, 1976).

3.5 MORPHOLOGY AND ULTRASTRUCTURE

The morphology and ultrastructure of MLO cells associated with plant diseases were established by electron microscopical studies of thin sections from plants or vectors naturally or experimentally infected with aster yellows type pathogens. Almost all of the papers published from 1968 until now are based on such investigations, which demonstrate that a high accumulation of MLO bodies can be detected in the sieve elements of infected plants. Sometimes the phloem cells are entirely filled with MLOs (Fig. 7).

High resolution electron microscopy of MLOs from several European diseases such as stolbur, clover dwarf, parastolbur, Crimean yellows, onion proliferation, aster yellows, nasturtium little leaf, wheat chlorosis, apple proliferation, and gladiolus virescence (all experimentally induced in periwinkle plants) reveals highly pleomorphic cells bounded by a single unit membrane about 95-100 Å thick (Ploaie, 1968a and b, 1969 a and b, 1970, 1973a; Ploaie et al., 1968; Ploaie and Maramorosch, 1969).

On the basis of their morphology, as they appear in thin section, MLO bodies have been grouped earlier in several categories: (1) very small particles measuring 50-100 nm in diameter that stain heavily and are highly electron opaque (Fig. 7, arrows); (2) 100-300 nm, spherical or ovoid bodies with visible unit membrane; (3) large, 400 nm-1 μm, ovoid bodies with electron-dense areas, nuclear-like net-like structures, and ribosomes; (4) filaments or segmented filaments 1-2 μm long and 0.08-0.3 μm wide; (5) irregular large bodies comprised of 2-3 cells on a common support, and (6) large cells having dense inclusions or numerous buds (Fig. 8b) on their surface (Ploaie, 1973a).

The simultaneous presence of small and large particles in the same phloem cell seemed to suggest different stages of development (Ploaie, 1968a) or different planes of sectioning of ovoid cells or filamentous branches. The shape and size of the bodies detected in diseased plants resemble either the genus *Mycoplasma*, the agents of psittacosis-lymphogranuloma venereum-trachoma group (PLT), or L-forms of bacteria (Ploaie, 1968a; Ploaie and Maramorosch, 1969).

The morphological aspects revealed by transmission electron microscopy of thin sections are insufficient for a correct appraisal of the shape of MLOs in phloem cells. The micrographs of MLO-infected plant tissue, freeze fractured and examined by scanning electron microscopy, showed MLOs to be represented by filamentous branching forms, dumbells, and spherical cells in the process of binary fission or budding (Hagis and Sinha, 1978; Marwitz and Petzold, 1978). This indicates that numerous spherical particles, which appear in thin sections, are in fact cross-sectioned filaments or branches. Consequently, greater attention should be given to the interpretation of MLO morphology from thin sections.

Structures resembling rod-shaped viruses, 31-33 by 85-88 nm, were also detected by us (Ploaie, 1968b, 1971b) free in the phloem cells of infected periwinkle plants or fixed on clover dwarf mycoplasma cells (Fig. 8a). These virus-like particles suggested an MLO viral infection and were almost identical morphologically with those isolated later from *Acholeplasma laidlawii* (Gourlay, 1971; Liss and Maniloff, 1971, Cole, 1978). Similar particles, but 27 ± 33 nm in diameter and 50-90 nm in length, were found associated with clover phyllody mycoplasma in both diseased plants and insect vectors (Gourret et al., 1973). Rod-like particles, 34 by 70 nm, were also described in association with aster yellows mycoplasma in aster plants (Allen, 1972), and in cultured aster yellows spiroplasma (Maramorosch, 1979).

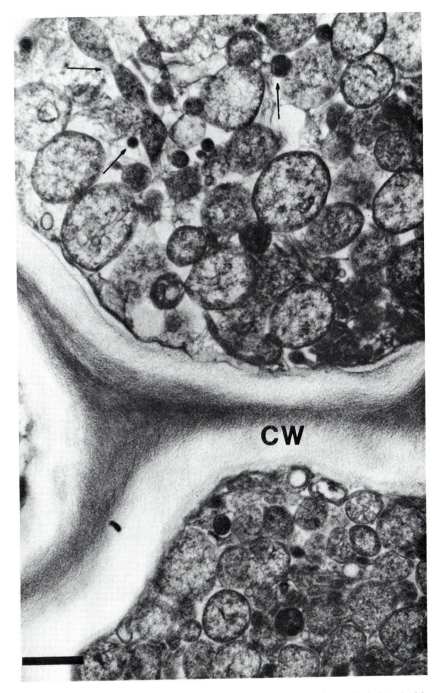

FIG. 7. Cross section of phloem elements of periwinkle plant experimentally infected with clover phyllody. Note the abundance of pleomorphic, mycoplasma-like bodies in the two cells separated by cell wall (CW). Very small electron dense bodies or filaments (arrows) are distributed between large bodies. × 34,000. Bar, 0.5 μm. (Original.)

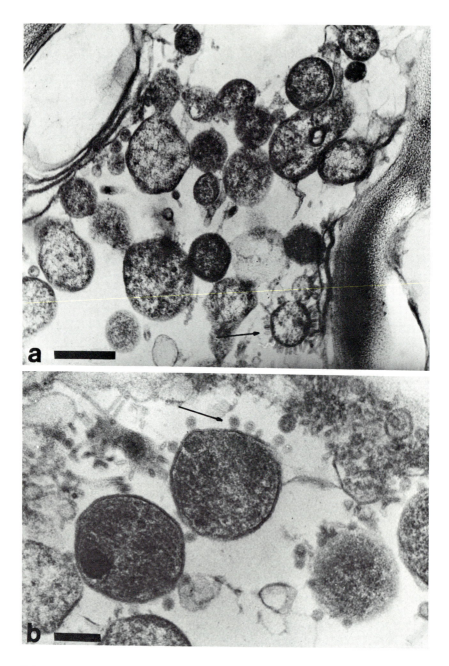

FIG. 8. Section through periwinkle phloem cells experimentally inoculated with clover dwarf mycoplasma. a) Rod-shaped virus-like particles are attached to membrane of myco-plasma-like organism (arrow). × 37,000. Bar, 0.5 μm. b) Section of bud-like structures formed on membrane of MLO cells. × 40,000. Bar, 0.3 μm. (Originals.)

3.6   CULTIVATION OF THE PATHOGENS AND REPRODUCTION OF DISEASE SYMPTOMS

The discovery of MLOs in diseased plants stimulated many laboratories all over the world to try to cultivate these microorganisms in artificial media and, subsequently, to reproduce disease symptoms in healthy plants, using cultured MLOs as inocula. Many plant pathologists have considered that MLOs must generally resemble animal mycoplasmas in their nutritional requirements. Therefore, numerous attempts have been made to grow the presumed mycoplasmas on artificial media similar to those sustaining animal mycoplasmas and supplemented with several components used in vertebrate and invertebrate cell culture (Jones et al., 1977).

From 1968 until now, dozens of reports from various laboratories in different countries have been published on successful as well as unsuccessful attempts at culturing MLOs and demonstrating the infectivity to plants (Müller et al., 1975; Elmendorf, 1977; Whitcomb and Tully, 1979). Unfortunately, none of the so-called mycoplasma isolates could be cultivated from the same diseased plants in other laboratories. Were the reported isolates secondary parasites of diseased plants or primary, causative agents of plant diseases? Were they contaminants from the culture media or from the soil? These questions have not been answered to the satisfaction of the scientific community. There is general agreement that the isolated species of Mycoplasma and Acholeplasma were not primary agents of yellows diseases. The claims of mycoplasma cultivation in Europe, from 1969 to the present, with the notable exception of the classic work on the non-European S. citri (Saglio et al., 1973; Daniels et al., 1973) have not been accepted by mycoplasmologists in other parts of the world. The main reason for this rejection is the failure to deposit the cultures in international type culture collections and the failure to disseminate them to experts, so as to enable others to verify the reports.

During the past decade several claims have been made that non-spiral mycoplasmas have been isolated from diseased plants, cultured on cell-free media, and re-inoculated into plants directly or via insect vectors, reproducing the original disease. In some instances no disease was reproduced but the claim was made that the cultured mycoplasmas were indeed plant pathogens. Unfortunatley the authors of such claims were unable or unwilling to submit their isolates to the scrutiny of recognized mycoplasma experts and thus failed to comply with the established rules of the International Organization of Mycoplasmologists. The rules require that new isolates must be deposited in one of the mycoplasma collections, such as the WHO/FAO Mycoplasma Center in Aarhus, Denmark, where an independent comparison can be made with existing species and strains of Mollicutes. Until the international rules are followed, all unsubstantiated claims have to be ignored by the scientific community. Readers are referred to a recent review of this subject (Whitcomb and Tully, 1979).

3.7 MIXED INFECTIONS: MLO AND RICKETTSIA

In the past 10 years, structures resembling rickettsia have been detected in association with several plant diseases assumed to be caused by viruses. This new aspect of plant pathology was discussed in detail by Davis and Whitcomb (1971), Whitcomb (1973), Maramorosch (1974), and Maramorosch et al., 1975.

Investigations carried out in Europe led to several important findings. First, there is the simultaneous presence of MLO and rickettsia-like organisms (RLOs) in the same cell. Thus, in diseased *Daucus carota* L. plants with symptoms of proliferation and in the vector of this disease agent, the psyllid *Trioza nigricornis* Forst, both MLOs and RLOs were detected in the same cells by electron microscopy (Giannotti et al., 1974a). This double infection was also observed in the cells of dodder, *C. subinclusa*, used for the transmission of this proliferation disease agent (Giannotti et al., 1974b). A second instance of this type of mixed infection (MLO/RLO) in clover (*T. repens* L.) with symptoms of dwarfing (Vago and Giannotti, 1977).

RLOs were also found in "Golden Delicious" apple trees, with symptoms of proliferation, mainly in the parenchyma, cambium, and tracheidal cells (Petzold et al., 1973). Structures resembling rickettsia have occurred in phloem cells of wheat plants (*T. durum* L) ever since 1971 (Ploaie, 1973b), in the phloem and parenchymatous cells of yellows diseased grapevines in Germany, France and Greece, and in the pseudocoelum of *Xiphinema index* through which the disease was successfully transmitted to grapevine cuttings (Rumbos, 1978).

A new disease, rosette disease of sugar beets, also associated with an RLO, was detected in roots, hypocotyls, epycotyls and dwarfed leaves. This RLO, transmitted by *Piesma quadratum* Fieb., reportedly can be cultivated in chick embryos, similarly to human pathogenic rickettsia (Green, 1978; Hasse and Schmutterer, 1978; Nienhaus et al., 1978).

The etiological role of RLOs requires more investigation. The presence of MLOs and RLOs in the same plant or vector cells (Maillet, 1970) may suggest the existence of a small bacterium and its L-form. Structures resembling MLOs and RLOs or PLT organisms appeared several times when we attempted the cultivation under controlled conditions of the agent of onion proliferation on a medium for *M. agalactiae*, after filtration through a 0.45 μm filter.

3.8 CONCLUSIONS

Proliferative diseases in Europe are similar to those known in America, Asia or Australia. In all these geographical areas MLOs were detected in infected plants. The expansion of these diseases depends on the vectors and donor (reservoir) plants. Their occurrence in herbaceous plants has a cyclic character and is related to the natural foci of the disease (insects and wild perennial plants). Destroying the natural foci is the best control method. Quarantine measures against stolbur and pear decline, apple proliferation, and apple chat fruit have already been recommended by the European Organization of Plant Protection.

3.9 REFERENCES

Albouy, J., Cousin, M. T., and Grison, C. (1967). Étude comparée de 3 maladies à virus: souche "Californienne" de l'aster du glaieul, phyllodie du trèfle et stolbur de la tomate sur *Vinca rosea. Ann Épiphyties* 18: 157-171.

Aldrovandi, U. (1642). Monstruorum historia. N. Tebaldini, Bologna.

Allen, T. C. (1972). Bacilliform particles within aster infected with a western strain of aster yellows. *Virology* 47: 491-493.

Amici, A., Belli, G., Corbetta, G., and Osler, E. (1973). II. Indagini al microscopio elettronico su piante infestanti della risaia. *Il Riso* 22: 111-118.

Bălășcuță, N., Ghiorghiu, E., and Ploaie, P. G. (1979). Necroza liniară a floemului și xilemului de păr și gutui, un simptom transmisibil prin altoire. *Analele Inst. cercet. protectia plantelor* 15.

Beakbane, A. B., Mishra, M. D., Posnette, A. F., and Slater, C. H. W. (1971). Mycoplasma-like organisms associated with chat fruit and rubbery wood disease of apple, *Malus domestica* Borch., compared with those of strawberry with green petal disease. *J. gen. Microbiol.* 66: 55-62.

Begtrup, J. (1975). Mycoplasma-like organisms in *Helenium* sp. *Phytopath. Z.* 82: 356-358.

Begtrup, J., and Thomsen, A. (1975). Mycoplasma-like organisms in phloem elements of *Cirsium, Stelaria* and *Epilobium. Phytopath. Z.* 83: 119-126.

Belli, G., Amici, A., and Osler, R. (1972). The mycoplasmas as a cause of plant disease with reference to the Italian situation. *Meded. Landbouwwetensch. Rijksuniv.* Gent 37: 441-449.

Bjørn, S. (1973). Mykoplasmasykdommer pa planten. Blyttia 31: 39-50.

Black, L. (1943). Some properties of aster yellows virus. *Phytopathology* 33: 2.

Black, L. M. (1955). Concepts and problems concerning purification of labile insect-transmitted plant viruses. *Phytopathology* 45: 208-216.

Blattný, C. (1958). The principal question of stolbur. In "Stolbur and similar virus diseases causing seedlessness of plants," Proc. Conf. on stolbur, Smolenice (E. Spaldon and C. Blattný, eds.), pp. 37-54. Slovak Academy of Sciences.

Blattný, C. (1960). Contribuții la cunoașterea circuitului unor virusuri in natura. *Studii si cercet. de biol., Seria biologie vegetală* 13: 7-34.

Blattný, C., and Váňa, V. (1974). Pear decline accompanied with mycoplasmalike organisms in Czechoslovakia. *Biol. Plantarum* 16: 474-475.

Bojňanský, V., and Kosljarová, V. (1964). Gradation of stolbur and possibilities of its prognosis in the present. *In* "Plant Virology," Proc. 5th Conf. Czechosl. Plant Virologists, Prague 1962 (C. Blattný, ed.), pp. 206-212. Academia Publishing House, Prague.

Böning, K. (1958). Auftreten einer stolburähnlichen Kartoffelkrankheit in Niederbayern in den Jahren 1927-1928. *Nach. Bl. des Dtsch. Pfl. Sch. Dienst* 10: 129-133.

Bos, L. (1957). Plant teratology and plant pathology. *Tijdschr. PlZiekt.* 63: 222-231.

Bos, L. (1960). Witches' broom virus disease of *Vaccinium myrtillus* in the Netherlands. *Tijdschr. PlZiekt.* 66: 259-263.

Bos, L. (1970). Mycoplasma's een nieuw hoofdstuk in de plantziek tenkunde? *Gewasbescherming* 1: 45-54.

Bos, L., and Grancini, P. (1965). Some experiments and considerations on the identification of witches' broom viruses, especially in clovers in the Netherlands and Italy. *Neth. J. Plant Path.* 71, Suppl. 1: 1-20.

Borges, M. De Lourdes, V., and David-Ferreira, J. F. (1968). Presence of mycoplasma in *Lycopersicon esculentum* Mill. with "Mal Azul." *Boletim. Soc. Broteriana*, Sér. 2, 42: 321-333.

Borges, M. De Lourdes, V., and Martins, M. M. (1972). A regressão-floral-do-tabaco. Presença de mycoplasmas e açao de antibióticos. *Agronomia lusit.* 33: 443-453.

Bovey, R. (1957). Une anomalie des fleurs du trèfle causée par un virus transmis par des Cicadells. *Rev. romande Agric. Vitic. Arboric.* 13: 106-108.

Carle, P. (1965). Fluoroscopie des symptòmes hystologiques de flavescence dorée de la vigne. Application à la détection rapide des lésions précoces sur cépage sensible (Baco 22A). *Ann. Épiphyties* 15: 113-118.

Carr, A. J. H., and Large, E. C. (1963). Surveys of phyllody in white clover seed crops, 1959-1962. *Plant Pathology* 12: 121-126.

Casper, R. (1969). Mycoplasmen als Erreger von Pflanzenkrankheiten. *Nach. Bl. des Dtsch. Pfl. Sch. Dienst* (Braunschweig) 21: 177-182.

Caudwell, A. (1964). Identification d'une nouvelle maladie à virus de la vigne, la "flavescence dorée." Étude des phénomènes de localisation des symptômes et de rétablissement. *Ann. Épiphyties* 15: no. hors sér. 1-193.

Caudwell, A. (1965). Note sur un virus provoquant des proliférations des fleurs des cuscutes. *Ann. Épiphyties* 15: 77-81.

Caudwell, A. (1978). Étiologie des jaunisses des plants. *Phytoma* 30: 5-9.

Caudwell, A., Kuszla, C., Bachelier, J. C., and Larrue, J. (1970). Transmission de la flavescence dorée de la vigne aux plantes herbacées par l'allongement de temps d'utilisation de la cicadelle *Scaphoideus littoralis* Ball et l'étude de la survie sur un grand nombre d'espèces végétales. *Ann. Phytopathol.* 2: 415-428.

Caudwell, A., Giannotti, J., Kuszala, C., and Larrue, J. (1971). Étude du rôle de particules de type "mycoplasme" dans l'étiologie de la flavescence dorée de la vigne. Examen cytologique des plants malades et de cicadelles infectieuses. *Ann. Phytopathol.* 3: 107-123.

Chen, T. A., and Su, H. J., Raju, B. C., and Huang, W. C. (1977). A new spiroplasma isolated from Bermuda grass (Cynodon dactylon L. Pers. *Proc. Am. Phytopathol. Soc.* 4: 152.

Chod, J., Polák, J., Novák, M., and Kříž, J. (1974). Verkommen von mycoplasmaähnlichen Organismen in Hopfenblättern mit nekrotischen Kräuselmosaik. *Phytopath. Z.* 80: 54-59.

Ciccarone, A. (1951). Sintomi do "virescenza ipertrofica" (big-bud) del pomodori nei pressi di Roma. Nota preliminare. *Boll. Staz. Patol. Veg.* Roma, Ser. 3, 7: 193.

Clark, T. B. (1977). *Spiroplasma* sp. A new pathogen in honey bees. *J. Invert. Pathol.* 29: 112-113.

Cole, R. M. (1978). Mycoplasma and spiroplasma viruses ultrastructure. *In* "The Mycoplasmatales-Cell Biology," Academic Press, New York (in press).

Cole, R. M., Tully, J. G., Popkin, T. J., and Bové, J. M. (1973). Morphology, ultrastructure and bacteriophage infection of the helical mycoplasma-like organism (*Spiroplasma citri* gen. nov. sp. nov.) cultured from "stubborn" disease of citrus. *J. Bacteriol.* 115; 367-386.

Cousin, M. T., and Grison, C. (1966a). Quelques observations et essais concernant le stolbur de la tomate. *Ann. Épiphyties* 17: 95-111.

Cousin, M. T., and Grison, C. (1966b). Premières observations concernant une fluorescence anormale dans le liber interne de plusieurs solanées infectées par le virus du stolbur et d'une apocynacée atteinte de phyllodie. *Ann. Épiphyties* 17: 93-98.

Cousin, M. T., Grison, C., and Decharme, M. (1968). Étude comparée de plusieurs types de flétrissements de Solanacées. polyphagie du stolbur. *Ann. Épiphyties* 19: 121-140.

Cousin, M. T., Moreau, J. P., Faivre-Amiot, A., and Staron, T. (1970). Mise en évidence de mycoplasmes dans des fraisiers atteint de la maladie des "pétales vert" recemment signalée en France. *C. R. Acad. Sc.*, Sér. D 270: 2000-2001.

Cousin, M. T., Moreau, J. P., Kartha, K. K., Staron, T., and Faivre-Amiot, A. (1971). Étude ultrastructurale des mycoplasmes infectant les tubes criblés de lavandins "abrial" atteints de dépérissement jaune. *Ann. Phytopathol.* 3: 243-250.

Dale, J. L., and Kim, K. S. (1969). Mycoplasma-like bodies in dodder parasitzing aster yellows-infected plants. *Phytopathology* 59: 1765-1766.

Daniels, M. J., Markham, P. G., Meddins, B. M., Plaskitt, A. K., Townsend, R., and Bar-Joseph, M. (1973). Axenic culture of plant pathogenic spiroplasma. *Nature* (London) 244: 523-524.

Daniels, M. J., and Meddins, B. M. (1974). The pathogenicity of *Spiroplasma citri. In* "Les Mycoplasmes de l'Homme, des Animaux, des Végétaux et de Insectes" (J. M. Bové and J. F. Duplan, eds.), pp. 195-200. INSEREM.

Davis, R. E. (1978a). Spiroplasma associated with flowers of the tulip tree (*Liriodendron tulipifera* L.). *Can. J. Microbiol.* 24: 954-959.

Davis, R. E. (1978b). Spiroplasmas from flowers of *Bidens pilosa* L. and honey bees in Florida: Relationship to honey bees spiroplasma AS 576 from Maryland. *Phytopathol. News* 12: PO-7.

Davis, R. E., and Whitcomb, R. F. (1971). Mycoplasmas, rickettsiae and chlamydiae: possible relation to yellows diseases and other disorders on plants and insects. *Ann. Rev. Phytopathol.* 9: 119-154.

Davis, R. E., and Worley, J. F. (1973). Spiroplasma: motile helical microorganism associated with corn stunt disease. *Phytopathology* 63: 403-408.

Davis, R. E., Worley, J. F., and Basciano, L. K. (1977). Association of spiroplasma and mycoplasma-like organisms with flower of tulip tree (*Liriodendron tulipifera* L.) *Proc. Am. Phytopathol. Soc.* 4: 185-186.

Davis, R. E., Worley, J. F., Clark, T. B., and Moseley, M. (1976). New spiroplasma in diseased honey bee (*Apis mellifera* L.): Isolation, pure culture, and partial characterization *in vitro. Proc. Am. Phytopathol. Soc.* 3: 304.

Davis, R. E., Worley, J. F., Whitcomb, R. F., Ishijima, T., and Steere, R. L. (1972). Helical filaments produced by a mycoplasma-like organism assocaited with corn stunt disease. *Science* 176: 521-523.

Delacroix, G. (1908). Maladies des plantes cultivées, vol. I. Berlliére, Paris. 22 pp.

Delattre, R., Giannotti, J., and Czarnecky, D. (1974). Maladies du cotonnier et de la vigne liés au sol et associées à des cochenilles endogées. Présence de mycoplasmes et étude comparatives des souches *in vitro. C. R. Acad. Sc.,* Sér. D 279: 315-318.

Delay, C., and Darmanaden, J. (1973). Association entre les caractères morphologiques de la variété horticole "minor" d'*Opuntia subulata* et la présence de microorganismes de type mycoplasme dans le phloème. Étude expérimentale et infrastructurale. *Annales Sc. Nat., Botanique et Biologie Végétale* 14: 408-459.

Diener, T. O., and Raymer, W. B. (1967). Potato spindle tuber virus: a plant virus with properties of a free nucleic acid. *Science* 158(3799): 378-381.

Dihoru, G., and Dihoru, A. (1973). Fitomorfe anormale semnalate în flora României. *Acta Botanica Horti. Bucurestiensis 1972-1973*: 449-453.

Doi, Y., Teranaka, M., Yora, K., and Asuyama, N. (1967). Mycoplasma- or PLT group-like microorganisms found in the phloem elements of plants infected with mulberry dwarf, potato witches' broom, aster yellows or pawlownia witches' broom. *Ann. Phytopathol. Soc. Japan* 33: 257-266.

Elmendorf, E. (1977). Newly discovered prokaryotic plant pathogens. A bibliography. *R 2901 Research Bull. College Agric.,* pp. 1-85. Univ. Wisconsin, Madison.

Fedotina, V. L. (1973). Mikoplazmopodobnye tela pri stolburnom zabolevanie tomatov. *Arch. Phytopathol. u. Pflanzenschutz* 9: 273-279.

Fedotina, V. L. (1977). Virus und mycoplasmaähnliche Organismen in Zelle von Hafer der von der Pseudorosettenkrankheit befallen ist. *Arch. Phytopathol. u. Pflanzenschutz* 13: 177-191.

Frazier, N. W., and Posnette, A. F. (1957). Transmission and host-range studies of strawberry green-petal virus. *Ann. Appl. Biol.* 45: 580-588.

Freitag, J. H., Aldrich, T. M., and Drake, R. M. (1962). The control of the spread of aster yellows virus to celery. *Meded. Landbouwhogeschool en de Opzoekingsstations van de Staat*, Gent 27: 1047-1052.

Fudl-Allah, A. E.-S. A., and Calavan, E. C. (1973). Effect of temperature and pH on growth *in vitro* of mycoplasmalike organism associated with stubborn disease of citrus. *Phyto-pathology* 63: 256-259.

Fulton, R. W. (1964). Transmission of plant viruses by grafting, dodder, seed and mechanical inoculation. *In* "Plant Virology" (M. K. Corbett and D. Sisler, eds.), pp. 42-44. Univ. Florida Press, Gainesville.

Gáborjanýi, R., and Sziráki, F. (1978). Cytokinins in white clover plants infected by clover phyllody mycoplasma. *Acta Phytopathol. Hung.* 12: 161-175.

Ghimpu, V. (1931). Contribution à la teratologie des *Nicotiana*. *Rev. Pathol. Veg.* 18: 289-295.

Ghiorghiu, A., Ionescu-Matiu, E., and Ploaie, P. G. (1966). Influence de l'infection virale sur la quantité et la qualité de la teneur en alcaloides de *Vinca rosea* L. (*Catharanthus roseus* G. Don). *Rev. Roum. Biochim.* 3: 221-227.

Giannotti, J. (1969). Transmission of clover phyllody by a new leafhopper vector, *Euscelidius variegatus*. *Pl. Dis. Reptr.* 53: 173.

Giannotti, J., Devauchelle, G., and Vago, C. (1968a). Microorganismes de type mycoplasmes chez une cicadelle et une plante infectées par la phyllodie. *C. R. Acad. Sc.*, Sér. D 266: 2168-2170.

Giannotti, J., Morvan, G., and Vago, C. (1968b). Micro-organismes de type mycoplasme dans les cellules libériennes de *Malus sylvestris* L. atteint de la maladie des proliférations. *C. R. Acad. Sc.*, Sér. D 267: 76-77.

Giannotti, J., Marchoux, G., and Devauchelle, G. (1969a). Observations de microorganismes de type mycoplasme dans les cellules libériennes de *Cuscuta subinclusa* L., plante vectrice du stolbur de la tomate. *Ann. Phytopathol.* 1: 445-455.

Giannotti, J., Marchoux, G., and Devauchelle, G. (1969b). Remarques à propos de l'observation de microorganismes de type mycoplasme chez les solanées maraichères. *Ann. Phytopathol.* 1: 419-431.

Giannotti, J., Sassine, J., Czarnecky, D., and Tournier, J. (1972). Caractérisation sérologique et biologique de trois mycoplasmes de plantes correspondant à trois maladies différentes. *Parasitica* 28: 78-88.

Giannotti, J., Louis, C., Leclant, F., Marchoux, G., and Vago, C. (1974a). Infection à mycoplasmes et à micro-organismes d'allure rickettsienne chez une plant atteinte de prolifération et chez le psylle vecteur de la maladie. *C. R. Acad. Sc.*, Sér. D 278: 465-471.

Giannotti, J., Marchoux, G., Devauchelle, G., and Louis, C. (1974b). Double infection cellulaire à mycoplasmes et à germes rickettsoides chez la plante parasite *Cuscuta subinclusa* L. *C. R. Acad. Sc.*, Sér. D 278: 751-753.

Gimenez, E. H. (1975). Detection de "stubborn" en citricos españolas. Método parr su diagnóstico. *Anales de la Real Academia de Farmacia* 2: 181-206.

Gourlay, R. N., Bruce, J., and Garwes, D. J. (1971). Characterization of Mycoplasmatales virus laidlawii 1. *Nature New Biology* 229: 118-119.

Gourret, J. P. (1970). Ultrastructure et micro-écologie des mycoplasmes de phloème dans trois maladis des pétales verts. Étude de lésion cellulaires. *J. Microscopie* 9: 807-822.

Gourret, J. P., Maillet, P. L., and Gouranton, J. (1973). Virus-like particles associated with the mycoplasmas of clover phyllody in the plant and in the insect vector. *J. Gen. Microbiol.* 74: 241-249.

Granados, R. R. (1965). Strains of aster yellows virus and their transmission by the six-spotted leafhopper, *Macrostelles fascifrons* (Stål). *Ph. D. Thesis*. Univ. of Wisconsin, Madison.

Granados, R. R., Maramorosch, K., and Shikata, E. (1968). Mycoplasma: suspected etiologic agent of corn stunt. *Proc. Nat. Acad. Sci. USA* 60: 841-844.

Green, S. K. (1978). Association of rickettsialike organisms with rosette disease of sugar beets. *3rd Int. Congress of Plant Pathol.*, München 16-23 Aug. Abstracts of papers, p. 79.

Grunewaldt-Stöcker, G., and Nienhaus, F. (1977). Mycoplasma-ähn-liche Organismen als Krankheitserreger in Pflanzen. *Acta Phytomedica* (Helft 5), pp. 1-115. Verlag Paul Parey, Berlin.

Haggis, G. H., and Sinha, R. C. (1978). Scanning electron microscopy of mycoplasma-like organisms after freeze fracture of plant tissues affected with clover phyllody and aster yellows. *Phytopathology* 68: 677-680.

Hampton, R. O. (1972). Mycoplasmas as plant pathogens: perspectives and principles. *Ann. Rev. Plant. Physiol.* 23: 384-418.

Harrison, B. D., and Roberts, J. M. (1969). Association of mycoplasma-like bodies with po-tato witches' broom disease from Scotland. *Ann. Appl. Biol.* 63: 347-349.

Hasse, V., and Schmutterer, H. (1978). Untersuchungen über den einfluss des Pathogenes der latenten Rosettenkrankheit der Beta-Rüben auf der Vektor *Piesma quadratum* Fieb. *Z. Pflanzenkrankh. u. Pflanzenschutz* 8: 735-744.

Heinze, L., and Kunze, L. (1955). Die europäische Astern gelbsucht und ihre Übertragung durch Zwergzikaden. *Nach. Bl. des Dtsch. Pfl. Sch. Dienst* (Braunschweig) 7: 161-164.

Hibben, C. R., and Wolansky, B. (1971). Dodder transmission of a mycoplasma from ash trees with yellows type symptoms. *Phytopathology* 60: 1295.

Hiruki, C., and Dijkstra, J. (1973a). Light and electron microscopy of *Vinca* plants infected with mycoplasma-like bodies of sandal spike disease. *Neth. J. Pl. Path. 79: 207-217.*
with mycoplasma-like bodies of sandal spike disease. *Neth. J. Pl. Path.* 79: 207-217.

Hiruki, C., and Shukla, P. (1973b). Mycoplasma-like bodies associated with witches' broom of bleeding heart. *Phytopathology* 63: 88-91.

Horváth, J. (1970). Properties of mycoplasmas causing yellowing type plant diseases and the plant mycoplasmoses. *Növénytermelés* 19: 327-337.

Hosford, R. M. Jr. (1968). Transmission of plant viruses by dodder. *Botanical Review* 33: 387-406.

Hull, R. (1971). Mycoplasma-like organisms in plants. *Rev. Plant Pathol.* 50: 121-130.

Hull, R. (1972). Mycoplasma and plant diseases. *Pans* 18: 154-164.

Ionică, M., Ploaie, P. G., Tîrcă, M., Găină, M., and Popa, I. (1978). Cloroza și uscarea pru-nului, o nouă boală asociată cu micoplasma în România. *Analele Inst. cercet. protecția plantelor* 14: 11-16.

Ishie, T., Doi, Y., Yora, K., and Asuyama, H. (1967). Suppressive effects of antibiotics of tetracycline group on symptom development of mulberry dwarf disease. *Ann. Phyto-pathol. Soc. Japan* 33: 267-275.

Jaczewski, A. A. (1926). Ved'miny metly kartofela. *Mater. Mykol. i Fitopat. Ross.* 5: 117-128.

Johnson, F. (1941). Transmission of plant viruses by dodder. *Phytopathology* 13: 649-656.

Jones, A. L., Whitcomb, R. F., Williamson, D. L., and Coan, M. E. (1977). Comparative growth and primary isolation of spiroplasmas in media based on insect tissue culture formulations. *Phytopathology* 67: 738-746.

Kaminska, M. (1971). Comparative anatomy of healthy and proliferation-infected apple tree. *In* "Plant Virology," Proc. 7th Conf. Czechosl. Plant Virologists, High Tatras (V. Bojňanský, ed.), pp. 335-341. Slovak Acad. Sciences, Bratislava.

Kegler, H., Müller, H. M., Kleinhempel, H., and Verderevskaja, T. D. (1973). Untersuchungen über der Kirschenverfall und die Hexenbesenkrankheit der Heidelbeere. *Nachrbl. f. Pfl. Schutzdienst in der DDR* 27(1): 5-8.

Kleinhempel, H., and Müller, H. M. (1978). Mycoplasmen-Erreger von Pflanzenkrankheiten. *Urania* DDR 14: 36-39.

Kleinhempel, H., Lehmann, W., and Spaar, D. (1971). Mycoplasmaähnliche Organismen als mögliche Erreger von Pflanzenkrankheiten. *Fortschrittsberichte für die Landwirtschaft und Nahrungsgüterwirtschaft* 9: 1-61.

Kleinhempel, H., Müller, H. M., and Spaar, D. (1972). Isolierung und Kultivierung von Mycoplasmatales aus Weissklee mit Blütenvergrünungssymptomen. *Arch. Pflanzenschutz* 8: 361-370.

Kondo, F., McIntosh, A. H., Padhi, S. B., and Maramorosch, K. (1976). Electron microscopy of a new plant-pathogenic spiroplasma isolated from *Opuntia. Proc. 34th Ann. Electron Microscopy Soc. Amer.*, Miami Beach, Florida (C. W. Bailey, ed.), pp. 56-57. Claiton Publ. Div. Baton Rouge, LA.

Kostoff, D. (1933). Virus disease causing sterility. *Phytopath. Z.* 5: 593-602.

Kovachevsky, I. V., Arabadžiev, D., Martinov, S., Elencov, F., Markov, M., Petkov, M., Christova, E., and Danova, D. (1964). Untersuchungen über die Stolbur-Krankheit in Bulgarien. Bulgarischen Akad. der Wissenschaften, pp. 1-129.

Krczal, H. (1960). Eine vom Weissklee aug *Fragaria vesca* L. übertragbare virose. *Z. Pflkrakh. PflSchutz.* 67: 599-602.

Kreitlow, K. W. (1963). Phyllody virus infection permits rooting of clover flower heads. *Pl. Dis. Reptr.* 47: 453-454.

Krujilin, S. A. (1953). Kak borotsia s uviadaniem i stolburom rastenii. *Sad and ogorod* 2: 63-67.

Kunkel, L. O. (1926). Studies on aster yellows. *Amer. J. Bot.* 13: 646-705.

Kunkel, L. O. (1931). Studies on aster yellows in some new host plants. *Contr. Boyce Thompson Inst. Pl. Res.* 3: 85-123.

Kunkel, L. O. (1953). Aster yellows. *In* "Plant Diseases." U.S. Dept. Agr. Yearbook of Agr. 1953. pp. 642-645.

Lackey, C. F., and Bennett, C. W. (1949). Pathological effects of curly-top virus on dodder. *Phytopathology* 39: 860.

Leclant, F. (1968). Premières observations sur *Hyalesthes obsoletus* Sig. dans le midi de la France. *Ann. Épiphyties* 19: 111-113.

Lee, P. E., and Chykowski, L. N. (1963). Infectivity of aster-yellows virus preparations after differential centrifugation of extracts from viruliferous leafhoppers. *Virology* 21: 667-669.

Leeuw, G. T. N. (1977). Mycoplasma's in planten. *Natur en Techniek* 45: 76-89.

Lehmann, W. (1973). *Euscelis plebejus* (Fallén) und *Macrosteles cristatus* Ribaut als Überträger von Pflanzenkrankheiten von vermutter Mycoplasma-Ätiologie. *Arch. Phytopathol. u. Pflanzenschutz* 9: 363-370.

Lemcke, R. M. (1972). Osmolar concentration and fixation of Mycoplasmas. *J. Bacteriol.* 110: 1154-1162.

Lesemann, D., Casper, R. (1970). Mycoplasma-like bodies in Kaktee mit Hexenbesenwuchs. *Phytopath. Z.* 67: 175-179.

Liss, A., and Maniloff, J. (1971). Isolation of Mycoplasmatales viruses and characterization of MVL 1, MVL 52, and MVG 51. *Science* 173: 725-727.

Lombardo, G., Bassi, M., and Gerola, F. M. (1970). Mycoplasma development and cell alterations in white clover affected by clover dwarf. An electron microscopy study. *Protoplasma* 70: 61-71.

Luckwill, L. C. (1957). Studies of fruit development in relation to plant hormones. IV Acidic auxins and growth inhibitors in leaves and fruits of apple. *J. Hort. Sci.* 32: 18-33.

Maillet, P. L. (1970). Infection simultanée par des particules de type PLT (Rickettsiales) et de type PPLO (Mycoplasmatales) chez un insecte vector de la phyllodie du trèfle *Euscelis lineolatus* Brullé (Homoptéra: Jassidae). *J. Microscopie* 9: 827-832.

Maillet, P. L., and Gouranton, J. (1971). Étude de cycle biologique du mycoplasme de la phyllodie du trèfle dans l'insecte vecteur *Euscelis lineolatus* Brullé (Homoptera: Jassidae). *J. Microscopie* 11: 143-162.

Maillet, P. L., Gourret, J. P., and Hamon, C. (1968). Sur la présence de particules de type mycoplasme dans le liber de plantes atteintes de maladies du type "jaunisse" (aster yellows, phyllodie du trèfle, stolbur de la tomate) et sur la parentéultrastructurale de ces particles avec celles trouvées chez divers insectes homoptères. *C. R. Acad. Sc.,* Sér. D 266: 2309-2311.

Maramorosch, K. (1956). Semiautomatic equipment for injecting insects with measured amount of liquids containing viruses or toxic subtances. *Phytopathology* 46: 188-190.

Maramorosch, K. (1960). Leafhopper-transmitted plant viruses. *Protoplasma* 2: 457-266.

Maramorosch, K. (1974). Mycoplasmas and rickettsiae in relation to plant diseases. *Ann. Rev. Microbiol.* 28: 301-324.

Maramorosch, K. (1976). Plant mycoplasma diseases. In: *Encyclopedia of Plant Physiology.* P. H. Williams and R. Haytefuss, eds. Springer: 150-171.

Maramorosch, K. (1976). Plant mycoplasma diseases. *In* "Encyclopedia of Plant Physiology" (P. H. Williams and R. Haytefuss, eds.). Springer: 150-171.

Maramorosch, K. (1979). Aster yellows spiroplasma ATCC 29747. Abstracts 79th Annu. Meeting Amer. Soc. Microbiol.: 85.

Maramorosch, K., and Jernberg, N. (1970). An adjustable multiple-insect holder for micro-injection. *J. Econ. Entomol.* 63: 1216-1218.

Maramorosch, K., and Kondo, I. (1978). Aster yellows spiroplasma: infectivity and association with a rod-shaped virus. *Zbl. Bakteriol. Parasitenk. Infektionskrankh. und Hyg.* A241: 196.

Maramorosch, K., Shikata, E., and Granados, R. R. (1968). Structures resembling mycoplasma in diseased plants and in insect vectors. *Trans. NY Acad. Sci.,* Ser. 2, 30: 841-855.

Maramorosch, K., Granados, R. R., and Hirumi, H. (1970). Mycoplasma diseases of plants and insects. *Adv. Virus Res.* 16: 135-193.

Maramorosch, K., Hirumi, H., Kimura, M., and Bird, J. (1975). Mollicutes and rickettsialike plant disease agents (Zoophytomicrobes) in insects. *Ann. N. Y. Acad. Sci.* 266: 276-292.

Marchoux, G., Giannotti, J., and Laterrot, H. (1969). Le stolbur P une novelle maladie de type jaunisse chez la tomate. Symptôme et examen cytologique des tissus au microscope électronique. *Ann. Phytopathol.* 1: 633-640.

Markham, P. G., Townsend, R., Bar-Joseph, M., Daniels, M. J., Plaskitt, A., and Meddins, B. M. (1974). Spiroplasmas are the causal agents of citrus little-leaf disease. *Ann. Appl. Biol.* 78: 49-57.

Marwitz, R., and Petzold, H. (1972). Nachweis mycoplasmaählicher Organismen in *Dactylorhiza majalis* (Rchb.) Hunt et Summerh. mit Vergilbungssymptomen. *Phytopath. Z.* 75: 360-364.

Marwitz, R., and Petzold, H. (1978). Examination of mycoplasma-like organisms in yellows diseased plants by scanning electron microscope. *3rd Int. Congress of Plant Pathol.,* München 16-23 Aug. Abstracts of papers, p. 78.

Marwitz, R., Petzold, H., and Özel, M. (1974). Untersuchungen zur Übertragbarkeit des möglichen Erregers der Triebsuchtkrankheit auf einen krautigen Wirt. *Phytopath. Z.* 81: 85-91.

Michailova, P. V. (1935). Pathologico-anatomical changes in the tomato incident to development of woodiness of the fruit. *Phytopathology* 25: 539-558.

Milkus, B. N. (1974). Mycoplasma or chlamydiaelike bodies in grape affected by Marbour. *Acta Phytopath. Hung.* 9: 385-388.

Mišiga, S. (1961). Prenos niektorých žltačkových virusov kukučinom. *Biológia* (Bratislava) 16: 340-350.

Moreau, J. P., Cousin, M. T., and Lacote, J. P. (1968). Rôle de jasside *Aprodes bicinctus* Schrk. (Homoptères -Auchénorrhynques) dans la transmission de la jaunisse du trèfle blanc. *Ann. Épiphyties* 19: 103-110.

Morvan, G., Giannotti, J., and Marchoux, G. (1973). Untersuchungen über die Aetiologie des chlorotischen Blattrolls des Aprikosenbaums. Nachweis von Mycoplasmen. *Phytopath. Z.* 76: 33-38.

Müller, H. M., Kleinhempel, H., Rabitschuch, A. W., Spaar, D., and Müller, J. H. (1975). Mykoplasmen in der Phytopathologie. Bibliographisches Verzeichnis der Literatur für die Jahre 1967-1974. (852 Titel). Inst. für Landwirtschaftliche Information und Dokumentation, Berlin.

Murant, A. R., and Roberts, I. M. (1971). Mycoplasma-like bodies associated with Rubus stunt disease. *Ann Appl. Biol.* 67-385-393.

Musil, M. (1965). Übertragung der gelbsuchtviren durch die zwergzikade *Euscelis plebejus* Fallen). *Biologické práce* 11: 1-85.

Musil, M., and Valenta, V. (1958). Prenos stolburu a pribuzných vírusov pomocou niektorých cikáď. *Biológia* (Bratislava) 13: 133-136.

Nasu, S., Sugiura, M., Wakimoto, T., and Iida, T. T. (1967). On the pathogen of rice yellow dwarf virus. *Ann. Phytopathol. Soc. Japan* 33: 343-344.

Neklyudova, E. T., and Dikii, S. P. (1973). Tarnished plant bugs as vectors of stolbur of Solanaceae. *Ref. Rev. Pl. Pathol.*: 59: 1866.

Nichiporich, W. (1965). The aerial migration of the six-spotted leafhopper and the spread of the virus disease aster-yellows. *Int. J. Biochim. Biomet.* 9: 219-227.

Nichols, C. W., and Shalla, Th. A. (1960). Le dépérissement de poiries et sa propagation apparent en direction du Sud, le long de la côte Pacifique de l'Amérique du Nord. *FAO Plant Prot. Bull.* 9: 39-42.

Nienhaus, F., Rumbos, I., and Green, S. (1978). Richettsialike organisms isolated from plants, cultivated in chick embryo. *3rd Int. Congress of Plant Pathol.*, München, 16-23 Aug. Abstracts of papers, p. 62.

Orad, A. G., and San Roman, F. P. (1954). Conditions which determine spindling sprout of potato in Spania. *Proc. 2nd Confer. Potato Virus Disease*, Lisse-Wageningen, pp. 160-170.

Panjan, M. (1950). Ispitivanje stolbura Solanacea in năcin suzbijanja. *Zaăstita bilja* 2: 49-58.

Pellegrini, S., Belli, G., and Gerola, E. M. (1969). Mycoplasma-like bodies in rice plants infected with a yellows-type disease. *Gior. Bot. Ital.* 103: 395-399.

Penzing, O. (1921). Pflanzen-Teratologie. 2nd ed. Berlin.

Petre, Z., and Ploaie, P. G. (1971). Proliferarea şi aspermia la ceapă, o nouă boală produsă de microplasma in Romanânia. *Analele Inst. Cercet. Protectia Plantelor* 9: 13-17.

Petzold, H., Marwitz, R., and Kunze, L. (1973). Elektronenmikroskopische untersuchungen über intrazellulâre rickettsienähnliche Bakterien in triebsuchtkranken Äpfeln. *Phytopath. Z.* 78: 170-181.

Pilipenko, K. D. (1959). Stolbur tomatiev na Ukraini i Moldavii. *In* "Virishi hvorobi seliskogospodarskih roslin na Ukraini" (S. M. Moscoveţ, ed.), pp. 74-86. Vidabnitvo Ukrainskoi Akad. Seliskogospodarskih Nauk, Kiev.

Plavšić, B., Buturović, D., Krivokapíć, K., and Erić, Ž. (1978). Some characteristics of Mycoplasma-like infection and the effects of kinetin on the MLO-infected plants. *3rd Int. Congress of Plant Pathol.*, München 16-23 Aug. Abstracts of papers, p. 81.

Ploaie, P. G. (1960). Contribuţii la studiul transmiterii virusului stolburului prin cicada *Hyalesthes obsoletus* Sign. *Studii şi cercetări de biologie, Seria biologie vegetală* 12: 497-505.

Ploaie, P. G. (1966). Cercetări asupra virusului filodiei trifoiului (clover phyllody virus) izolat în România. *Studii şi cercetări de biologie, Seria botanică* 18: 569-577.

Ploaie, P. G. (1967). Electron microscopic evidence of virus-like particles in both plant and vector infected with clover phyllody virus. *In* "Plant Virology." Proc. 6th Conf. Czechosl. Plant Virologists, Olomouc (C. Blattný, ed.), pp. 134-17. Academia Publishing, Prague.

Ploaie, P. G. (1968a). Mycoplasma-or PLT group-like organism in *Vinca rosea* infected with European yellows-type disease. *Meded. Rijksfakulteit Landbouwwetenschappen Gent* 33: 1223-1231.

Ploaie, P. G. (1968b). Ultrastructura agenţilor de tipul micoplasma (mycoplasma-like agents) asociaţi cu bolile Europene de tipul clorozelor. *Microbiologia*, Vol. I. Conferinţa Naţională de microbiologie generală şi aplicată, Bucereşti 4-7 Dec., pp. 85-88.

Ploaie, P. G. (1968c). Probleme actuale privind prezența agenților de tipul micoplasma (pleuropneumonia-like organisms) sau psitacoză-limfogranulomă-trahomă în bolile plantelor. *Microbiologia*, Vol. I. Conferința Națională de microbiologie generală și aplicată, București 4-7 Dec., pp. 77-82.

Ploaie, P. G. (1969a). Cercetări asupra agenților de tipul clorozelor izolați în România. *Teză de doctorat*, Institutul de biologie Traian Săvulescu, București, pp. 1-280.

Ploaie, P. G. (1969b). Aster yellows agent in Romania. *Rev. Roum. Biol.-Botanique* 14: 335-339.

Ploaie, P. G. (1970). Association of mycoplasma-like organisms with clover phyllody in Romania. *Rev. Roum. Biol.-Botanique* 15: 443-445.

Ploaie, P. G. (1971a). Dezvoltarea bolilor și prevenirea lor în condiții de seră și solarii. *Sinteza 782*. Centrul de informare și documentare pentru agricultură și silvicultură, pp. 1-135.

Ploaie, P. G. (1971b). Particles resembling viruses associated with mycoplasma-like organism in plants. *Rev. Roum. Biol.-Botanique* 16: 3-6.

Ploaie, P. G. (1971c). Cîteva probleme actuale privind particularitățile și etiologia unor boli proliferative de la plante izolate în România. *Studii și cercetări de biologie, Seria botanică* 23-181-192.

Ploaie, P. G. (1971d). Micoplasma în bolile plantelor. Progrese și perspective. *Studii și cercetări de biologe, Seria botanică* 23: 87-96.

Ploaie, P. G. (1972). Boli de tip proliferativ ale plantelor. *Sinteza 781*. Centrul de informare și documentare pentru agricultură și silvicultură, pp. 1-80.

Ploaie, P. G. (1973a) Micoplasma și bolile proliferative ale plantelor. Editura "Ceres," București, pp. 1-178.

Ploaie, P. G. (1973b). Rickettsia-like bodies associated with plant cells. *Rev. Roum. Virol.* 10: 319-320.

Ploaie, P. G. (1976). Observation of molecular planes in protein crystals of plant cells. *Rev. Roum. Biol.-Biol. veǵ.* 21: 3-5.

Ploaie, P. G. (1977). Mikoplazmi i spiropazmi-vozbuditeli nekotorîh zabolevanii rastenii. *Soviet Economiceskoi Vzaimopomoșci, Koordinationnii tentr.* Poznan, pp. 1-75.

Ploaie, P. G. (1978). Present problems in research of Mycoplasma-like organisms in association with plant diseases in Romania. *3rd Int. Congress of Plant Pathol.*, München 16-23 Aug. Abstracts of papers, p. 64.

Ploaie, P. G., and Ionică, M. (1968). Micoplasma (PPLO) in glanda salivară a vectorului stolburului *Hyalesthes obsoletus* Sign. *Conferinta Natională de microbiologie generală și aplicată*. Rezumatul comunicărilor. București 4-7 Dec., pp. 15-16.

Ploaie, P. G., and Maramorosch, K. (1969). Electron microscopic demonstration of particles resembling mycoplasma or psittacosis-lymphogranuloma-trachoma group in plants infected with European yellows-type diseases. *Phytopathology* 59: 536-544.

Ploaie, P. G., Granados, R. R., and Maramorosch, K. (1968). Mycoplasma-like structures in periwinkle plants with Crimean yellows, European clover dwarf, stolbur and parastolbur. *Phytopathology* 58: 1063.

Ploaie, P. G., Ionică, M., Petre, Z., and Tapu, Z. (1975). Cloroza gramineelor in România. Date experimentale privind natura agentului cauzal. *Analele Inst. cercet. Protecția plantelor* 11: 15-22.

Ploaie, P. G., Petre, Z., and Ionică, M. (1977). Identificarea unor boli din grupul proliferării, clorozelor și nanismului la plante asociate cu micoplasma în România. *Analele Inst. cercet. protectia plantelor* 12: 17-25.

Protsenko, A. E. (1958a). O vzaimootnoshenii virusov stolbura (*Leptomotropus solani* Ryzkov) i zheltukhi astr (*Leptomotropus callistephy* Ryschkov). *Vopr. Virusol.* 3: 292-296.

Protsenko, A. E. (1958b). Electronnoskopia fitopatoghenih virusov I. Virusi zheltukhi astr. *Microbiologhia* 27: 128-130.

Protsenko, A. E., and Surgucheva, N. A. (1973). Mycoplasma-like bodies in *Erysimum* plantain tissue and their morphology. *Selskokhoz. Biol.* 8: 629-622.

Quoirin, M., Vanderveken, J., and Cousin, M. T. (1970). Des microorganismes de type "mycoplasme" associés à des maladies de plantes. *Annales de Gembloux* (Brussels) 76: 93-107.

Rasa, E. A., and Esau, K. (1961). Anatomic effects of curly top and aster yellows viruses on tomato. *Hilgardia* 30: 469-529.

Razviazkina, G. M. (1950). O rasprostranenii virusa stolbura v priorde. *Mikrobiologhia* 19: 256-259.

Refati, E. (1967). La "moria" du poirier en Italie. *FAO Plant. Prot. Bull.* 12: 6-12.

Rodeia, N., and Borges, M., De Lurdes, V. (1973). Alterations in phosphorus and anthocyanins related to the presence of mycoplasmas in tomato plants. *Portugaliae Acta Biologica*, Serie A 13: 72-78.

Rozsnyay, J. (1963). La culture des apricots en Hongrie et les problémes. *Ann. Épiphyties* 14: 119-161.

Rui, D. (1950). Una malattia inedita: la virosi a scopazzi del melo. *Humus* 6: 7-10.

Rumbos, I. (1978). Rickettsialike organisms and nematode fauna associated with yellows disease of grapevines in Germany, France and Greece. *3rd Int. Congress of Plant Pathol.*, München 16-23 Aug. Abstracts of papers, p. 65.

Ryshkov, V. L., Koratshevsky, I., and Michailova, P. (1933). Über die Fruchtverholzung bei Tomaten. *Z. Pfl. Krankh.* 43: 496-498.

Saglio, P., L'Hospital, M., Laflèche, D., Dupont, G., Bové, J. M., Tully, J. G., and Freundt, E. A. (1973). *Spiroplasma citri* gen. nov., and sp. nov: a mycoplasma-like organism associated with "stubborn" disease of citrus. *Int. J. Syst. Bacteriol.* 23: 191-204.

Samundzheva, E. M. (1949). Nekótorye dannye po izucheniya perenoschika stolbura tomatov v Gruzii. *In* "Work of the Institute of Plant Protection." Acad. Sci. of the Giorgian SSR 6: 154-160.

Sánchez-Capuchino, J. A., Llácer, G., Casanova, R., Forner, J. B., and Bono, R. (1976). Epidemiological studies on fruit tree mycoplasma disease in the eastern region of Spain. *Acta Horticulture* 67: 109-111.

Sǎvulescu, A., and Ploaie, P. G. (1961). Phyllodia trifoiului în R. P. R. oovirozǎ din grupa stolburului? *Communicǎrile Academiei R. P. R.* 11: 1357-1363.

Sǎvulescu, A., and Ploaie, P. G. (1964). Studies on the incidence of the stolbur virus disease in relation to the vector, host plant and ecological conditions. *In* "Plant Virology." Proc. 5th Conf. Czechosl. Plant Virologists, Prague, 1962 (C. Blattný, ed.), pp. 195-202. Academia Publishing, Prague.

Sǎvulescu, A., and Ploaie, P. G. (1967a). Comparative studies on the stolbur and clover phyllody viruses isolated in R. S. Romania. *In* "Plant Virology." Proc. 6th Conf. Czechosl. Plant Virologists, Olomouc (C. Blattný, ed.), pp. 236-240. Academia Publishing, Prague.

Sǎvulescu, A., and Ploaie, P. G. (1967b). Virogeographische Studien über das Kleeverlaubungsvirus und seine Vektoren. *Phytopath. Z.* 58: 315-322.

Sǎvulescu, A., and Pop, I. (1956). Contribuţii la studiul stolburului în România. *Bul. stiintific, Secţia de biologie agric.* 8: 723-737.

Seemüller, E. (1976). Investigations to demonstrate mycoplasma-like organisms in diseased plants by fluorescence microscopy. *Acta Horticulture* 67: 109-111.

Seidl, V., and Komárková, V. (1976). Vliv mycoplasmózni proliferance (metlovitosti) jabloňe na tvozbu a kličvost semen. *Acta Inst. Bot. Acad. Sci. Slov.* Bl: 247-256.

Selydiyko, Y. U., and Reifman, V. G. (1978). Viroidî- novîi class patoghenov. Izdatelystvo "Nauka." Moscva, pp. 3-86.

Semancik, J. S., Weathers, L. G. (1972). Exocortis virus: an infectious free nucleic acid plant virus with unusual properties. *Virology* 47: 456-466.

Shikata, E., Maramorosch, K., and Ling, K. C. (1969). Presumptive mycoplasma etiology of yellows diseases. *FAO Plant Prot. Bull.* 10: 121-128.

Silvere, A. P. (1970). Mycoplasma-like organisms in association with black currant reversion. 10th Int. Congress of Microbiology, Mexico City. Abstracts of papers. 222.

Slogteren, D. H. M. van (1974). Mycoplasmas in bulb crops, particularly in gladiolus and hyacynth. *Antonie van Leeuwenhoek J. Microbiol. and Serol.* 40: 314.

Slogteren, D. H. M. van, and Muller, P. J. (1972). "Lissers" a yellows disease in hyacinths, apparently caused by a mycoplasma. *Meded. Fakulteit Landbouwwetenschappen* Gent 37: 449-457.

Smith, R. E. (1902). Growing China aster. *Hatch. Exp. Sta. Massachusetts Agr. Col. Bul.* 79: 1-26.

Spaldon, E. (1958). Opening address. *In* "Stolbur and similar virus diseases causing seedlessness of plants." Proc. Conf. on stolbur, Smolenice, 1956 (E. Spaldon and C. Blattný, eds.), pp. 25-33. Slovak Acad. Publishing House, Bratislava.

Steere, R. L. (1967). Gel filtration of aster-yellows virus. *Phytopathology* 57: 832-833.

Su, H. J., Lei, J. O., and Chen, T. A. (1978). Spiroplasmas isolated from green leaf bug (*Trigonotylus ruficornis*) and rice plants. *3rd Int. Congress Plant Pathol.*, München 16-23 Aug. Abstracts of papers, p. 61.

Sukhov, K. S., and Vovk, A. M. (1946). Cikadka *Hyalesthes obsoletus* Sign., perenoschik stolbur paslyonovykh. *Doklady Acad. Nauk SSSR* 53:153-156.

Sukhov, K. S., and Vovk, A. M. (1949). Stolbur paslyonovykh. Moscow-Leningrad.

Surgucheva, N. A., and Protsenko, A. E. (1971). Mycoplasma-like bodies in cells of mint affected with witches' broom. (russ). *Dokl. Acad. Nauk SSSR*, Ser. Biol. 200: 1447-1448.

Suringar, W. F. R. (1882). Observations sur une monstruosité de *Sisymbrium alliaria* avec phyllodie des carpelles et des ovules. Ass. Franç. Avancem. Sci., pp. 444-449.

Tahon, J. (1963). Decouverte en Belgique de plantes de *Trifolium repens* atteintes de phyllodie. *Parasitica* 19: 172-175.

Tapio, E. (1970). Virus diseases of legumes in Finland and in the Scandinavian countries. *Ann. Agric. Fenn.* 9: 1-97.

Todd, J. M. (1958). Witches' broom of potatoes and similar diseases in Great Britain. *In* "Stolbur and similar virus diseases cuasing seedlessness of plants." Proc. Conf. on stolbur, Smolenice, 1956 (E. Spaldon, and C. Blattný, eds.), pp. 77-101. Slovak Acad. Publishing House, Bratislava.

Ulrychová, M., and Limberk, J. (1964). Some metabolic disturbances in tomato plants infected with potato witches' broom. *Biologia plantarum* (Praha) 6: 291-298.

Ulrychová, M., and Limberk, J. (1967). Metabolic disturbances in *Nicotiana glauca*, a symptomless carrier of potato witches' broom. *Biologia plantarum* (Praha) 9: 56-60.

Ulrychová, M., and Sosnova, V. (1970). Effect on phosphorus deficiency on anthocyanin contenr in tomato plants. *Biologia plantarum* (Praha) 12: 231-235.

Vago, C., and Giannotti, J. (1971). Evolution et état actuel des recherches sur le mycoplasmes des végétaux et des vertebres. *Bull. de la Soc. de Sc. Veterinaires et le Médicine Comparée de Lyon* 73: 69-82.

Vago, C., and Giannotti, J. (1972). Les mxcoplasme chez les végétaux et chez les vecteurs. *Physiol. Vég.* 10: 87-101.

Vago, C., and Giannotti, J. (1977). Infection comlexe a virus mycoplasmes et rickettsoides a l'échelle d'un meme cellule chez des plantes malades. *Travaux dédiés àG. Viennot-Bourgin*, pp. 393-403.

Valenta, V. (1956). Virus metlovitosti zemiakov v Československu (prispevok k oktoźke "Severneho" stolburu. *Biológia (Bratislava)* 11: 449-456.

Valenta, V. (1958). A new yellows virus causing flower proliferations in dodder, *Cuscuta campestris* Yunck. *Phytopathol. Z.* 33: 316-318.

Valenta, V., and Musil, M. (1963). Investigations on European yellows-type viruses. II. The clover dwarf and parastolbur viruses. *Phytopathol. Z.* 47: 38-65.

Valenta, V., Musil, M., and Mišiga, S. (1961). Investigations on European yellows-type viruses.

I. The stolbur virus. *Phytopathol. Z.* 42: 1-38.

Vovk, A. M., and Nikiforova, S. G. (1955). Issledovanije virusa stolburu v elektronomikroscope. *Dokladi Acad. Nauk SSSR* 102: 839-840.

Vries, H. de (1896). Een epidemie van vergroeningen. *Bot. Jaarb.* Gent 8: 66-91.

Waters, H., Eden-Green, S. J., and Dabek, A. J. (1978). Coconut lethal yellowing and diseases of other plants associated with mycoplasma-like organisms in Jamaica. *3rd Int. Congress Plant Pathol.*, München 16-12 Aug. Abstracts of papers, p. 79.

Westphal, E., and Heitz, B. (1971). Mis en évidence de mycoplasmes dans le phlòeme de Linum austriacum L. atteint de virescence. *C. R. Acad. Sc.*, Sér D 272: 2552-2554.

Whitcomb, R. F. (1973). Diversity of procaryotic plant pathogens. *Proc. North Central Branch E. S. A.* 28: 38-60.

Whitcomb, R. F., and Davis, R. E. (1970). Mycoplasma and phytarboviruses as plant pathogens persistently transmitted by insects. *Ann. Rev. Entomol.* 15: 405-464.

Whitcomb, R. F., and Tully, J. G., eds (1979). The Mycoplasmas III. Academic Press, New York.

Chapter 4

EPIDEMIOLOGY OF DISEASES CAUSED BY LEAFHOPPER-BORNE PATHOGENS*

Lloyd N. Chiykowski

Chemistry and Biology Research Institute
Agriculture Canada
Ottawa, Ontario, Canada

Contribution No. 1129 Chemistry and Biology Research Institute.

4.1 INTRODUCTION

The epidemiology of diseases caused by leafhopper-borne plant pathogens involves four components, namely, the pathogen, the leafhopper, the plant, and the environment (Maramorosch and Harris, 1979). Thus, a plant pathogen is transmitted by the leafhopper to a plant within a certain environment. While this situation may appear to be relatively simple, it is complicated by the fact that each component can be affected by a variety of factors. The role of the pathogen can be affected by its availability, quantity, location and strains. The insect is influenced by such factors as population size, number of generations, longevity, diapause, dispersal patterns, feeding behavior and reaction to the pathogen. The performance of the plant may be influenced by its susceptibility to the pathogen or strains of pathogen, susceptibility to the insect, location, and stage of growth. Environmental factors may be in the form of temperature, moisture, air currents, predators or various cultural practices. The study of epidemiology of diseases caused by leafhopper-borne pathogens then becomes a highly complex problem involving the interaction of factors within components and between components. Discovery that a new disease is associated with a leafhopper-borne pathogen is generally just the beginning of a long and arduous task toward understanding its epidemiology. In this chapter, an attempt has been made to outline the roles played by the four components and to discuss the influence of various factors on their performance.

4.2 IDENTIFICATION OF DISEASES

In order to understand the epidemiology of a disease one must first be able to recognize and distinguish it from other similar diseases. This has often proven to be a difficult and frustrating task because of the similarities that exist among some diseases and the lack of suitable criteria for their identification.

Although symtomatology is not the most reliable tool for such identification, it is still widely used mainly because it is often the only one available. Used in combination with selected plant hosts, its reliability can be greatly enhanced. Musil (1967), for example, used symptom differences in *Trifolium repens* L., *Vinca rosea* L., *Chrysanthemum carinatum* Schousb., and *Senecio vulgaris* L., to show the presence of six different yellows type diseases in Czechoslovakia. Three California strains of aster yellows were differentiated by Freitag (1963, 1964) on the basis of symptoms developing on *Nicotiana rustica* L., *Plantago major* L., *V. rosea* and *Callistephus chinensis* Nees.

One of the characteristic symptoms commonly used to identify a disease as belonging to the yellows group is a phylloid or virescent flower produced by the infected plant. A review of the literature indicates that there are a number of these so-called yellows diseases which produce flowers that are reduced in size and somewhat faded in color but never phylloid or virescent. This reduced flower characteristic, best seen on infected periwinkle, *V. rosea*, has been associated with blueberry stunt (Hutchinson *et al.*, 1960), clover witches' broom (Posnette and Ellenberger, 1963), peach X-disease (Gilmer *et al.*, 1966), metastolbur (Musil, 1967), flavescence doreé of grape (Caudwell *et al.*, 1973). Thus, yellows diseases can be divided into two groups, based on flower symptoms and future studies may reveal what biological or morphological properties the pathogens within each group have in common.

Although considerably more work is required a combination of symptomatology with other characteristics such as host range, vector species, and vector-pathogen relationships, generally produces more reliable results. For several years, the only leafhopper transmitted pathogen recognized in Canada was aster yellows agent of which there were believed to be two strains. On the basis of the above characteristics, four additional diseases and one additional strain of aster yellows agent have been identified (Chiykowski, 1962a, 1965a, 1969; Raine, 1967; Westdal and Richardson, 1969). Similarly, Lindsten *et al.*, (1970) distinguished three new cereal diseases in Sweden and, on the basis of symptoms and pathogen transmission characteristics, identified them as being closely related or identical to oat blue dwarf, aster yellows and wheat dwarf diseases. Vector species, host range, host symptoms and reaction to heat *in vivo* were used by Posnette and Ellenberger (1963) to distinguish strawberry green petal, clover witches' broom, stolbur and Delphinium yellows from one another and from aster yellows, tomato big bud and cranberry falseblossom. Vector transmission and symptom expression have been successfully used in some cases to show that diseases occurring in different hosts and under different names may be caused by the same pathogen. Strawberry green petal, for example, was found to be caused by the pathogen responsible for clover phyllody (Frazier and Posnette, 1957; Chiykowski, 1962b) and, in Upper Volta, the phyllody diseases of cotton and sesame were similarly shown to be the result of infection by the same pathogen (Desmidts and Laboucheix, 1974).

Corn stunt disease serves as a good example of the confusion which can occur in disease identification based largely on symptomatology and its effect on understanding the epidemiology of a disease. For a number of years the disease was considered to be associated with several pathogen strains, differing somewhat in disease symptomatology, vector species, transmission behavior and geographic distribution. Based on the results of serology, rate-zonal density gradient centrifugation, vector transmission and light and electron microscopy, four disease agents have now been identified as being responsible for the corn stunt disease complex. The reader is referred to a recent review of Nault and Bradfute (1979) for a detailed account on corn stunt.

A disease complex which must still be sorted out occurs in southeast Asia and is comprised of a number of diseases characterized by stunting and yellow or orange discoloration in rice. Tungro (Rivera and Ou, 1965), penyakit merah (Ou *et al.*, 1965), penyakit habang (Saito *et al.*, 1975), yellow-orange leaf (Wathanakul and Weerapat, 1969) and Waika (Hirao and Inoue, 1978) diseases display a similar symptomatology and are transmitted in a semipersistent manner by *Nephotettix* species (Ling, 1966; Ting and Paramsothy, 1970; Singh, 1969; Saito *et al.*, 1975; Hino *et al.*, 1974; Hirao and Inoue, 1978). Electron microscopy has associated two very different virus particles with these diseases, one isometric and the other bacilliform (Galvez, 1967; Galvez *et al.*, 1971; Saito *et al.*, 1971, 1975, 1976; Nishi *et al.*, 1975). Further confusion is provided by the presence of both particle types in plants infected with some of the diseases (Saito *et al.*, 1976; Hibino *et al.*, 1978). Unlike the corn stunt complex in the United States, this one involves diseases from several countries, making direct comparisons difficult.

4.3 PATHOGENIC AGENTS

Circumstantial evidence such as filterability, graft transmissibility, and multiplication in some leafhopper vectors led earlier scientists to believe that all leafhopper-borne plant pathogens were viruses. This belief was further strengthened when the viral etiology of such diseases as wound tumor, potato yellow dwarf and rice dwarf was demonstrated (Bils and Hall, 1962; Black *et al.*, 1965; Fukushi and Shikata, 1963; Fukushi *et al.*, 1962). Today, three basic leafhopper-borne plant pathogen types are recognized, viruses, mollicutes and rickettsialike organisms.

4.3.1 Viruses

Considerable diversity exists in particle size, morphology and chemical composition among viruses transitted by leafhoppers. Rice dwarf virus, for example, is a large polyhedral particle (70 nm) made up of 32 capsomeres and containing double stranded ribonucleic acid (RNA) (Fukushi *et al.*, 1962; Fukushi and Shikata, 1963; Miura *et al.*, 1966; Kimura and Shikata, 1968). Although wound tumor virus is in the same size range and also contains double stranded RNA, it is made up of 92 capsomeres (Bils and Hall, 1962; Black and Markham, 1963). Oat blue dwarf and maize chlorotic dwarf viruses are small isometric particles, measuring ca. 30 nm in diameter, which contain single stranded RNA (Zeyen and Banttari, 1972; Pring *et al.*, 1973; Bradfute *et al.*, 1972a, 1972b; Gingery, 1976). A third and even smaller size range is recorded for maize streak virus (Boch *et al.*, 1974). These hexagonal particles, which probably contain RNA, measure ca. 20 nm in diameter and usually occur in pairs. Rounding out the diversity of the leafhopper-borne viruses are those having a bacilliform particle. Included in this group are potato yellow dwarf, wheat striate mosaic and rice transitory yellowing (MacLeod *et al.*, 1966; Lee, 1967; Shikata and Chen, 1969). Although they are generally similar in morphology some differences have been noted by Chen and Shikata (1971).

4.3.2 Mollicutes

The plant pathogenic nature of organisms belonging to the class Mollicutes was first shown by Doi *et al.* (1967) and Ishiie *et al.* (1967) thus ending years of frustration in the search for the causal agents of yellows type diseases. Mycoplasma-like organisms (MLOs) have now been implicated as the etiologic agents of numerous leafhopper associated plant diseases (Whitcomb and Davis, 1970; Maramorosch *et al.*, 1970; Maramorosch, 1974; 1976). The extreme pleomorphism displayed by these procaryotic organisms has, in most cases, prevented morphological differentiation of mycoplasma species causing different diseases. However, the corn stunt pathogen was shown to exhibit a filamentous, helical morphology during certain stages of its development (Davis *et al.*, 1972; Davis and Worley, 1973). The name "spiroplasma" was proposed for this organism and this has now been adopted as the genus name for other similar mollicutes (Saglio *et al.*, 1973; Davis, 1977). Successful *in vivo* culturing of spiroplasmas has provided proof of their pathogenicity and their relationship to such diseases as citrus stubborn (Saglio *et al.*, 1973; Daniels *et al.*, 1973; Markham and Townsend, 1974) corn stunt (Chen and Liao, 1975; Williamson and Whitcomb, 1975) and aster yellows (Maramorosch, 1979).

4.3.3 Rickettsialike Organisms

Etiologic studies stimulated by the discovery or mycoplasma as new agents of plant diseases led to the subsequent recognition of yet another group of plant pathogens, the rickettsialike organisms (RLOs). These organisms may be xylem inhabiting as in the case of Pierce's disease of grape (Hopkins and Mollenhauer, 1973; Goheen *et al.*, 1973) or limited to the phloem as in clover club leaf (Windsor and Black, 1973) and rugose leaf curl (Behncken and Gownlock, 1976). Although both types of RLOs show the ultrastructure of gram negative bacteria, they differ from one another in certain vector-pathogen relationships, with the phloem RLOs appearing to be more intimately associated with their leafhopper vectors (Black, 1948, 1950; Grylls, 1954; Purcell and Finlay, 1979). These organisms and the diseases they cause have been comprehensively covered in recent reviews by Hopkins (1977) and Purcell (1979).

4.4 MODES OF TRANSMISSION

4.4.1 Persistent

A majority of the plant pathogenic agents transmitted by leafhoppers are of the persistent type, enabling the insect to continue transmitting for a considerable time after having acquired the agent from a diseased source. Within this group, however, one encounters a variety of vector-pathogen relationships (Harris, 1979). A number of pathogens have been shown to be circulative and to multiply within their vector hosts. This type of relationship has been demonstrated for viruses (Black and Brakke, 1952; Sinha, 1965; Sinha and Chiykowksi, 1969), mycoplasmalike organisms (Black, 1941a; Maramorosch, 1952; Sinha and Chiykowski, 1967, 1968; Whitcomb et al., 1966, 1967) and rickettsialike organisms (Black, 1950). The association of curly top virus with its vector exemplifies another type of persistent transmission. Although insects lose infectivity over a period of time (Severin, 1939) some may remain infective for more than 100 days (Freitag, 1936). This virus has been shown to be circulative in the insect (Bennett and Wallace, 1938) but multiplication has not been demonstrated. A third type of persistent association is that found in the Pierce's disease bacterium and its vector. Once infective, adult leafhoppers remain so for their lifetime (Severin, 1949; Purcell, 1979). Although nymphs can acquire and transmit, recent work has shown that they failed to transmit after molting (Purcell, 1978; Purcell and Finlay, 1979). Lack of a pathogen latent period in the leafhopper and loss of infectivity after molting led the workers to conclude that the Pierce's disease bacterium is noncirculative and is retained anterior to the midgut in infective leafhoppers.

4.4.2 Transitory

Only a few plant diseases are caused by pathogens that have a "semipersistent" or transitory (Ling and Tiongco, 1979) relationship with their leafhopper vectors and all are known to have a virus etiology. Ling (1966) was first to demonstrate this type of association with the transmission of rice tungro virus by *Nephotettix virescens* (Distant). Retention of infectivity was dependent on temperature, ranging from 6 days at 32°C to 22 days at 13°C (Ling and Tiongco, 1975, 1979). Leafhoppers that acquired virus in the nymphal stage lost their infectivity following a molt. Infectivity could be restored with reacquisition feeding. Recently, similar transitory relationships have been shown for maize chlorotic dwarf virus (Nault *et al.*, 1973; Nault and Bradfute, 1979) and rice waika virus (Hirao *et al.*, 1974; Yokoyama *et al.*, 1974). The infectivity retention period of waika virus varied with vector species from 72 hr in *N. virescens* to 36 hr in *Nephotettix nigropictus* (Stål).

4.4.3 Transovarial

Relatively few plant pathogenic agents have been shown to be transmitted by infective female leafhoppers to their progeny. Rice dwarf virus, the first virus demonstrated to be transovarially transmitted, could be passed through the eggs of *Nephotettix cincticeps* (Uhler) to as high as 85% of the progeny of six succeeding generations (Fukushi, 1933, 1939, 1940). In other vector species the proportion of infective progeny was considerably lower (Shinkai, 1958, 1962; Nasu, 1963). An electron microscopic study of the transovarial passage of rice dwarf virus in *N. cincticeps* by Nasu (1965) showed that the virus is passed to the egg at the yolk-forming stage as the mycetocytes of the ovarioles, containing both symbiotes and virus particles, invade the oocytes. Two other leafhopper-borne viruses, wound tumor and potato yellow dwarf, have been shown to be transmitted in the same manner but infectivity of the progeny is extremely low, being 1.8% and 0.8%, respectively (Black, 1953).

The phenomenon of transovarial transmission is found also in the plant pathogenic rickettsialike organisms. Black (1948) demonstrated that clover club leaf RLO was passed through the egg of its vector *Agalliopsis novella* (Say) to a high proportion of its progeny. In subsequent tests, infectivity was passed to the progeny of 21 consecutive generations (Black, 1950). Rugose leaf curl agent, also a phloem-limited RLO, was transmitted to 42% of the progeny of *Austroagallia torrida* Evans (Grylls, 1954).

The only case of transovarial transmission of a plant mycoplasmalike organism is that of clover phyllody MLO reported by Frazier and Posnette (1957) and Posnette and Ellenberger (1963). However, consistent failure by these workers to obtain transmission with progeny of infective *Euscelis incisus* (Kirsch.) (= *E. plebejus*) females reared on oats and transmission only by progeny reared on wheat, a species of which is now known to be susceptible to clover phyllody (Chiykowski, 1967), suggests that the transovarial transmission status of this disease agent should be re-examined.

4.5 THE INSECT

4.5.1 Biology

Since the leafhopper is the prime means of spreading the pathogenic agents, knowledge of its biology is essential to understanding the epidemiology of a disease. Much is known about the bionomics of a number of nonvector leafhopper species, especially *Eucelis* spp. (Müller, 1979). Unfortunately, however, because of the time and effort required to obtain such data, the biology, ecology and behavior of many vectors, other than their ability to transmit, are relatively unknown or poorly understood.

One of the vector species which has been extensively studied and which serves as an example of the problems involved is *N. cincticeps*, the principal vector of rice dwarf virus. Life cycle studies have revealed several important factors governing insect development and population size (Kuno, 1968, 1973; Kiritani *et al.*, 1970; Hokyo, 1972). The number of generations per year may vary, depending on the geographical location, and number and timing of rice crops. Fecundity of females in the field is considerably below their potential and was attributed partly to the short longevity of the adults due to predation and dispersal as a result of crowding. Parasites and predators can cause egg mortality ranging from 10 to 90%, depending on geographical area. Nymphal mortality can range from 25 to 28%, having a tendency to rise towards autumn, similar to that found in eggs. According to Kiritani *et al.* (1972), spiders are the most important factor responsible for such nymph mortality. On the basis of life tables Kiritani *et al.* (1970) and Hokyo (1972) have concluded that *N. cincticeps* has several mechanisms of population regulation that come into operation at different densities. Escape of the population from one mechanism automatically brings into action another density dependent mechanism. The density is regulated within a moderate level and the energy for further population growth is converted to the activity for dispersal to neighboring less crowded areas. Sasaba (1974) has shown that intraspecific competition reduces the number of survivors. Density of adults effects the number of eggs laid and survival rate from egg to adult emergence. Longevity of emerged adults increased with decreasing parental or nymphal density. Mathematical model systems have been formulated to describe the life cycle of *N. cincticeps*, taking into consideration such biological factors as parasitism, predation, intraspecific competition and dispersal (Sasaba and Kiritani, 1975; Hokyo, 1976).

Extensive studies of the beet leafhopper, *Neoaliturus tenellus* (Baker) (= *Circulifer tenellus*) have revealed a similar complex biological pattern of development. The insect was shown to have great adaptability and a tremendous potential for increase. Although it is essentially a desert insect, it has many variations in pattern of breeding and dispersal in different areas depending on environmental conditions. Factors affecting the development of *N. tenellus* have been fully dealt with in studies by Knowlton (1932), Severin (1933), Wallace and Murphy (1938) and Lawson *et al.* (1951) and in reviews by Bennett (1967, 1971).

The information gained from biological studies often can be applied to the rearing of vector species under artificial conditions (Mitsuhashi, 1979) thereby making available disease-free populations for transmission work. *Aphrodes bicinctus* (Schrank), for example, a univoltine species in the Ottawa area was shown to lay diapausing eggs during the late summer. Exposure of these eggs to low temperature for several weeks followed by a return to normal growing conditions resulted in the emergence of nymphs (Chiykowski, 1970). Over the past few years this method has provided several thousand insects for experimental purposes. Cold treatment of *Elymana sulphurella* (Zett.) eggs has given similar results (Chiykowski, unpublished). Some species, such as *Colladonus clitellarius* (Say), may require vastly different plant species for oviposition and nymphal development (George and Davidson, 1959). Others may be greatly influenced by temperature as *E. inimica* (Westdal and Richardson, 1966; Coupe and Schultz, 1968) or by day-length as in the case of *Scaphytopius delongi* Young (Swenson, 1971) and *N. cincticeps* (Oya, 1978).

The taxonomic relationships of leafhopper vectors have been recently reviewed by Nielson (1979).

4.5.2 Movement

The extent to which leafhopper-borne plant pathogens spread is dependent upon the type and extent of insect movement. These movements may be of a local type as between plants in a field, or of a dispersive type as from breeding areas or overwintering sites to crops, or of a migratory type in which the leafhoppers may move considerable distances varying from several miles to several hundred miles. Factors responsible for these movements are many and, as pointed out by Carter (1961), may involve such biotic factors as the normal life history of the insect, its host range and host preferences, the availability and condition of these hosts and their status as disease reservoirs and physical factors of the environment.

Much of our knowledge about local movement of leafhoppers in the field has been gained by extrapolation from results obtained in studies conducted under artificial conditions. When observed on caged plants, leafhoppers will remain in one place for hours if not disturbed. This sedentary behavior, however, can be affected by several factors. Ling (1975a) showed that watering of seedlings in the greenhouse caused insects to move and resulted in an increase in transmission of rice tungro disease. Resistant cultivars of rice caused leafhoppers to move more often than did susceptible ones (Ling and Carbonell, 1975). Similarly, more movement was observed on nonpreferred than on preferred hosts of the beet leafhopper (Thomas, 1972). Light intensity and temperature were also shown to be important factors in influencing movement (Kuwahara, 1974). In the field, such environmental factors as light, rain, and wind or physical disturbances such as passage of cultivating equipment or irrigation would have similar disturbing effects. Rose (1974) for example, suggested that increased infection rates of maize streak virus in plots of maize where infected plants were removed daily were presumably the results of the disturbance of vectors which then flew to surrounding healthy plants.

In contrast to these local movements are those in which insects move considerable distances. The movement of the beet leafhopper from local or distant over-wintering and breeding areas has been well documented (Knowlton, 1932; Wallace and Murphy, 1938; Dorst and Davis, 1937; Lawson et al., 1951; Fulton and Romney, 1940) and reviewed (Bennett, 1967, 1971; Delong, 1971). Distances covered by migrating leafhoppers vary considerably. In some parts of the western United States, migrations of 30 to 60 miles are commonplace while in others, these distances may be close to 400 miles. The routes followed by the leafhopper in its migration are determined by temperature and winds that blow at times most favorable for flight. There is no evidence for any directional in-stinct or other form of guidance. Any consistency in routes followed is the re-sult of correlations between weather factors and only indirectly a product of biological causes (Lawson et al., 1951).

The dispersal behavior of three species of *Cicadulina* which transmit maize streak virus has been investigated in considerable detail in the high plateau region of Rhodesia and has been recently reviewed (Rose, 1978). Tethered flight exper-iments indicate that populations of these leafhoppers are composed of two morphs that are distinguishable by their mean flight duration (Rose, 1972b). The proportion of long and short-distance fliers in a population varies between species and seasonally. Flight ability is inherited. Mean distances flown have been estimated as 4 to 12 m by short-distance fliers and 1.8 km downwind by long-distance fliers at average wind speeds of 13.1 km/hr. At this wind speed it is estimated that some hoppers will travel at least 118 km in one flight (Rose, 1973). The main flight period occurs at the end of the wet season when popula-tions in natural grasses are at their highest and extends through March to Sep-tember. There was a correlation between total number of *Cicadulina* captured July through September and the amount of rain falling during the end of the previous wet season (Rose, 1972a).

Another important disease whose epidemiology in some areas is affected by long distance migration of a leafhopper vector is aster yellows. Although the principle vector, *Macrosteles fascifrons* (Stål) overwinters in the egg stage in the northern part of the United States and Canada, Drake and Chapman (1965) showed that, in many years, the first insects observed in the spring in Wisconsin were in the adult stage. Subsequent studies showed that *M. fascifrons* migrates northward several hundred miles each spring, affecting most northern states in the midwest United States (Chiykowski and Chapman, 1965) and the Prairie Provinces of Canada (Westdal et al., 1961). These spring migrant populations were found to contain a higher percentage of infective individuals than did pop-ulations that developed locally later in the season (Chiykowski, 1958). Meade and Peterson (1964) concluded that leafhoppers originating outside the state of Minnesota were more important as initial sources of aster yellows inoculum than were populations produced locally. Areas affected by the migration are ordinar-ily subject to infection much earlier than those involving local populations and hence control procedures would be different (Chapman, 1973). A high proportion of the leafhoppers migrating into Manitoba also were shown to be carrying oat blue dwarf virus (Westdal, 1968).

4.5.3 Specificity and Versatility of Vectors

Leafhoppers can be specific and yet versatile in their transmission of plant disease agents. While this may sound paradoxical there are many examples that this family of insects is both and it is this "paradox" that complicates our attempts to understand the epidemiology of certain diseases. Using the term "specificity" in its broadest sense, one could say that a specific relationship exists between leafhoppers and the pathogens they transmit since pathogens transmitted by leafhoppers, with rare exceptions, are not transmitted by any other group of insects. The transmission of Pierce's disease agent of grape by species of Cercopidae (Severin, 1950a) and stolbur agent by a species of Cixiidae (Valenta *et al.*, 1961), as well as by leafhoppers, are two exceptions that have been recorded.

Subfamily specificity is exhibited in the transmission of several plant disease agents. The most illustrative example of this type of association is that of Pierce's disease agent of grape. All 24 known leafhopper vectors of this pathogen belong to Cicadellinae (Nielson, 1968). Similarly, all vectors of phony peach agent are found in the same subfamily. Transmission of potato yellow dwarf and wound tumor viruses is limited to members of Agalliinae (Black, 1944). Deltocephalinae, which contains the largest number of vector species, offers other examples of subfamily specificity (Nielson, 1968).

Genus specificity, where two or more vectors are known, is frequently encountered in leafhopper transmission. Rice yellow dwarf virus, for example, is transmitted by four species, all members of *Nephotettix* (Hirao and Inoue, 1978). Only members of this genus are also responsible for vectoring rice transitory yellowing (Chiu *et al.*, 1968; Hsieh *et al.*, 1970), and rice waika viruses (Hirao *et al.*, 1974; Kimura *et al.*, 1975; Satomi *et al.*, 1975; Inoue, 1977). Maize streak virus transmission is limited to *Cicadulina* and Rose (1978) speculates that most species of this genus are capable of transmission.

Species specificity involving one disease agent and one vector species which transmits no other pathogen exists in a limited number of associations. Although Nielson (1968) lists 10 such cases, there is good indication that this number may be high. For example, among those listed is golden flavescence pathogen of grape and the vector *Scaphoideus titanus* Ball (= *S. littoralis*) in France (Schvester *et al.*, 1961). Successful transmission of the pathogen to herbaceous hosts (Caudwell *et al.*, 1969) enabled the testing of additional species and led to the discovery of two new vectors (Caudwell *et al.*, 1972). Further studies on other such specific associations will probably further reduce their number.

Pathogen specificity, in which a leafhopper species transmits only one pathogen although that pathogen may have several vector species, is another type of specificity found in leafhopper transmitted disease agents. There are some 80 leafhopper species listed by Nielson (1968) which fall into this grouping. Many of the vectors of Pierce's disease agent and western aster yellows agent are known to transmit only those respective pathogens, and some species of *Cicadulina* are limited to the transmission of maize streak virus. Other examples include *S. titanus* for golden flavescence pathogen of grape (Schvester *et al.*, 1961) and *Scaphytopius nitridus* DeL. for citrus stubborn agent (Kaloostian *et al.*, 1975). As in the case of other specificity groupings, these numbers may be exaggerated for lack of sufficient testing. For example, aster yellows agent was, for a number of years, considered to be the only pathogen transmitted by *M. fascifrons*. This species is now known to transmit clover phyllody (Chiykowski, 1962a) and clover proliferation mollicutes (Chiykowski, 1965a) and oat blue dwarf virus (Banttari and Moore, 1962). Similarly, *Dalbulus maidis* (DeL. and Wal.) and *Dalbulus elimatus* (Ball) were believed to transmit only corn stunt spiroplasma but are now known to transmit maize bushy stunt MLO and maize rayado fino virus, as well (Nault and Bradfute, 1979).

While continued studies on transmission have failed to disprove that certain degrees or catagories of specificity do exist in some leafhopper-borne pathogens, they have illustrated the considerable diversity of transmission efficiencies and abilities which some leafhopper species exhibit. An excellent example of this diversity is that shown by *N. nigropictus*. This species is capable of transmitting rice dwarf virus (RDV), a large polyhedral particle (Nasu, 1963; Fukushi and Shikata, 1963), rice tungro virus (RTV), a small spherical particle (Ling, 1970; Galvez, 1968) and rice transitory yellowing virus (RTYV), a bacilliform particle (Chiu *et al.*, 1965; Shikata and Chen, 1969). It may transmit in a persistent manner as with RDV and RTYV (Nasu, 1963; Chiu *et al.*, 1965) or in a semipersistent manner as with RTV (Ling, 1970). It is also capable of transovarially transmitting RDV to its progeny (Nasu, 1963). In addition, *N. nigropictus* is a vector of rice yellow dwarf MLO (Ouchi and Suenaga, 1963; Nasu *et al.*, 1967; Chen and Liu, 1974). Similarly, *A. novella* can transmit potato yellow dwarf virus, a bacilliform particle, wound tumor virus, a large polyhedral particle and the clover club leaf pathogen, a rickettsialike organism (Black, 1944). Other examples of leafhopper transmission of different etiologic agents can be found in the transmission of oat blue dwarf virus and aster yellows MLO by *M. fascifrons* (Frederiksen, 1964), wheat striate mosaic virus and aster yellows MLO by both *E. inimica* and *E. sulphurella* (Slykhuis, 1953; Chiykowski, 1963; Sinha, 1970; Chiykowski and Sinha, 1969) and corn stunt spiroplasma, maize bushy stunt MLO and maize rayado fino virus by *D. maidis* and *D. elimatus* (Nault and Bradfute, 1979).

4.6 THE PATHOGEN

4.6.1 Host Range

There is considerable variation among pathogens as to the plant species which they are capable of infecting. Some appear to be highly specific, affecting only one or two species or only species within one family, while others may infect large numbers of species in many families, including herbaceous and woody plants. Maize, for example, is the only known host for maize bushy stunt pathogen (Nault and Bradfute, 1979) and maize and teosinte are the only ones known for maize rayado fino virus (Gamez, 1973). However, Gamez (1977) suggests that certain yet undetermined species of wild grass probably serve as hosts both for the latter pathogen and the vector between maize planting seasons in Latin America. His suggestion is based on the fact that the growth range of teosinte is restricted to northern Guatemala and Mexico and yet, in Costa Rica, the virus and leafhoppers appear very rapidly with new corn plantings beginning at the first of the rainy season. Corn stunt agent was considered to be limited to the genus *Zea* until Markham *et al.* (1977) transmitted the pathogen to two dicotyledonous plant species using injected leafhoppers of the species *E. variegatus*. The known host range of maize streak virus includes only Gramineae but it would appear that studies on host range have been limited to only this family (Rose, 1978). All known hosts of maize chlorotic dwarf virus occur in the family Gramineae and Nault *et al.* (1976) postulate that this might be partly due to the fact that vector species are restricted to grasses, thus preventing plant species outside this family from being tested. This then raises the question of how much of the plant specificity is due to the pathogen itself and how much is dependent on plant specificity of the leafhopper.

At the other extreme we find curly top virus which has a host range involving some 300 plant species (Thornberry, 1966; Thomas, 1969). Within this host range are plant species which are susceptible to certain strains of the virus and immune to others (Giddings, 1938). Aster yellows MLO also has an extensive and diverse host range made up of over 300 plant species belonging to 50 or more families (Kunkel, 1953; Banttari, 1966; Westdal and Richardson, 1969). Clover phyllody MLO was shown to have a potentially large host range in North America (Chiykowski, 1967, 1974). Seventy-nine species in 22 families of plants have been experimentally infected and there is little doubt that the list will increase with further testing since up to 40% of the species tested in some families were found to be susceptible. The pathogen causing X-disease also displays considerable diversity in its host range, infecting both woody and herbaceous plant species (Jensen, 1971a). Although only one herbaceous species has been shown to be naturally infected (Gilmer, 1960) it serves to awaken the researcher to the potential role of such plants in the spread of tree diseases.

4.6.2 Strains

The epidemiology of diseases caused by leafhopper-borne pathogens is affected by the occurrence of different pathogen strains which may vary in virulence, host range and transmissibility. Nearly 50 years ago, two strains of aster yellows MLO were recognized on the basis of host range, geographical distribution and vector specificity (Kunkel, 1932; Severin, 1934a). Today, the situation is somewhat complicated by changes in geographical distribution of the original strains (Magie *et al.*, 1952; Maramorosch, 1963; George and Richardson, 1957; Chiykowski, 1958), and the occurrence of additional strains (Freitag, 1963, 1967; Granados, 1965; Granados and Chapman, 1968; Westdal and Richardson, 1969; Westdal, 1969). A diagnostic feature which has remained unchanged is the transmission of the eastern strain by only the aster leafhopper, *M. fascifrons.* One of the more potentially important strains reported was isolated from infected barley in Manitoba (Westdal and Richardson, 1969). This noncelery infecting strain was shown to readily infect hosts considered to be immune to aster yellows pathogen such as oats, wild oats, rye and *Triticale*. Of 69 isolates recovered from migrant *M. fascifrons* in Manitoba, 32 readily infected oats. All attempts to infect oats with noncelery- and celery-infecting strains which had been collected from the field several years earlier and maintained in the greenhouse, failed. Strains of aster yellows MLO which vary in their infectivity to oats have also been isolated in Ontario (Chiykowski, unpublished). One of these strains, resembling eastern aster yellows agent in resulting symptomatology has been shown to readily infect celery but the average length of time required for symptoms to appear was 115.7 days (Chiykowski, 1978).

Variations in virulence of curly top virus, as evidenced by symptoms produced in sugar beet, were originally attributed to lack of uniformity in resistance or susceptibility in the host plant or to attenuation of the virus (Carsner, 1925). Since then curly top virus has been shown to consist of a number of strains that range from mild to severe (Giddings, 1938, 1950; Thomas 1969). Recent studies have shown that current strains occurring in beet fields and in weeds in the foothills of the San Joaquin Valley in California are far more virulent than those found in the 1950s and the 1960s (Magyarosy and Duffus, 1977). The significant fact is that these severe strains are capable of infecting and causing losses in curly top resistant sugar beet cultivars even when infection occurs up to 10 weeks after seeding (Duffus and Skoyen, 1977). Maize mosaic virus has also been shown to exist as several strains, each varying in virulence and host range (Storey, 1936; Boch, 1974; Rose 1978). Such diseases as streak of sugarcane and guinea grass, for example, are now known to be the result of infection with host-adapted strains of maize mosaic virus.

The existence of strains of X-disease agent has been noted in the literature almost since the time the disease was first reported. Lack of suitable criteria for the designation of strains has led to some uncertainty as to their number, authenticity and importance in the epidemiology of X-disease. Hildebrand (1953) compared X-disease from the east with that from the west and, on the basis of symptomatology and epidemiology, concluded that they were distinctly different but probably related. At the same time, he suggested that further studies were required to clarify the situation. Three strains of X-disease pathogen in eastern North America were erected on the basis of differences in symptomatology in chokecherry by Gilmer *et al.* (1954) but their significance in field infection was not indicated. Four strains of western X-disease pathogen were recognized on the basis of symptomatology and host range by Jensen (1956) in California. One of these, peach yellow leafroll strain, was earlier considered to be a distinct disease pathogen but studies on transmission, symptom expression, hosts and cross protection led to the conclusion that it was a severe strain of X-disease pathogen.

Although some leafhopper-borne pathogens exist as strains, one must exercise caution in the designation of strains. Nault and Bradfute (1979) have clearly described the confusion created with the corn stunt disease complex by the discovery or description of so-called pathogen "strains" which have now been shown to be different unrelated pathogens. In Australia, Helson (1951) suggested that lucerne witches' broom and tomato big bud diseases were caused by the same or closely related strains of virus while Hutton and Grylls (1956) concluded that legume little leaf was probably another strain of the big bud-witches' broom complex. Bowyer (1974) has shown that tomato big bud, legume little leaf and a 'mild' form of lucerne witches' broom diseases could be distinguished by vector transmission, symptomatology and incubation periods in differential hosts and concluded that the three diseases were caused by different mycoplasmalike organisms. Three strains of potato witches' broom pathogen which caused identical symptoms in potato but could be distinguished on tomato were described in Canada by Wright (1954, 1957). Subsequently Raine (1967) showed that the strains were actually different pathogens and suggested that although they were distinct they were probably related.

4.6.3 Source

Leafhoppers moving into a crop after acquiring the pathogen from outside sources such as infected weeds or other crop plants are generally the prime source of infection. The amount of disease that develops in that crop will depend on size of the insect population and percentage of infective individuals in the population. Thus, a very large population containing even a low number of infective individuals could result in a relatively high incidence of disease. Similarly, a small population with a high percentage of infective insects could produce considerable infection. Needless to say, a combination of a large population with a high percentage of infective leafhoppers would produce catastrophic results. In describing a severe curly top disease outbreak in western Idaho, Mumford and Peay (1970) stated that the percentage of infective leafhoppers in nearby breeding areas averaged 37% and that large populations of these insects had moved into newly planted sugar beet fields. During the 10 years prior to the outbreak the infectivity of such populations had averaged only 7%. Wallace and Murphy (1938) have also pointed out variations in infectivity from year to year ranging from 4% in one year to 67% in another. Rose (1974), working with maize streak in Rhodesia observed that when there are few migrant leafhoppers the incidence of disease increased arithmetically through transient settling and feeding of previously infective adults. By comparison, large migrating populations resulted in an exponential increase.

Although the sources of the insects moving into the crop may be located considerable distances away, they are more commonly found relatively near the affected crops. The distribution pattern of a disease in a crop is usually indicative of the location of the source of inoculum. Hewitt *et al.* (1946) observed that the incidence of alfalfa dwarf in new plantings was highest in areas of the field nearest older diseased plantings. They also observed that Pierce's disease of grape was usually more prevalent in vineyards of districts where considerable alfalfa was grown and in portions of vineyards adjacent to alfalfa, an observation which led to the discovery that both diseases were caused by the same pathogen. In some instances, 100% of the vines were infected in the first 4 to 6 rows adjacent to alfalfa fields. Freitag and Frazier (1954) found that vector species collected in a number of habitats such as vineyards, irrigated pastures, alfalfa fields, along roadsides, ditches and natural breeding areas such as uncultivated range, meadows, pastures, bogs and riverbanks were naturally infective with Pierce's disease agent. The trend of natural infectivity during the year indicated a high percentage of infective leafhoppers in March, a dropping off in April and summer and a gradual increase toward the fall, reaching its highest peak during October. Purcell (1974), in mapping vineyards in the Napa Valley in California found that the percentage of diseased vines was highest nearest natural riparian vegetation, declining along a steep gradient with the distance (up to 400 to 500 feet) from natural areas of the vector and then levelling off. The source of inoculum for phyllody of cotton in Upper Volta apparently is in infected weeds, particularly *Sidia cordifolia* L. (Desmidts *et al.*, 1973). The hibernating vector leafhopper, *Orosium cellulosus* Lindberg, moves to this preferred weed host, many plants of which are infected, and produces its first generation. This generation, when mature, provides the infective migrants which move into the cotton fields.

The spread of X-disease in peach orchards in eastern North America appears to be largely due to an inoculum source in areas surrounding the orchard. Stoddard (1947) and Hildebrand (1953) demonstrated that the removal of infected wild chokecherry from areas adjoining orchards markedly reduced the spread of X-disease. No evidence could be found for peach to peach spread. Lukens *et al.* (1971) reported that where infected chokecherry had not been destroyed, 72% of initially healthy peach trees became infected over a three-year period while only 13% became infected when chokecherry was removed. The epidemiology of western X-disease which involves several pathogen strains, apparently differs somewhat from that of eastern X-disease (Jensen, 1971a). There is no correlation between the presence of infected chokecherry and western X-disease incidence in orchards. The known vectors of western X pathogen are capable of transmitting from peach to peach and cherry as well as from cherry to cherry and peach (Wolfe *et al.*, 1951; Anthon and Wolfe, 1951; Kaloostian, 1951, and Wolfe, 1955). Jensen (1971a) experimentally infected 12 species of herbaceous plants with western X although the disease has not been shown to occur in such hosts in nature. He pointed out that because most of the vector species feed on both herbaceous and tree hosts, the former could provide a source of disease inoculum. This could also be true for eastern X since Gilmer (1960) demonstrated the presence of what appears to be X-disease in naturally infected milkweed. An herbaceous source of inoculum would explain the presence of eastern X in a number of orchards in the Niagara Peninsula of Ontario where no infected chokecherry could be found (Conners and Savile, 1943; T. R. Davidson, personal comm.).

Secondary spread, where newly infected plants within a crop serve as a source of inoculum for further infection in that crop, has not been widely reported for leafhopper-borne diseases. Prerequisites for such a situation to occur in annual crop plants would require the pathogen to have a short incubation period in the inoculated plant and the leafhopper to be able to acquire and transmit the pathogen in a relatively short space of time. Ideally suited to this type of situation is sugar beet curly top virus and its vector *N. tenellus* since the insect can acquire and transmit the virus within a matter of hours of feeding on an infected plant (Bennett and Wallace, 1938). Wallace and Murphy (1938) showed that spring brood leafhoppers moving into beet fields often contained only low percentages of insects infective with curly top virus. Repeated testing of populations in these fields throughout the summer showed an increase in percent infectivity sometimes reaching 100% by harvest time. Landis *et al.* (1970) attributed the 100% infection in some beet fields in Washington as being at least partly due to the progeny of the spring broods that remains in the beet fields throughout the growing period.

The viruses transmitted in a semipersistant manner also appear well suited for significant amounts of secondary spread. Narayanasamy (1972) found that 40% of the rice plants inoculated at 10 days of age with rice tungro virus were able to serve as virus sources for leafhoppers 2 days after inoculation although no symptoms were visible. Even plants inoculated at 90 days of age became infected and provided a source of virus. Similarly, maize chlorotic dwarf virus could be recovered by leafhoppers 2-4 days in advance of symptom appearance (Nault and Bradfute, 1979). Discovery of a new strain of rice waika virus which is much more virulent than the common strain suggests that secondary spread could occur with this disease as well (Inoue, 1978). With inoculation and acquisition thresholds of 30 min, no latent period in the insect and an incubation period of 10-14 days in the plant, this strain of waika virus possesses the necessary qualities of rapid spread within a crop. Epidemics of maize streak virus in Rhodesia are apparently caused by spread from infected to healthy plants within crops since only a low proportion of immigrants alighting during the flight season transmit the virus (Rose, 1974).

Secondary spread apparently is not important in the epidemiology of Pierce's disease agent in grape although conditions for such spread appear to be ideal. Within hours of feeding on infected plants, the leafhoppers are capable of transmitting the pathogen to new plants. The insects can acquire or transmit the pathogen at anytime and grapevines remain susceptible to infection and provide a source of inoculum throughout the growing season. The work of Purcell (1974, 1975, 1979) indicates that secondary spread does occur in vineyards but that infections occurring late in the season do not persist through the winter dormant season. The role of secondary spread in other perennial corps such as tree fruits is often poorly understood because of the lack of sufficient information. A least five leafhopper vectors have been experimentally shown capable of transmitting western X-disease pathogen from peach to peach (Wolfe *et al.*, 1951; Anthon and Wolfe, 1951; Wolfe 1955; Jensen, 1957a) and although this type of transmission, no doubt, occurs in nature, the extent to which it occurs is unknown. In contrast, only one case of leafhopper acquisition from infected peach trees has been reported for eastern X-disease pathogen (Gilmer *et al.*, 1966) and circumstantial evidence suggests that peach trees play only a minor role, if any, as a source of inoculum. The role of infected peach in the epidemiology of phony peach disease is also not clearly understood. Although leafhoppers were shown capable of acquiring the pathogen from this host, considerably more success was obtained from plum (Turner and Pollard, 1959). Evidence that roguing of diseased peach trees in an orchard reduces the spread of phony peach disease is not considered to be conclusive and would suggest that peach plays a minor role in the spread of this disease (Purcell, 1979).

4.6.4   Adverse effects on the vector

Although several plant pathogenic organisms are known to have a close biological relationship with their leafhopper vector species, only three have been shown to adversely affect the biology of the insect. The first such effects observed were those reported by Jensen (1959) in which the western X-disease agent was shown to cause premature death of its vector, *Colladonus montanus* (Van D.). Preliminary results indicated that longevity of another vector, *Colladonus geminatus* (Van D.) was also reduced but not to the same extent as *C. montanus* Extensive cytological studies of infective *C. montanus* have shown that the X-disease agent produces severe cytopathic effects in various tissues and that these may be responsible for the premature death of the insects (Whitcomb and Jensen, 1968; Whitcomb *et al.*, 1967, 1968a, b). Heat treatment of infective leafhoppers allowed them to live longer than untreated infective leafhoppers suggesting arrested or retarded development of the pathogen by heat (Jensen, 1968). Fecundity of *C. montanus* was impaired by the pathogen when infection was acquired either by feeding on diseased plants or by injection with inoculum (Jensen, 1971b). Reduction in fecundity was greater in insects acquiring from plants than through injection and this was interpreted as an effect of the diseased plant.

Rice dwarf virus, which is transovarially transmitted to some of its progeny, also has been shown to adversely affect its vector. Progeny of *Inazuma dorsalis* Motsch. which receive virus through the egg die prematurely (Shinkai, 1962).

Nasu (1963) observed that congenitally infected leafhoppers had a high mortality, a shorter diapause and fecundity reduced by 28-68%. Nakasuji and Kiritani (1970) found the virus to adversely affect the developmental period and survival rate of nymphs, longevity of adults and the preoviposition and fecundity of female adults of *N. cincticeps*. The greatest decrease in fecundity was observed in the third or summer generation and was attributed, at least partly, to the enhanced effect of high temperature.

Corn stunt spiroplasma was also shown to cause premature death in its vector *D. elimatus* (Granados and Meehan, 1975). Mean survival of transmitters was considerably reduced regardless of whether the insects acquired the pathogen through feeding on diseased plants or injection. Although the pathogen also proved lethal to *D. maidis*, the effect occurred about two weeks later than in *D. elimatus*. Mean egg production per female was considerably lower in infected than in healthy *D. elimatus* leafhoppers.

The role which these adverse effects play in the epidemiology of the above diseases has only been evaluated for rice dwarf. Here, the reduction in longevity and fecundity is considered to be highly beneficial in reducing the percentage of infected insects in the field (Nakasuji and Kiritani, 1971, 1972, 1977). In the case of X-disease and corn stunt pathogens, this role is unknown because of the general lack of field data on the vectors on such aspects as biology, natural infectivity and transmission efficiency. For example, most of the studies dealing with the harmful effects of X-disease pathogen were conducted on *C. montanus*, a vector which is very inefficient in transmitting the pathogen to peach (Jensen, 1957b) while the effect on more efficient vectors is unknown. In fact, the importance of the various vector species in the epidemiology of this disease has not been fully evaluated.

Shikata (1979) recently reviewed the cytopathological changes that occur in leafhopper vectors of several circulative-propagative plant viruses.

4.7 THE PLANT

The role of the plant in the epidemiology of leafhopper-borne plant pathogens is threefold. It may be involved in the biology of the vector, it may be the host in which the damage by the pathogen occurs or it may act as the source or reservoir for disease spread. Since the role of the plant cannot be readily separated from that of the other components some of its effects on epidemiology have been included in discussions in other sections of this review. This section then will deal with plant effects which have not been dealt with or which have only been briefly mentioned elsewhere.

Oman (1949), in discussing the role of plants in the bionomics of leafhoppers, refers to two categories of plants. Those on which eggs are deposited and on which nymphal development takes place are termed "host plants." Those that are utilized as sources of food by the adults but are not normally used for oviposition are referred to as "food plants." The results of Pitre (1967) with *D. maidis* are illustrative of this dual role by the plant. At least 33 different species will serve as food plants for this leafhopper but only 20 species of Gramineae were found suitable for oviposition. Although eggs hatched on 16 of these, nymphs completed their development only on corn. The limited host range of the two most efficient vectors of corn stunt and maize bushy stunt pathogens, *D. maidis* and *D. elimatus*, severely restricts not only their ability to overwinter in areas where corn is not grown continuously but also limits the potential of these pathogens to adapt to new hosts (Nault and Bradfute, 1979). In contrast, the vectors of curly top pathogen, *N. tenellus* (Severin, 1930; Wallace and Murphy, 1938; Nielson, 1968) and *Neoaliturus opacipennis* (Lethierry) (Kaur *et al.*, 1971), are capable of completing their development as well as feeding on a number of plant species. Purcell (1976), studying the leafhopper *G. circellata*, found that while a wide variety of plants served as food plants for adults, relatively few were condusive to nymph development. This species also displayed seasonal changes in plant preferences. Although such knowledge about plant species utilized by leafhoppers is important in studying epidemiology, it is often sparse or lacking for many vector species.

Biological studies on plant preferences by leafhoppers have led to some important findings useful in reducing disease losses. Wilcox and Beckwith (1933), for example, found a negative correlation between the rate of spread of cranberry false blossom and the relative attractiveness of cranberry varieties as food plants for the vector. This suggested that "field resistance" to the disease actually consisted in resistance to attack by the insect. In testing plant species and varieties for attractiveness to the vector, Wilcox (1951) found that care must be exercised in selecting test insects since leafhoppers showed a preference for the kinds of plants on which they formerly lived and fed. *N. virescens*, one of the most important of the rice leafhoppers, has received considerable attention in host preference studies. Pathak *et al.* (1969) found two varieties of rice that were highly resistant to insect damage even when large numbers of leafhoppers were used. Nymphs caged on resistant plants had a low survival rate and grew more slowly suggesting that varieties resistant to *N. virescens* may contain some toxic materials or lack nutrients vital to that insect. Later screening studies, utilizing a large number of rice varieties and selections revealed that several were highly resistant (Cheng and Pathak, 1972). Resistant varieties generally were not preferred by adults or nymphs and insects caged on these plants suffered high mortality. On highly resistant varieties, only 0-3% of the first instars reached the adult stage. Adults confined on susceptible varieties lived 4 to 8 times longer and laid 15 times more eggs than on resistant varieties. Insects made more probing punctures in resistant than in susceptible varieties but did little feeding. Ling and Carbonell (1975) found that more insects moved and each individual insect moved more times on an insect resistant than on an insect susceptible variety of rice. They postulated that field resistance to tungro disease in rice varieties could be related to the nonpreference of the insect for the rice variety, resulting in shorter visits to plants and hence less chance of transmission.

Although a plant species may not be a preferred food plant of a leafhopper, it may still become severely infected with a pathogen transmitted by that vector species. Curly top incidence often approaches 100% in fields of tomato (Martin, 1970), peppers (Dana and Dodge, 1947), bean (Schultz and Dean, 1947) and species of cucurbits (Dana, 1938) as a result of transient movement of the leafhopper. Thomas (1972) examined the mode of expression of plant preference and found that the leafhopper apparently finds preferred food plants by random movement from plant to plant and does not discriminate until after spending 25 to 50 min on a plant. The insect feeds on both preferred and nonpreferred plants before discriminating and then expresses preference by departing from nonpreferred and remaining on preferred plants. Even in a nonpreferred species such as tomato there may be a further degree of nonpreference shown which may be responsible for some of the resistance to curly top infection (Thomas and Martin, 1971). Transmission of curly top agent to tomato was found to be equal to that of sugar beet, a preferred host, during the first hour of insect confinement. During the next 3 hours transmission to tomato was higher than that to sugar beet. Thereafter transmission to sugar beet increased while that to tomato decreased and almost stopped at 8 hr. This pattern of transmission appeared to reflect changes in feeding behavior and health of the vector when confined on the two plant species because leafhoppers confined on tomato began dying after 12 to 16 hr (Thomas and Boll, 1977). The incidence of a corn stunting disease in Mississippi was apparently related not only to the size of *G. nigrifrons* populations but also to the availability to the vector of preferred food plants (Pitre and Boyd, 1970). Although more leafhoppers were collected in weedy than in weed free plots, the latter had a higher disease incidence suggesting that the vector showed preferential feeding on vegetation other than corn which is not a preferred food or host plant.

The reaction of plant species to infection by specific pathogens may vary considerably, ranging from reduction in yield to death of the plant. The degree of injury depends upon age of the plant at time of infection and upon the species or variety used. Scott *et al.* (1977) found that yields from corn infected with a corn stunt complex were reduced by more than 50%. Diseased plants showing symptoms 52 days after planting produced no grain while plants showing symptoms as late as 107 days after planting produced a normal yield. Losses in susceptible hybrids were higher than in resistant ones because they showed not only a higher number of infected plants but also an earlier expression of disease symptoms. The effect of infection by curly top virus on the production of sugar beet seed was found to depend on time of infection (Hills and Brubaker, 1968). The greatest reduction in the percentage of germinating seed occurred when curly top infective leafhoppers invaded the fields just as plants were approaching the bloom stage. The growth stage of peach trees was found to affect the transmission of phony peach disease pathogen (Turner and Pollard, 1959; Kaloostian and Pollard, 1962). Trees inoculated early in the season during active growth became infected more readily than those inoculated later in the season and susceptibility decreased as the season progressed. Activation of new growth by pruning also increased chances of infection.

Various varietal responses to infection have been observed with a number of leafhopper-borne pathogens. In Costa Rica, reduction in mature ear weight due to maize rayado fino on individual plants of Central American varieties ranged from 40 to 50% but reached 100% in introduced foreign or newly developed maize varieties (Gamez, 1977). Disease incidence in locally adapted varieties usually showed 0 to 20% infection but in some of the more susceptible genotypes 100% infection was observed. Considerable variation was found in the reaction of strawberry cultivars to green petal disease pathogen transmitted by the leafhopper *A. bicinctus* under greenhouse conditions. The results suggested that disease incidence in the field is dependent on cultivar preference by the vector as well as on cultivar susceptibility to the causal agent (Chiykowski and Craig, 1975). Slykhuis (1961, 1962, 1963) found marked differences in the reaction of grasses and grains to infection by wheat striate mosaic virus. All durum wheats were highly susceptible as were several hard red spring and winter wheats but a few of the latter were highly resistant or immune. Most varieties of oats and barley and 10 species of grasses were moderately susceptible. The disease was difficult to recognize in the field in many varieties because of the moderate symptoms. Gorter (1953, 1959) found the hybrids of Peruvian x Hickory tended to tolerate maize streak virus but showed no marked resistance to infection. In studying the resistance of tomato cultivars to curly top infection, Thomas and Martin (1971) concluded that resistance was not due to the ability of plants to recover after being infected but rather resulted primarily from characteristics which reduce the chances of being infected. They also observed that the lines tested were less resistant in the greenhouse than in the field.

Some plant species display no symptoms at all or only mild symptoms when infected, a situation which presents some difficulty in recognizing potential pathogen sources in the field. Turner and Pollard (1959) state that for many years, phony peach was recognized only on peach. After a method of artificial transmission of phony peach pathogen was discovered, numerous species of wild plum as well as apricots and almonds were found to be susceptible. Although most species of *Prunus* showed symptoms, several species of infected plums were found to be symptomless. Among the latter was the wild plum *Prunus angustifolia* Marsh., which was found to be generally distributed and infected throughout the peach growing areas where the disease was active. Plum was also shown to be a better source of inoculum for leafhopper transmission than was peach. Freitag (1951), in studying the host range of Pierce's disease of grape, found 36 species of plants in 18 families to be naturally infected. Although symptoms developed on a few species a majority of the plants were symptomless carriers.

Several studies have been carried out on the mechanism of feeding in an attempt to explain the transmitting ability of leafhoppers on different plant species. For a leafhopper to be an efficient vector of a phloem- or xylem- restricted pathogen its mouthparts must reach these tissues in a reasonably high percentage of its feeds. Day *et al.* (1952) found that most leafhopper species tested fed in both parenchyma and vascular bundles and exhibited varying degrees of preference for phloem tissue. The anatomy of some plant species was found to influence feeding and prevented some phloem-feeding leafhoppers from reaching the phloem. The authors concluded that leafhoppers find the tissues upon which they feed by random probing with the stylets. Moreau and Boulay (1967), compared three vector species on different plant species and observed that the number of feeding tracks, their ramifications and their success in reaching the phleom are dependent on the adaptation of the insect to the plant species. In a preferred plant species the feeding tracks were observed to be less branched, fewer in number and better directed to the phloem. The feeding habits of each developmental stage of *N. cincticeps* were histologically compared by Naito (1976). He reported that only a few salivary sheaths of 1st instar nymphs reached the large vascular bundles and attributed this to the length of the stylets. More feeding tracks of nymphs terminated in the xylem than in the phloem and the author postulated that this might have some relation to their nutritional requirement during the growing stages. According to Sogawa (1973), the feeding process can be divided into two main behavioral phases, probing and sucking. He postulates that these phases are controlled by two types of chemical stimuli found in the plant; a probing stimulant that activates stylet insertion but suppresses ingestion and a sucking stimulant that activates ingestion. When the stylet reaches and enters the vascular tissues the insects cease probing and salivation of sheath material and begin ingestion. Once having decided the sucking site, the length of ingestion time becomes dependent on the chemical quality of the plant sap.

4.8   THE ENVIRONMENT

The effects of some environmental factors have already been mentioned at various places in this chapter because they are so closely linked with other epidemiological components that one cannot be discussed without the other. Thus, insect movement, especially of a migratory type, is dependent on air currents and temperature; vector biology is influenced by temperature, moisture and daylength. The discussion in this section will be limited to additional factors of the environment which can have profound effects on disease incidence. Some, such as cultural practices, are controlled by man while others, like weather are not although they can be utilized in predicting disease behavior.

4.8.1   Cultural Practices

Serious disease outbreaks can sometimes be traced to changes in cultural practices. The introduction of early planted rice in parts of Japan to avoid typhoon damage was apparently at least partly responsible for an increase in losses due to rice dwarf virus (Nakasuji, 1974; Nakasuji and Kiritani, 1977). Although there had been earlier outbreaks of the disease, these were characterized by a mosaic pattern of infected areas without regional continuity and lasting only one or two years. Around 1955 the epidemics became characterized by continuity in time and space over an extended area. Introduction of early planted rice into an area normally used for middle season rice production resulted in mix cropping which provided a continuous supply of rice plants for leafhopper and disease development. This was evidenced by a 5-fold increase in annual catches of the vector, *N. cincticeps*. In early planted rice areas, adults of the overwintering generation transmit rice dwarf to seedlings in the nursery which then provide a disease source for nymphs developing on these plants, resulting in a rapid increase in infected insects. In an area of purely middle season rice, the percentage of infected insects in the first generation is low because the insects are forced to develop on weeds that are generally immune to infection. If an area consists of only early season rice, the prevalence of rice dwarf decreases because of a decrease in the abundance of the vector due to the lack of host plants for insects of the fourth and fifth generations. Thus, it is a combination or mixture of early and middle season crops that results in severe outbreaks of the disease. Similar changes in cultural practices are also apparently responsible for severe outbreaks of penyakit merah disease in rice in North Krian, Malaysia. According to Lim (1972) the introduction of double cropping together with variations in planting have created conditions favorable for outbreaks of this disease. These cropping procedures make available young plants over an extended period, thereby increasing leafhopper populations and disease inoculum.

An increase in the amount of fallow paddy acreage may also be responsible for increases in rice dwarf incidence according to Nakasuji (1974). With the increase in early season rice and reduction in acreage of winter barley and wheat crops, large areas of paddy field are left fallow. These fallow fields give rise to gramineous weeds which give rise to overwintering and first generation leafhoppers. Winter plowing of these fallow fields greatly reduced the incidence of rice dwarf the following year and demonstrated the importance of cultural practices in enhancing disease control.

The introduction of new crops into areas where they were not previously grown may produce unexpected outbreaks of new diseases. Following the initiation of a peach industry in North America, a host of problems were encountered as a result of indigenous diseases, such as peach yellows, X-disease and phony peach, to which peach was highly susceptible. Similar problems followed the introduction of corn to Africa and sugar beet to parts of South America (Bennett, 1967). Pierce's disease of grape presents a situation in which both the crop and the pathogen were introduced into a new area. The grape industry, consisting largely of European vines was well established in California when Pierce's disease pathogen was introduced into the state probably in grape wood from the Gulf Plain area of the U. S. (Hewitt, 1958; Hopkins *et al.*, 1974). Although certain

Vitis species in this area where the disease is indigenous are highly tolerant, attempts to grow European type and American bunch grapes have not been successful.

Irrigation as a cultural practice can play an important role in disease epidemiology. In Rhodesia, Rose (1974, 1978) found that irrigated cereal crops and pastures, grown during the dry season, provided inoculum sources for maize streak virus and feeding and breeding areas for leafhopper vectors. These irrigated crops, surrounded by dry areas, remain attractive for several months to insects which are continually arriving, settling temporarily, laying eggs and then leaving. During this period a high proportion of plants may become infected and the leafhopper nymphs produced on such plants become infective. As they mature, the adult insects move and spread the disease to later planted irrigated crops in adjoining areas. Frazier and Posnette (1957) reported an unusually high incidence of green petal in new strawberry plantings grown under irrigation. As a result of a dry autumn, vector leafhoppers apparently moved in greater numbers to irrigated fields, resulting in a disease incidence that was twice that observed in nonirrigated fields.

Crop selection and location may influence disease incidence. Alfalfa growing adjacent to vineyards may serve as a source of both vectors and disease inoculum for Pierce's disease of grape. Older stands of alfalfa can also act as reservoirs of infection for new alfalfa plantings (Hewitt *et al.*, 1946) The location of such perennial crops as clovers adjacent to strawberry plantings may provide reservoirs for green petal infection since clovers are susceptible to the pathogen and are also breeding hosts for some vectors (Chiykowski, 1962b). The incidence of maize streak disease can be reduced by the use of strips of bare land about 10 m wide between plantings of maize which act as barriers to leafhopper movement (Gorter, 1953; Rose, 1978). Their effectiveness, however, is greatly reduced when new plantings are located downwind from previously planted, infested maize.

Time of planting is an important factor in avoiding severe losses from some diseases. Wallace and Murphy (1938) considered the stage of beet development at time of infestation with leafhoppers to be the single most important factor in the epidemiology of curly top disease. A delay in time of infestation by even a few days was decidedly advantageous if it came after the plants had developed beyond the cotyledon stage. Delayed planting of sugar beets in the San Joaquin Valley until May, greatly increased incidence of curly top to the extent that a profitable root yield could not be achieved (Ritenour *et al.*, 1970). Curly top losses in other susceptible crops such as cantaloupe and watermelon were also found to be more severe if infection occurred at early stages of plant development (Hills and Taylor, 1954; Hills, 1958). Differences in incidence of penyakit merah in rice cultivars in Malaysia, first attributed to reaction to the pathogen, were later shown to be the result of planting time. Cultivars showing lower infection rates were being used in a double cropping system and were being planted earlier thus escaping infection (Lim, 1972). Martinez-Lopez (1977) reported that incidence of maize rayado fino in experimental plots in Colombia could be reduced from 80% to less than 2% by controlling planting dates and using crop rotation.

The use of chemicals in cultural practices may produce some unexpected epidemiological results. Synthetic insecticides have been listed as being at least partly responsible for the increase in rice dwarf incidence in Japan (Nakasuji, 1974). These insecticides were introduced around 1955 to control the rice stem borer and about this time rice dwarf began to increase. Light trap records of many regions showed remarkable increases in abundance of *N. cincticeps* in every region regardless of the cropping procedure (Kiritani, 1972). The reason for the increase in leafhoppers was attributed to the fact that the insecticides were killing spiders

et al., 1962; Kiritani *et al.*, 1971). Kiritani *et al.* (1972) estimated that 4.4% to 100% of the nymphs and 9.0% to 100% of the adults of *N. cincticeps* were preyed upon by spiders. The situation was further aggravated when the leafhopper developed resistance to organophosphate and carbamate insecticides in many regions in southern Japan (Hama and Iwata, 1973; Hama, 1976; Hama *et al.*, 1977; Miyata and Saito, 1976). A change in control practices several years ago resulted in an increase in the incidence of rubus stunt (Fluiter and Van der Meer, 1955). In this case a tar oil winterspray had been used specifically to control *Lampronia rubiella* Bjerk on cultivated *Rubus* spp. An unknown beneficial side effect was the destruction of eggs of *Macropsis fuscula* Zett., the vector of rubus stunt pathogen. Abandonment of the tar oil spray in favor of DDT resulted in increased leafhopper populations and disease incidence.

4.8.2 Climatic conditions

Specific climatic conditions have been correlated with the outbreak of many leafhopper-borne pathogens. Wright (1954) observed that incidence of potato witches' broom was higher during summers when the weather was hot and dry. The drying up of grasses, native vegetation and alfalfa stands presumably caused the insect vectors to move from preferred hosts to potatoes resulting in a higher incidence of disease in this crop. A geographic study of stolbur incidence by Savulescu and Ploaie (1962) showed that severe outbreaks followed periods of drought which lasted several years and high temperatures, conditions favoring high insect populations. Conversely low incidence followed heavy rainfall and cool temperatures, conditions unfavorable to the vector. A warm early spring was found to produce spring broods of beet leafhopper which were highly viruliferous with curly top virus (Wallace and Murphy, 1938) while dry autumn and dry early winters reduced spread of this disease by overwintering adults (Severin, 1939). Years of low incidence of cotton phyllody were correlated with low moisture in May and June in Upper Volta (Desmidts *et al.*, 1973). This lack of moisture retarded the growth of weed source plants and resulted in later infection and development of phyllody in cotton fields.

The correlation of climatic conditions to particular phases of epidemiology can be used in predicting disease outbreaks. Chapman (1973), in describing the migration of *M. fascifrons* in the central United States, pointed out the dependence of the migration on climatic conditions in the south central states. Early growing conditions and warm southerly winds signal an early leafhopper migration often accompanied by a severe outbreak of aster yellows disease. Conversely cool, wet weather in the migration source areas may delay or even prevent such a movement. Rose (1972a, 1978) showed that temperature and moisture determined the developmental time, length of flight season and size of flight populations of *Cicadulina* spp. and, hence, determined the incidence of maize streak infection. The correlation of large spring flights with high autumn rainfall provided several months warning that severe streak infection could occur the following year.

4.9   LEAFHOPPER TRANSMITTED PATHOGENS AND THEIR VECTORS

In an historical account of leafhopper-borne pathogens Nielson (1968) states that prior to 1920, only three vector species and two viruses were known. By 1965 this number had increased dramatically to 114 species of leafhoppers transmitting 65 "virus and virus strains." The list has continued to grow and now includes some 135 vector species and 86 pathogens and strains (Table I; Nielson, 1979). New vectors such as *E. sulphurella* and *Athysanus argentarius* Metc., have been discovered for the long established aster yellows agent. Vectors have been found for such pathogens as cotton phyllody and soyabean witches' broom pathogens which were not known to be leafhopper transmitted. In other cases, previously known vectors like *A. bicinctus* and *N. pallida* have been shown to transmit new pathogens. With some 6,000 leafhopper species recognized in the world (DeLong, 1971) and the knowledge that less than 2% of these are known vectors, one can expect the vector list to grow.

Some of the pathogens listed may eventually be found to be synonyms or strains of other pathogens when suitable identification criteria are developed for establishing such relationships. Almond leaf scorch, for example, does not appear in the list because recent findings suggest that it is caused by the pathogen responsible for Pierce's disease of grape (Mircetich *et al.*, 1976; Lowe *et al.*, 1976). Other diseases may in fact be complexes made up of several different diseases as was demonstrated for corn stunt (Nault and Bradfute, 1979). Mycoplasmalike organisms have been associated with a number of diseases in various parts of the world for which vector associations are not yet known. Since most of the diseases shown to have MLO etiologies have eventually been found to have leafhopper vector associations, some of these MLOs, no doubt, will eventually be shown to be associated with this group of insects.

TABLE I. Leafhopper Transmitted Plant Pathogens

Pathogen	Vector	Reference
Alfalfa dwarf (see Pierce's disease)		
Aster yellows (M)[b]		
North American		
Eastern strain	*Macrosteles fascifrons* (Stål)	Kunkel 1924
Western strain	*Acinopterus angulatus* Lawson	Severin & Frazier 1945
	Aphrodes bicinctus (Schrank)	Chiykowski 1977
	Athysanus argentarius Metc.	Chiykowski 1979
	Chlorotettix similis DeL.	Severin 1947
	Colladonus flavocapitatus (Van D.)	Severin 1947
	Colladonus geminatus (Van D.)	Severin 1934b
	Colladonus holmesi Bliven	Severin 1947
	Colladonus intricatus (Ball)	Severin 1947
	Colladonus kirkaldyi (Ball)	Severin 1947
	Colladonus montanus (Van D.)	Severin 1934b
	Colladonus rupinatus (Ball)	Severin 1947
	Elymana sulphurella (Zett.) (=*virescens*)	Chiykowski & Sinha 1969
	Endria inimica (Say)	Chiykowski 1963
	Euscelidius variegatus (Kbm.)	Severin 1947
	Gyponana angulata (Spang.) (=*hasta* of Nielson 1968)	Severin 1946
	Idiodonus sp.	Severin 1948
	Macrosteles fascifrons (Stål)	Severin 1929
	Paraphlepsius apertinus (Osb. & Lath.)	Severin 1945
	Scaphytopius acutus (Say)	Chiykowski 1962a
	Sacphytopius delongi Young	Severin 1947
	Scaphytopius irroratus (Van D.)	Severin 1947

132

Disease	Vector	Reference
	Texananus incurvatus (Osb. & Lath.)	Severin 1950b
	Texananus lathropi (Baker)	Severin 1945
	Texananus latipex DeL.	Severin 1945
	Texananus oregonus (Ball)	Severin 1945
	Texananus pergradus DeL.	Severin 1945
	Texananus spatulatus (Van D.)	Severin 1945
Eurpoean	*Aphrodes bicinctus* (Schrank)	Heinze & Kunze 1955
	Macrosteles laevis Rib.	Heinze & Kunze 1955
Japanese	*Macrosteles quadripunctulatus*(Kbm.)	Sukhov & Vovk 1945
Blueberry stunt (M)	*Scleroracus flavopictus* (Ish.)	Fukuski & Nemoto 1953
Brinjal little leaf (M)	*Scaphytopius magdalensis* (Prov.)	Tomlinsen *et al.* 1950
	Empoasca devastans Dist.	Thomas & Krishnaswami 1939
	Hishimonus phycitis (Dist.)	Thomas & Krishnaswami 1939
Cereal Chlorotic mottle (V)	*Nesoclutha pallida* (Evans) (=*obscura*)	Greber 1977
Chloris striate (U)	*Nesoclutha pallida* (Evans) (=*obscura*)	Grylls 1963
Citrus stubborn (M)	*Euscelis incisus* (Kbm.) (=*plebejus* (Fall.))	Daniels *et al.* 1973
	Scaphytopius nitridus (DeL.)	Kaloostian *et al.* 1975
	Neoaliturus tenellus (Baker) (=*Circulifer tenellus*)	Oldfield *et al.* 1976
Clover club leaf (R)	*Agalliopsis ancistra* Oman (=*novella* Say)	Black 1944
Clover dwarf (M)	*Aphrodes bicinctus* (Schrank)	Musil 1960
	Euscelis incisus (Kbm.) (=*plebejus* (Fall.))	Musil & Valenta 1958
	Macrosteles laevis (Rib.)	Musil & Valenta 1958
Clover phyllody (M)	*Anoscopus albifrons* (Linné) (=*Aphrodes albifrons*)	Evenhuis 1958
	Aphrodes bicinctus (Schrank)	Evenhuis 1958

TABLE I. Leafhopper Transmitted Plant Pathogens

Pathogen	Vector	Reference
	Euscelis incisus (Kbm.) (=*plebejus* (Fall.))	Frasier & Posnette 1956
	Euscelis lineolata Brullé	Frazier & Posnette 1956
	Macrosteles fascifrons (Stål)	Chiykowski 1962a
	Macrosteles viridigriseus (Edw.)	Frazier & Posnette 1956
	Paraphlepsuis irroratus (Say)	Chiykowski 1965b
	Scaphytopius acutus (Say)	Chiykowski 1962a
	Scleroracus balli Med.	Raine 1967
	Scleroracus dasidus Med.	Raine 1967
	Speudotettix subfusculus (Fall.)	East Malling Res. Stn. 1960
Clover proliferation (M)	*Macrosteles fascifrons* (Stål)	Chiykowski 1965a
Clover yellow edge (M)	*Aphrodes bicinctus* (Schrank)	Chiykowski 1969
Clover wound tumor (V)	*Agalliopsis ancistra* Oman (=*novella* Say)	Black 1944
	Agallia constricta Van D.	Black 1944
	Agallia quadripunctata (Prov.)	Niederhauser & Cervantes 1950
Corn stunt (M)	*Dalbulus elimatus* (Ball)	Kunkel 1946
	Dalbulus maidis (DeL.)	Granados *et al.* 1966
	Graminella nigrifrons (Forbes)	Laboucheix *et al.* 1972
Cotton phyllody (M)	*Orosius cellulosus* (Lind.)	Dobroscky 1929
Cranberry false blossom (M)	*Scleroracus vaccinii* (Van D.)	
Curly top		
Sugar beet		
North American (V)	*Neoaliturus tenellus* (Baker) (=*Circulifer tenellus*)	Shaw 1910
Turkish (U)	*Neoaliturus opacipennis* (Leth.) (=*Circulifer opacipennis*)	Bennett & Tanrisever 1958
Argentine (U)	*Agalliana ensigera* Oman	Fawcett 1927
Tomato (U)		
braziliensis strain	*Agallia albidula* Uhler	Sauer 1946

solanacearum strain	*Agalliana ensigera* Oman	Costa 1952
	Agalliana sticticollis (Stål)	Costa 1952
Delphinium yellows (U)	*Macrosteles sexnotatus* (Fall.)	Posnette & Ellenberger 1963
Eastern wheat striate (V)	*Cicadulina mbila* (Naudé)	Nagaich & Sinha 1974
Elm phloem necrosis (M)	*Scaphoideus luteolus* (Van D.)	Baker 1948
Enanismo of small grains (M)	*Cicadulina pastusae* Rut. & Del.	Galvez *et al.* 1963
Grapevine golden flavescence (M) (flavescence dorée)	*Euscelidius variegatus* (Kbm.)	Caudwell *et al.* 1972
	Euscelis incisus (Kbm.) (*=plebejus*)	Caudwell *et al.* 1972
	Scaphoideus titanus Ball (*=littoralis*)	Schvester *et al.* 1961
Legume little leaf (M)	*Orosius argentatus* (Evans)	Hutton & Grylls 1956
Groundnut mosaic I (U)	*Orosius argentatus* (Evans)	Bergman 1956
Little peach (M)	*Macropsis trimaculata* Fitch	Manns & Manns 1935
Maize bushy stunt (M)	*Baldulus tripsaci* Kr. & Whit.	Granados & Whitcomb 1971
	Dalbulus elimatus (Ball)	Maramorosch 1958
	Dalbulus maidis (DeL.)	Maramorosch 1955
	Graminella nigrifrons (Forbes)	Granados *et al.* 1966
	Graminella sonora (Ball) (*=Deltocephalus sonorus*)	Granados *et al.* 1968
Maize chlorotic dwarf (V)	*Graminella nigrifrons* (Forbes)	Nault *et al.* 1973
	Graminella sonora (Ball) (*=Deltocephalus sonorus*)	Nault *et al.* 1973
Maize mottle (U)	*Cicadulina bipunctata bipunctata* (Mel.) (*=bipunctella zeae* China)	Storey 1937
	Cicadulina mbila (Naudé)	Storey 1937
	Cicadulina storeyi China	Storey 1937
Maize rayado fino (V)	*Baldulus tripsaci* Kr. & Whit.	Nault & Bradfute 1977

TABLE I. Leafhopper Transmitted Plant Pathogens

Pathogen	Vector	Reference
Maize streak (V)	*Dalbulus elimatus* (Ball)	Nault & Bradfute 1977
	Dalbulus maidis (DeL.)	Gámez 1969
	Graminella nigrifrons (Forbes)	Nault & Bradfute 1977
	Cicadulina bipunctata bipunctata (Mel.) (=bipunctella zeae China)	Storey 1936
	Cicadulina latens Fennah	Fennah 1959
	Cicadulina mbila (Naudé)	Storey 1924
	Cicadulina parazeae Ghauri	Rose 1962
	Cicadulina storeyi China	Storey 1936
	Hishimonus sellatus (Uhler)	Sakai 1937
Mulberry dwarf (M)		
Oat blue dwarf (V)		
North American	*Macrosteles fascifrons* (Stål)	Banttari & Moore 1962
Sweden	*Macrosteles laevis* (Rib.)	Lindsten *et al*. 1970
Papaya bunchy top (M)	*Empoasca papayae* Oman	Adsuar 1946
Parastolbur (U)	*Euscelis incisus* (Kbm.) (=plebejus (Fall.))	Musil 1962
	Macropsis trimaculata Fitch	Kunkel 1933
Peach yellows (M)		
Peach X (M)		
Eastern	*Colladonus clitellarius* (Say)	Thornberry 1954
	Fieberiella florii (Stål)	Gilmer & McEwen 1958
	Gyponana lamina DeL.	Gilmer & McEwen 1958
	Norvellina seminuda (Say)	Gilmer *et al*. 1966
	Orientus ishidae (Mats.)	Rosenberger & Jones 1978
	Paraphlepsius irroratus (Say)	Gilmer *et al*. 1966
	Scaphytopius acutus (Say)	Hildebrand 1953
	Scaphoideus sp.	Rosenberger & Jones 1978
Western	*Colladonus geminatus* (Van D.)	Wolfe *et al*. 1950
	Colladonus montanus (Van D.)	Wolfe 1955
	Euscelidius variegatus (Kbm.)	Jensen 1969

136

Disease	Vector species	Reference
	Fiebgeriella florii (Stål)	Anthon & Wolfe 1951
	Graphocephala confluens (Uhler)	Anthon & Wolfe 1951
	(=*Keonolla confluens* (Uhler)	
	Osbornellus borealis DeL. & Mohr	
Penyakit merah (V)	*Scaphytopius acutus* (Say)	Jensen 1957b
	Nephotettix virescens (Dist.)	Anthon & Wolfe 1951
	(=*impicticeps* Ishihara)	Ou *et al.* 1965
Phony Peach (R)	*Cuerna costalis* (Fab.)	Turner 1949
	Draeculacephala portola Ball	Turner & Pollard 1955
	Graphocephala versuta (Say)	Turner 1949
	Homalodisca coagulata (Say)	Turner & Pollard 1955
	Homalodisca insolita (Walker)	Turner & Pollard 1955
	Oncometopia nigricans (Walker)	Pollard (see Nielson 1968)
	Oncometopia orbona (Fab.)	Turner 1949
Pierce's disease of grape (R)	*Carneocephala flaviceps* (Riley)	Stoner 1953
	Carneocephala fulgida Nott.	Hewitt *et al.* 1942a
	Carneocephala triguttata Nott.	Frazier 1944
	Cuerna costalis (Fab.)	Kaloostian *et al.* 1962
	Cuerna occidentalis Oman & Beamer	Frazier 1944
	Cuerna yuccae Oman & Beamer	Freitag *et al.* 1952
	Draeculacephala crassicornis Van D.	Freitag *et al.* 1952
	Draeculacephala minerva Ball	Hewitt *et al.* 1942a
	Draeculacephala novaeboracensis (Fitch)	Freitag *et al.* 1952
	Draeculacephala portola Ball	Freitag *et al.* 1952
	Friscanus friscanus (Ball)	Frazier & Freitag 1946
	Graphocephala atropunctata (Sign.)	Hewitt *et al.* 1942b
	(=*Hordnia circellata* (Baker)	
	Graphocephala cythura (Baker)	Freitag *et al.* 1952
	Helochara communis Fitch	Frazier 1944
	Homalodisca coagulata (Say)	Kaloostian *et al.* 1962
	Homalodisca lacerta (Fowler)	Freitag *et al.* 1952

TABLE I. Leafhopper Transmitted Plant Pathogens

Pathogen	Vector	Reference
	Graphocephala confluens pacifica DeL. & Sev. (=*Keonolla confluens pacifica* DeL. & Sev.)	Frazier 1944
	Graphocephala dolobrata (Ball) (=*Keonolla dolobrata* Ball) (=*Neokolla hieroglyphica* of authors)	Frazier 1944
	Neokolla severini DeL.	Frazier 1944
	Oncometopia orbona (Fab.)	Kaloostian *et al.* 1962
	Pagaronia confusa Oman	Frazier & Freitag 1946
	Pagaronia furcata Oman	Frazier & Freitag 1946
	Pagaronia tredecimpunctata Ball	Frazier & Freitag 1946
	Pagaronia triunata Ball	Frazier & Freitag 1946
	Orosius albicinctus Dist.	Nagaich & Giri 1971
Potato marginal flavescence (M)		
Potato yellow dwarf (V) New York strain	*Aceratagallia curvata* Oman	Black 1944
	Aceratagallia longula (Van D.)	Black 1944
	Aceratagallia obscura Oman	Black 1944
	Aceratagallia sanguinolenta (Prov.)	Black 1934
	Agallia quadripunctata (Prov.)	Black 1944
	Agalliopsis ancistra Oman (=*novella* (Say))	Black 1944
New Jersey Strain	*Agallia constricta* Van D.	Black 1941b
	Agallia quadripunctata (Prov.)	Balck 1944
	Agalliopsis ancistra Oman (=*novella* (Say))	Black 1944
Rice dwarf (V)	*Inazuma dorsalis* (Motsch.) (=*Recilia dorsalis* Motsch.)	Takata 1895
	Nephotettix cincticeps (Uhler)	Shiga Agri. Exp. Stn. 1899
	Nephotettix nigropictus (Stål) (=*apicalis* (Motsch.))	Nasu 1963

Disease	Vector	Reference
Rice orange leaf (M)	*Inazuma dorsalis* (Motsch.) (=*Recilia dorsalis* Motsch.)	Rivera *et al.* 1963
Rice transitory yellowing (V)	*Nephotettix cincticeps* (Uhler)	Chiu *et al.* 1968
	Nephotettix nigropictus (Stål) (=*apicalis* (Motsch.))	Chiu *et al.* 1965
	Nephotettix virescens (Dist.) (=*impicticeps* Ishihara)	Hsieh *et al.* 1970
Rice tungro (V)	*Nephotettix nigropictus* (Stål) (=*apicalis* (Motsch.))	Ling 1970
	Nephotettix virescens (Dist.) (=*impicticeps* Ishihara)	Rivera & Ou 1965
Rice waika (V)	*Nephotettix cincticeps* (Uhler)	Hirao *et al.* 1974
	Nephotettix malayanus Ish. & Kawase	Inoue 1977
	Nephotettix nigropictus (Stål) (=*apicalis* (Motsch.))	Satomi *et al.* 1975
	Nephotettix virescens (Dist.) (=*impicticeps* Ishihara)	Kimura *et al.* 1975
Rice yellow dwarf (M)	*Nephotettix cincticeps* (Uhler)	Iida & Shinkai 1950
	Nephotettix nigropictus (Stål) (=*apicalis* (Motsch.))	Ouchi & Suenaga 1963
	Nephotettix virescens (Dist.) (=*impicticeps* Ishihara)	Shinkai 1962
Rubus stunt (M)	*Macropsis fuscula* (Zett.)	Fluiter & Van der Meer 1953
		Fluiter & Van der Meer 1958
Rugose leaf curl (R)	*Austroagallia torrida* Evans	Grylls 1954
Safflower phyllody (M)	*Neoaliturus fenestratus* (H.-S.)	Klein 1970

TABLE I. Leafhopper Transmitted Plant Pathogens

Pathogen	Vector	Reference
Sandal spike (M)	(=Circulifer fenestratus) Coelidia indica Walker	Rangaswami & Griffith 1941
Sesamum phyllody (M)	Orosius albicinctus Dist.	Vasudeva & Sahambi 1955
Sowbane mosaic (U)	Neoaliturus tenellus (Baker) (=Circulifer tenellus)	Bennett & Costa 1961
Stolbur (M)	Aphrodes bicinctus (Schrank)	Brčák 1954
	Euscelis incisus (Kbm.) (=plebejus (Fall.))	Valenta et al. 1961
	Macrosteles cristata Rib.	Blattný et al. 1954
	Macrosteles laevis Rib.	Valenta 1958
	Leopotettix dilutior (Kbm.)	Posnette & Ellenberger 1953
Sugarbeet yellow vein (U)	Aceratagallia calcaris Oman	Staples et al. 1970
Sugarbeet yellow wilt (M)	Paratanus exitiosus (Beamer)	Bennett & Munck 1946
Sugarcane chlorotic streak (V)	Draeculacephala portola Ball	Abbott & Ingram 1942
Sugarcane white leaf (M)	Matsumuratettix hiroglyphicus (Mats.) (=Epitettix hiroglyphicus Mats.)	Matsumoto et al. 1969
Tobacco yellow dwarf (U)	Orosius argentatus (Evans)	Hill 1941
Tomato big bud (M)	Orosius argentatus (Evans)	Hill 1943
Tomato leaf crinkle (U)	Anaceratagallia venosa (Fourcroy)	Sukhov & Vovk 1947

Disease / host	Leafhopper	Reference
Wheat dwarf (U)	*Psammotettix alienus* (Dahlb.)	Vacke 1961
Wheat pale green dwarf (U)	*Psammotettix alienus* (Dahlb.)	Agarkov 1966
Wheat striate mosaic (V)	*Macrosteles laevis* Rib.	Vacke 1973
	Endria inimica (Say)	Slykhuis 1953
Winter wheat mosaic (V)	*Elymana sulphurella* (Zett.) (=*virescens* (F.))	Sinha 1970
	Psammotettix striatus (Linné)	Zazhurilo & Sitnikova 1939
Witches' broom of Alfalfa (U)	*Scaphytopius acutus* (Say)	Menzies 1944
Blueberry (U)	*Idiodonus cruentatus* Panz.	Blattný 1963-64
Clover (M)	*Euscelis lineolata* Brullé	Frazier & Posnette 1956
	Macrosteles viridigriseus (Edw.)	Frazier & Posnette 1957
Cryptotaenia japonica (M)	*Macrosteles orientalis* Wagner	Okuda & Nishimura 1974
Groundnut (U)	*Orosius argentatus* (Evans)	Bergman 1956
Jujube (M)	*Hishimonus discigutta* (Wlk.)	Kim 1966
	Nesophrosyne orientalis (Mats.)	Shinkai 1964
Legumes (U)	*Hishimonus concavus* Knight	Chou *et al.* 1975
Loofah (M)	*Orosius argentatus* (Evans)	Helson 1951
Lucerne (M)	*Scleroracus balli* Med.	Raine 1967
Potato (M)	*Scleroracus dasidus* Med.	Raine 1967
	Scleroracus flavopictus (Ish.)	Fukushi *et al.* 1955
Soyabean (U)	*Nesophrosyne orientalis* (Mats.)	Lo & Han 1969
Sweet potato (M)	*Nesophrosyne ryukyuensis* Ish.	Shinki 1964

[a] Leafhopper names used in this table represent the latest accepted scientific names as determined by Dr. K.G.A. Hamilton, Biosystematics Research Institute, Agriculture Canada.

[b] Letter in parenthesis signifies the etiologic agent as follows: M=mollicute; V=virus; R=rickettsialike organism; U=pathogen unknown.

4.10 EPILOGUE

Our ability to understand the epidemiology of diseases caused by leafhopper-borne pathogens is dependent on the extent of our knowledge of the components involved and the effects of various factors on their performance. The greater our knowledge, the better are our chances of successfully predicting disease outbreaks and controlling or reducing losses due to the disease. Considerable progress has been made in our overall understanding of epidemiology. Electron microscopy has revealed the existence of diverse types of plant pathogenic agents. Vector-pathogen studies have disclosed new modes of transmission and vector-pathogen associations. Field experiments and observations have increased our awareness of the many factors influencing the performance of the vector and pathogen in their natural surroundings. However, our understanding of specific epidemiologies of many diseases is poor, consisting of little more than the knowledge that a pathogen is transmitted by leafhoppers. One of the obstacles to obtaining the required information is the diversity of the components involved. Since few scientists are trained in all the disciplines required to carry on an integrated program involving insects, pathogens, plants and all the related aspects, our knowledge of a disease is usually fragmented and limited to those areas in which the worker has had previous training or experience. A group approach, involving scientists with different disciplinary backgrounds appears to be the most logical solution to such problems. We should also ensure that the educational programs presently being used to train scientists entering this field of work are properly geared toward the "discipline" of plant pathogen transmission by insects. Realization that the epidemiology of diseases caused by leafhopper-borne pathogens is a highly complex field, requiring a multiplicity of skills, is the first step towards solving the problem.

The reader is referred to "Leafhopper Vectors and Plant Disease Agents" (Maramorosch and Harris, 1979) for a comprehensive up-to-date treatise on the worldwide importance of leafhopper vectors and the latest findings in leafhopper vector research.

4.11 ACKNOWLEDGMENTS

I thank Drs. Y.C. Paliwal, E.F. Schneider, and R.C. Sinha for their helpful reviews of the manuscript and Dr. K.G.A. Hamilton for his taxonomic expertise in the preparation of the list of leafhopper vectors.

4.12 REFERENCES

Abbott, E. V., and Ingram, J. W. (1942). Transmission of chlorotic streak of sugar cane by the leafhopper *Draeculacephala portola*. *Phytopathology* 32: 99-100.
Adsuar, J. (1946). Transmission of papaya bunchy top by a leafhopper of the genus *Empoasca*. *Science* 103: 316.

Agarkov, V. A. (1966). Virusnaja blednozeljonaja karlikovost ozimoj psenicy v Chmel-nickoj oblasti (wheat pale-green dwarf virus disease in Chmelnicki county). *Biol. nauki, Nautsh. Dokl. Vysh. Shkoly* 1: 201-206.

Anthon, E. W., and Wolfe, H. R. (1951). Additional insect vectors of western X-disease. *Plant Dis. Rep.* 35: 345-346.

Baker, W. L. (1948). Transmission by leafhoppers of the virus causing phloem necrosis of American elm. *Science* 108: 307-308.

Bantarri, E. E. (1966). Grass hosts of aster yellows virus. *Plant Dis. Rep.* 50: 17-21.

Banttari, E. E., and Moore, M. B. (1962). Virus cause of blue dwarf of oats and its trans-mission to barley and flax. *Phytopathology* 52: 897-902.

Behncken, G. M., and Gowanlock, D. H. (1976). Association of a bacterium-like organism with rugose leaf curl disease of clovers. *Aust. J. Biol. Sci.* 29: 137-146.

Bennett, C. W. (1967). Epidemiology of leafhopper-transmitted viruses. *Ann. Rev. Phyto-pathol.* 5: 87-108.

Bennett, C. W. (1971). The curly top disease of sugar beet and other plants. *Phytopathol. Monogr. No. 7*, Am. Phytopathol. Soc., St. Paul, Minnesota. 81 pp.

Bennett, C. W., and Costa, A. S. (1961). Sowbane mosaic caused by a seed-transmitted virus. *Phytopathology* 51: 546-550.

Bennett, C. W., and Munck, C. (1946). Yellow wilt of sugar beet in Argentina. *J. Agr. Res.* 73: 45-64.

Bennett, C. W., and Tanrisever, A. (1958). Curly top disease in Turkey and its relationship to curly top in North America. *J. Am. Soc. Sugar Beet Technol.* 10: 189-211.

Bennett, C. W., and Wallace, H. E. (1938). Relation of the curly top virus to the vector, *Eutettix tenellus. J. Agr. Res.* 56: 31-51.

Bergman, R. H. H. (1956). Het mozaiek en de heksenbezemziekte van de aardnoot (*Arachis hypogaea* L.) in West Java en hun vector, de Jassidae *Orosius argentatus* (Evans). (Mosiac I and witches' broom disease of peanut in West Java and the vector). *Tijdschr. over Plantenziekten* 62: 291-304.

Bils, R. F., and Hall, C. E. (1962). Electron microscopy of wound-tumor virus. *Virology* 17: 123-130.

Black, L. M. (1934). The potato yellow dwarf disease. *Am. Potato J.* 11: 148-152.

Black, L. M. (1941a). Further evidence for multiplication of the aster-yellows virus in the aster leafhopper. *Phytopathology* 31: 120-135.

Black, L. M. (1941b). Specific transmission of varieties of potato yellow-dwarf virus by re-lated insects. *Am. Potato J.* 18: 231-233.

Black, L. M. (1944). Some viruses transmitted by agallian leafhoppers. *Proc. Am. Phil. Soc.* 88: 132-144.

Black, L. M. (1948). Transmission of clover club-leaf virus through the egg of its insect vector. *Phytopathology* 38: 2.

Black, L. M. (1950). A plant virus that multiplies in its insect vector. *Nature* 166: 852-853.

Black, L. M. (1953). Occasional transmission of some plant viruses through the eggs of their insect vectors. *Phytopathology* 43: 9-10.

Black, L. M., and Brakke, M. K. (1952). Multiplication of wound-tumor virus in an insect vector. *Phytopathology* 42: 269-273.

Black, L. M., and Markham, R. (1963). Base pairing in the ribonucleic acid of wound tumor virus. *Netherlands J. Plant Pathol.* 69: 215.

Black, L. M., Smith, K. M., Hills, G. J., and Markham, R. (1965). Ultrastructure of potato yellow-dwarf virus. *Virology* 27: 446-449.

Blattný, C., Jr. (1963-64). Die übertragung der Hexenbesenkrankheit der Heidelbeeren (*Vaccinium myrtillus* L.) durch die zikade *Idiodonus cruentatus* Pang. *Phytopath. Z.* 49: 203-205.

Blattný, C., Brčák, J., Pozdena, J., Dlabola, J., Limberk, J., and Bojňaský, V. (1954). Die Übertragung des Stolburvirus bei Tabak und Tomaten und seine virogeographischen Beziehungen. *Phytopath. Z.* 22: 381-416.

Boch, K. R. (1974). Maize streak virus. *Common. Mycol. Inst. Assoc. Appl. Biol.* No. 133, 4 pp.

Boch, K. R., Guthrie, E. J., and Woods, R. D. (1974). Purification of maize streak virus and its relationship to viruses associated with streak diseases of sugar cane and *Panicum maximum. Ann. Appl. Biol.* 77: 289-296.

Bowyer, J. W. (1974). Tomato big bud, legume little leaf, and lucerne witches' broom: three diseases associated with different mycoplasma-like organisms in Australia. *Aust. J. Agric. Res.* 25: 449-457.

Bradfute, O. E., Gingery, R. E., Gordon, D. T., and Nault, L. R. (1972a). Tissue ultrastructure, sedimentation and leafhopper transmission of a virus associated with a maize dwarfing disease. *J. Cell. Biol.* 55: 25a.

Bradfute, O. E., Louie, R., and Knoke, J. K. (1972b). Isometric virus-like particles in maize with stunt symtpoms. *Phytopathology* 62: 748.

Brčák, J. (1954). Novy přenašeč stolburu (bezesemennosti) rajčete a tabǎku-krisek *Aphrodes bicinictus* Schrk. *Zool. Ent. Listy* 3: 231-237.

Carsner, E. (1925). Attenuation of the virus of sugar beet curly-top. *Phytopathology* 15: 745-757.

Carter, W. (1961). Ecological aspects of plant virus transmission. *Ann. Rev. Entomol.* 6: 347-370.

Caudwell, A., Bachelier, J. C., Kuszala, C., et Larrue, J. (1969). Etude de la survie de la cicadelle *Scaphoideus littoralis* (Ball) sur les plantes herbacées, et utilisation de ces données pour transmettre la Flavescence dorée de la vigne à d'autres espèces végétales. *C. R. Acad. Sci. Paris* 269: 101-103.

Caudwell, A., Kuszala, C., Larrue, J., et Bachelier, J. C. (1972). Transmission de la flavescence dorée a la fève a la fève par des cicadelles des genres *Euscelis* et *Euscelidius* intervention possible de ces insectes. Dans l'épidémiologie du bois noir en Bourgogne. *Ann. Phytopathol. n. H. S..* 1972: 181-189.

Caudwell, A., Kuszala, C., and Larrue, J. (1973). Technique utilisables pour l'etude de la flavescence dorée de la vigne. *Riv. Pathol. Veg.* 9: 269-276.

Chapman, R. K. (1973). Integrated control of aster yellows. *Proc. N. Cent. Br. Entomol. Soc. Am.* 28: 71-92.

Chen, T. A., and Liao, C. H. (1975). Corn stunt spiroplasma: isolation, cultivation, and proof of pathogenicity. *Science* 188: 1015-1017.

Chen, M., and Liu, H. (1974). Electron microscopic study of rice yellow dwarf. *Plant Protec. Bull.* 16: 42-55.

Chen, M. J., and Shikata, E. (1971). Morphology and intracellular localization of rice transitory yellowing virus. *Virology* 46: 786-796.

Cheng, C. H., and Pathak, M. D. (1972). Resistance to *Nephotettix virescens* in rice varieties. *J. Econ. Entomol.* 65: 1148-1153.

Chiu, R. J., Jean, J. H., Chen, M. H., and Lo, T. C. (1968). Transmission of transitory yellowing virus of rice by two leafhoppers. *Phytopathology* 58: 740-745.

Chiu, R. J., Lo, T. C., Pi, C. L., and Chen, M. H. (1965). Transitory yellowing of rice and its transmission by the leafhopper *Nephotettix apicalis* (Motsch.). *Bot. Bull. Acad. Sinica* 6: 1-18.

Chiykowski, L. N. (1958). Studies on migration and control of the six-spotted leafhopper, *Macrosteles fascifrons* (Stål) in relation to transmission of aster yellows virus. Ph. D. dissertation, Univ. Wisconsin, Madison. 126 pp.

Chiykowski, L. N. (1962a). Clover phyllody virus in Canada and its transmission. *Can. J. Bot.* 40: 397-404.

Chiykowski, L. N. (1962b). Clover phyllody and strawberry green petal diseases, caused by the same virus in eastern Canada. *Can. J. Bot.* 40: 1615-1617.

Chiykowski, L. N. (1963). *Endria inimica* (Say), a new leafhopper vector of a celery-infecting strain of aster-yellows virus in barley and wheat. *Can. J. Bot.* 41: 669-672.

Chiykowski, L. N. (1965a). A yellows-type virus of alsike clover in Alberta. *Can. J. Bot.* 43: 527-536.

Chiykowski, L. N. (1965b). Transmission of clover phyllody virus by the leafhopper, *Paraphlepsius irroratus* (Say). *Can. Entomol.* 97: 1171-1173.

Chiykowski, L. N. (1967). Some host plants of a Canadian isolate of the clover phyllody virus. *Can. J. Plant Sci.* 47: 141-148.

Chiykowski, L. N. (1969). A leafhopper transmitted clover disease in the Ottawa area. *Can. Plant Dis. Surv.* 49: 16-19.

Chiykowski, L. N. (1970). Notes on the biology of the leafhopper *Aphrodes bicincta* (Homoptera: Cicadellidae) in the Ottawa area. *Can. Entomol.* 102: 750-758.

Chiykowski, L. N. (1974). Additional host plants of clover phyllody in Canada. *Can. J. Plant Sci.* 54: 755-763.

Chiykowski, L. N. (1976). Transmission characteristics and host range of the clover yellow edge agent. *Can. J. Bot.* 54: 1171-1179.

Chiykowski, L. N. (1977). Transmission of a celery-infecting strain of aster-yellows by the leafhopper *Aphrodes bicinctus*. *Phytopathology* 67: 522-524.

Chiykowski, L. N. (1978). Delayed expression of aster yellows symptoms in celery. *Can. J. Bot.* 56: 2987-2989.

Chiykowski, L. N. (1979). *Athysanus argentarius*, an introduced European leafhopper, as a vector of aster yellows in North America. *Can. J. Plant Path.* 1: 37-41.

Chiykowski, L. N., and Chapman, R. K. (1965). Migration of the six-spotted leafhopper *Macrosteles fascifrons* (Stål). Part 2. Migration of the six-spotted leafhopper in central North America. *Univ. Wisconsin Res. Bull.* 261: 21-45.

Chiykowski, L. N., and Craig, D. L. (1975). Reaction of strawberry cultivars to clover phyllody (green petal) agent transmitted by *Aphrodes bicincta*. *Can. Plant Dis. Surv.* 55: 66-68.

Chiykowski, L. N., and Sinha, R. C. (1969). Comparative efficiency of transmission of aster yellows by *Elymana virescens* and *Macrosteles fascifrons* and the relative concentration of the causal agent in the vectors. *J. Econ. Entomol.* 62: 883-886.

Chou, T. G., Yang, S. J., Huang, P. Y., and Chung, S. J. (1975). Studies on loofah (*Luffa cylindrica* Roem.) witches' broom in Taiwan. III. The leafhopper *Hishimonus concavus* Knight as a vector. *Plant Prot. Bull.* 17: 384-389.

Connors, I. L., and Savile, D. B. O. (1943). Twenty-second Ann. Rep. Can. Plant Dis. Surv., 1942, p. 85.

Costa, A. S. (1952). Further studies on tomato curly top in Brazil. *Phytopathology* 42: 393-403.

Coupe, T. R., and Schulz, J. T. (1968). The influence of controlled environments and grass hosts on the life cycle of *Endria inimica* (Homoptera: Cicadellidae). *Ann. Entomol. Soc. Am.* 61: 74-77.

Dana, B. F. (1938). Resistance and susceptibility to curly top in varieties of squash, *Cucurbita maxima*. *Phytopathology* 28: 649-659.

Dana, B. F., and Dodge, J. C. (1947). Curly top disease of vegetable crops in Washington State. *State Col. Wash. Ext. Bull.* No. 357, 8 pp.

Daniels, M. J., Markham, P. G., Meddins, B. M., Plaskitt, A. K., Townsend, R., and Bar-Joseph, M. (1973). Axenic culture of a plant pathogenic spiroplasma. *Nature* 244: 523-524.

Davis, R. E. (1977). Spiroplasma: Role in the diagnosis of corn stunt disease. *Proc. Maize Virus Dis. Colloq. Workshop*, 16-19 Aug, 1976, Ohio Agric. Res. and Develop. Ctr., Wooster. pp. 92-98.

Davis, R. E., and Worley, J. F. (1973). Spiroplasma: Motile, helical microorganisms associated with corn stunt disease. *Phytopathology* 63: 403-408.

Davis, R. E., Worley, J. F., Whitcomb, R. F., Ishijima, T., and Steere, R. L. (1972). Helical filaments associated with a mycoplasma-like organism in corn stunt-infected plants.

Phytopathology 62: 494.

Day, M. F., Irzykiewicz, H., and McKinnon, A. (1952). Observations on the feeding of the virus vector *Orosius argentatus* (Evans), and comparisons with certain other jassids. *Aust. J. Sci. Res.* 5: 128-142.

DeLong, D. M. (1971). The bionomics of leafhoppers. *Ann. Rev. Entomol.* 16: 179-210.

Desmidts, M., and Laboucheix, J. (1974). Relationship between cotton phyllody and a similar disease of sesame. *FAO Plant Protec. Bull.* 22: 19-20.

Desmidts, M. J., Laboucheix, J., et Offeren, Van A. (1973). Importance économique et épidémiologie de la phyllodie du cotonnier. *Cot. Feb. Trop.* 28: 473-482.

Dobroscky, I. D. (1929). Cranberry false-blossom disease spread by a leafhopper. *Science* 70: 635.

Doi, Y., Teranaka, M., Yora, K., and Asuyama, H. (1967). Mycoplasma- or PLT group-like microorganisms found in the phloem elements of plants infected with mulberry dwarf, potato witches' broom, aster yellows, or paulownia witches' broom. *Ann. Phytopathol. Soc. Japan* 33: 259-266.

Dorst, H. E., and Davis, E. W. (1937). Tracing long-distance movements of beet leafhoppers in the desert. *J. Econ. Entomol.* 30: 948-954.

Drake, D. C., and Chapman, R. K. (1965). Migration of the six-spotted leafhopper *Macrosteles fascifrons* (Stål). Part I. Evidence for long distance migration of the six-spotted leafhopper into Wisconsin. *Univ. Wisconsin Res. Bull.* 261: 5-20.

Duffus, J. E., and Skoyen, I. O. (1977). Relationship of age of plants and resistance to a severe isolate of the beet curly top virus. *Phytopathology* 67: 151-154.

East Malling Research Station (1960). Annual Report. Kent, England. 148 pp.

Evenhuis, H. H. (1958). De vectoren van het bloemvergroeningsvirus van klaver. *Tijdschr. Plantenziekten* 64: 335-336.

Fawcett, G. L. (1927). The curly top of sugar beet in the Argentine. *Phytopathology* 17: 407-408.

Fennah, R. G. (1959). A new species of *Cicadulina* (Homoptera: Cicadellidae) from East Africa. *Ann. Mag. Nat. Hist.* 2: 757-758.

Fluiter, H. J. De., and Van Der Meer, F. A. (1953). Rubus stunt, a leafhopper-borne virus disease. *Tijdschr. Plziekt.* 59: 195-197.

Fluiter, H. J. De., and Van Der Meer, F. A. (1955). Rubus stunt disease of raspberry, its spread and control in the Netherlands. *Med. Landbouw. Hogeschool Gent* 20: 419-434.

Fluiter, H. J. De., and Van Der Meer, F. A. (1958). The biology and control of *Macropsis fuscula* Zett., the vector of *Rubus* stunt virus. *Proc. 10th Int. Congr. Entomol.* 3: 341-345.

Frazier, N. W. (1944). Phylogenetic relationship of the nine leafhopper vectors of Pierce's disease of grape. *Phytopathology* 34: 1000-1001.

Frazier, N. W., and Freitag, J. H. (1946). Ten additional vectors of the virus causing Pierce's disease of grape. *Phytopathology* 36: 634-637.

Frazier, N. W., and Posnette, A. F. (1956). Leafhopper transmission of a clover virus causing green petal disease of strawberry. *Nature* 177: 1040-1041.

Frazier, N. W., and Posnette, A. F. (1957). Transmission and host range studies of strawberry green-petal virus. *Ann. Appl. Biol.* 45: 580-588.

Frederiksen, R. A. (1964). Simultaneous infection and transmission of two viruses in flax by *Macrosteles fascifrons*. *Phytopathology* 54: 1028-1030.

Freitag, J. H. (1936). Negative evidence on multiplication of curly-top virus in the beet leafhopper, *Eutettix tenellus*. *Hilgardia* 10: 305-342.

Freitag, J. H. (1951). Host range of the Pierce's disease virus of grapes as determined by insect transmission. *Phytopathology* 41: 920-934.

Freitag, J. H. (1963). Cross protection of three strains of the aster yellows virus in the leafhopper and in the plant. *Neth. J. Plant Path.* 69: 215.

Freitag, J. H. (1964). Interaction and mutual suppression among three strains of aster yellows virus. *Virology*, 24: 401-413.

Freitag, J. H. (1967). Interaction between strains of aster yellows virus in the six-spotted leafhopper *Macrosteles fascifrons. Phytopathology* 57: 1016-1024.

Freitag, J. H., and Frazier, N. W. (1954). Natural infectivity of leafhopper vectors of Pierce's disease virus of grape in California. *Phytopathology* 44: 7-11.

Freitag, J. H., Frazier, N. W., and Flock, R. A. (1952). Six new leafhopper vectors of Pierce's disease virus. *Phytopathology* 42: 533.

Fukushi, T. (1933). Transmission of the virus through the eggs of an insect vector. *Proc. Imp. Acad. (Tokyo)* 9: 457-460.

Fukushi, T. (1939). Retention of virus by its insect vectors through several generations. *Proc. Imp. Acad. (Tokyo)* 15: 142-145.

Fukushi, T. (1940). Further studies on the dwarf disease of rice plant. *J. Fac. Agr., Hokkaido Imp. Univ.* 45: 83-154.

Fukushi, T., and Nemoto, M. (1953). Insect vectors of aster yellows. *Virus* 3: 208.

Fukushi, T., and Shikata, E. (1963). Fine structure of rice dwarf virus. *Virology* 21: 500-503.

Fukushi, T., Shikata, E., Shioda, H., Sekiyama, E., Tanaka, L. Oshima, N., and Nishia, Y. (1955). Insect transmission of potato witches' broom in Japan. *Proc. Jap. Acad.* 31: 234-236.

Fukushi, T., Shikata, E., and Kimura, L. (1962). Some morphological characters of rice dwarf virus. *Virology*, 18: 192-205.

Fulton, R. A., and Romney, Van E. (1940). The chloroform-soluble components of beet leafhoppers as an indication of the distance they move in spring. *J. Agr. Res.* 61: 737-743.

Galvez, G. E. (1967). The purification of virus-like particles from rice tungro virus-infected plants. *Virology* 33: 357-359.

Galvez, G. E. (1968). Purification and characterization of rice tungro virus by analytical density-gradient centrifugation. *Virology* 35: 418-426.

Galvez, G. E., Thurston, H. D., and Bravo, G. (1963). Leafhopper transmission of enanismo of small grains. *Phytopathology* 53: 106-108.

Galvez, G. E., Shikata, E., and Miah, M. S. A. (1971). Transmission and electron microscopy of a rice tungro virus strain. *Phytopathol. Z.* 70: 53-61.

Gámez, R. (1969). A new leafhopper-borne virus of corn in Central America. *Plant Dis. Rep.* 53: 929.

Gámez, R. (1973). Transmission of rayado fino virus of maize (*Zea mays*) by *Dalbulus maidis. Ann. Appl. Biol.* 73: 285-292.

Gámez, R. (1977). Leafhopper-transmitted maize rayado fino virus in Central America. *Proc. Maize Virus Dis. Collog. Workshop*, 16-19 Aug. 1976. Ohio Agric. Res. and Develop. Ctr. Wooster. pp. 15-19.

George, J. A., and Davidson, T. R. (1959). Notes on life history and rearing of *Colladonus clitellarius* (Say) (Homoptera: Cicadellidae). *Can. Entomol.* 91: 376-379.

George, J. A., and Richardson, J. K. (1957). Aster yellows on celery in Ontario. *Can. J. Plant Sci.* 37: 132-135.

Giddings, N. J. (1938). Studies of selected strains of curly top virus. *J. Agr. Res.* 56: 883-894.

Giddings, N. J. (1950). Some interrelationships of virus strains in sugar-beet curly top. *Phytopathology* 40: 377-388.

Gilmer, R. M. (1960). Recovery of X-disease virus from naturally infected milkweeds. *Phytopathology* 50: 636.

Gilmer, R. M., and McEwen, F. L. (1958). Insect transmissions of X-disease virus. *Phytopathology* 48: 262.

Gilmer, R. M., Moore, J. D., and Keitt, G. W. (1954). X-disease virus: I. Host range and pathogenesis in chokecherry. *Phytopathology* 44: 180-185.

Gilmer, R. M., Palmiter, D. H., Schaefers, G. A., and McEwen, F. L. (1966). Insect transmission of X-disease virus of stone fruits in New York. *N. Y. State Agric. Exp. Stn. (Geneva) Bull.* No. 813, 22 pp.

Gingery, R. E. (1976). Properties of maize chlorotic dwarf virus and its ribonucleic acid. *Virology* 73: 311-318.

Goheen, A. C., Nyland, G., and Lowe, S. K. (1973). Association of a rickettsia-like organism with Pierce's disease of grapevines and alfalfa dwarf and heat therapy of the disease in grapevines. *Phytopathology* 63: 341-345.

Gorter, G. J. M. A. (1953). Studies on the spread and control of streak disease of maize. *Union S. Africa Sci. Bull.* No. 341, 20 pp.

Gorter, G. J. M. A. (1959). Breeding maize for resistance to streak. *Euphytica* 8: 234.

Granados, R. R. (1965). Strains of aster-yellows virus and their transmission by the six-spotted leafhopper, *Macrosteles fascifrons* (Stål). *Ph. D. dissertation*. University of Wisconsin, Madison. 150 pp.

Granados, R. R., and Chapman, R. K. (1968). Identification of some new aster yellows virus strains and their transmission by the aster leafhopper, *Macrosteles fascifrons. Phytopathology* 58: 1685-1692.

Granados, R. R., and Meehan, D. J. (1975). Pathogenicity of the corn stunt agent to an insect vector, *Dalbulus elimatus. J. Invert. Pathol.* 26: 313-320.

Granados, R. R., and Whitcomb, R. F. (1971). Transmission of corn stunt mycoplasma by leafhopper *Baldulus tripsaci. Phytopathology* 61: 240-241.

Granados, R. R., Maramorosch, K., Everett, T., and Pirone, T. P. (1966). Transmission of corn stunt virus by a new leafhopper vector, *Graminella nigrifrons* (Forbes). *Contrib. Boyce Thomp. Inst. Plant Res.* 23: 275-280.

Granados, R. R., Gustin, R. D., Maramorosch, K., and Stoner, W. N. (1968). Transmission of corn stunt virus by the leafhopper *Deltocephalus sonorus* Ball. *Contrib. Boyce Thomp. Inst. Plant Res.* 24: 57-60.

Greber, R. S. (1977). Cereal chlorotic mottle virus (CCMV), a rhabdovirus of Gramineae transmitted by the leafhopper *Nesoclutha pallida. Aust. Plant Pathol. Soc. Newsl.* 6: 17.

Grylls, N. E. (1954). Rugose leaf curl—a new virus disease transovarially transmitted by the leafhopper *Austroagallia torrida. Aust. J. Biol. Sci.* 7: 47-58.

Grylls, N. E. (1963). A striate mosaic virus disease of grasses and cereals in Australia, transmitted by the cicadellid *Nesoclutha obscura. Aust. J. Agr. Res.* 14: 143-153.

Hama, H. (1976). Modified and normal cholinesterases in the respective strains of carbamate-resistant and susceptible green rice leafhoppers, *Nephotettix cincticeps* Uhler (Hemiptera: Cicadellidae). *Appl. Entomol. Zool.* 11: 239-247.

Hama, H., and Iwata, T. (1973). Resistance to carbamate insecticides and its mechanism in the green rice leafhopper *Nephotettix cincticeps* Uhler. *Jap. J. Appl. Entomol. Zool.* 17: 154-161.

Hama, H., Iwata, T., Tomizawa, C., and Murai, T. (1977). Mechanism of resistance to malathion in the green rice leafhopper, *Nephotettix cincticepts* Uhler. *Botyu-Kagaku* 42: 188-197.

Harris, K. F. (1979). Leafhoppers and aphids as biological vectors: vector-virus relationships. *In* "Leafhopper Vectors and Plant Disease Agents" (K. Maramorosch and K. F. Harris, eds.), pp. 217-308. Academic Press, New York.

Heinze, K., and Kunze, L. (1955). Die europäische Asterngelbsucht und ihre übertragung durch zwergzikaden. *Nachrichtenbl. Deutsch. Pflanzenschutzd.* 7: 161-164.

Helson, G. A. H. (1951). The transmission of witches' broom virus disease of lucerne by the common brown leafhopper, *Orosius argentatus* (Evans). *Aust. J. Sci. Res.*, Ser. B, 4: 115-124.

Hewitt, W. B. (1958). The probable home of Pierce's disease virus. *Plant Dis. Rep.* 42: 211-215.

Hewitt, W. B., Frazier, N. W., Jacob, H. E., and Freitag, J. H. (1942a). Pierce's disease of grapevines. *Calif. Agr. Exp. Stn. Cir.* No. 353, 32 pp.

Hewitt, W. B., Frazier, N. W., and Houston, B. R. (1942b). Transmission of Pierce's disease of grapevines with a leafhopper. *Phytopathology* 32: 8.

Hewitt, W. B., Houston, B. R., Frazier, N. W., and Freitag, J. H. (1946). Leafhopper transmission of the virus causing Pierce's disease of grape and dwarf of alfalfa. *Phytopathology* 36: 117-128.

Hibino, H., Roechan, M., and Sudarisman, S. (1978). Association of two types of virus particles with penyakit habang (tungro disease) of rice in Indonesia. *Phytopathology* 68: 1412-1416.

Hildebrand, E. M. (1953). Yellow-red or X-disease of peach. *Cornell Univ. Agr. Exp. Stn. Memoir* No. 323, 54 pp.

Hill, A. V. (1941). Yellow dwarf of tobacco in Australia. II. Transmission by the jassid *Thamnotettix argentata* (Evans). *J. Coun. Sci. Industr. Res. Aust.* 14: 181-186.

Hill, A. V. (1943). Insect transmission and host plants of virescence (big bud of tomato). *J. Coun. Sci. Industr. Res. Aust.* 16: 85-90.

Hills, O. A. (1958). Effect of curly top-infective beet leafhopper on watermelon plants in different stages of development. *J. Econ. Entomol.* 51: 434-436.

Hills, O. A., and Brubaker, R. W. (1968). Comparison of the effect on beet seed production of spring and fall infestations of beet leafhoppers carrying curly top virus. *J. Am. Soc. Sugar Beet Technol.* 15: 214-220.

Hills, O. A., and Taylor, E. A. (1954). Effect of curly top-infective beet leafhoppers on cantaloup plants in varying stages of development. *J. Econ. Entomol.* 47: 44-46.

Hino, T., Wathanakul, L., Nabheerong, N., Surin, P., Chaimongkol, U., Disthaporn, S., Putta, M., Kerdochokchai, D., and Surin, A. (1974). Studies on rice yellow orange leaf virus disease in Thailand. *Trop. Agric. Res. Centre (Tokyo) Tech. Bull.* TARC No. 7, 67 pp.

Hirao, J., and Inoue, H. (1978). Transmission efficiency of rice waika virus by the green rice leafhoppers, *Nephotettix* spp. (Hemiptera: Cicadellidae). *Appl. Ent. Zool.* 13: 264-273.

Hirao, J., Satomi, H., and Okada, T. (1974). Transmission of the "Waisei" disease of rice plant by the green rice leafhopper, *Nephotettix cincticeps* Uhler. *Proc. Assoc. Pl. Prot. Kyushu* 20: 128-133.

Hokyo, N. (1972). Studies on the life history and the population dynamics of the green rice leafhopper, *Nephotettix cincticeps* Uhler. *Bull. Kyushu Agric. Exp. Stn.* 16: 282-382.

Hokyo, N. (1976). Population dynamics, forecasting and control of the green rice leafhopper, *Nephotettix cincticeps* Uhler. *Rev. Plant Protec. Res.* 8: 1-13.

Hopkins, D. L. (1977). Diseases caused by leafhopper-borne rickettsia-like bacteria. *Ann. Rev. Phytopathol.* 17: 277-294.

Hopkins, D. L., and Mollenhauer, H. H. (1973). Rickettsia-like bacterium associated with Pierce's disease of grapes. *Science* 179: 298-300.

Hopkins, D. L., Mollenhauer, H. H., and Mortensen, J. A. (1974). Tolerance to Pierce's disease and the associated rickettsia-like bacterium in muscadine grape. *J. Am. Soc. Hortic. Sci.* 99: 436-439.

Hsieh, S. P., Chiu, R. J., and Chen, C. C. (1970). Transmission of rice transitory yellowing virus by *Nephotettix impicticeps*. *Phytopathology* 60: 1534.

Hutchinson, M. T., Goheen, A. C., and Varney, E. H. (1960). Wild sources of blueberry stunt virus in New Jersey. *Phytopathology* 50: 308-312.

Hutton, E. M., and Grylls, N. E. (1956). Legume "little leaf," a virus disease of subtropical pasture species. *Aust. J. Agric. Res.* 7: 85-97.

Iida, T. T., and Shinkai, A. (1950). Transmission of rice yellow dwarf by green rice leafhopper. *Ann. Phytopathol. Soc. Japan* 14: 113-114.

Inoue, H. (1977). A new leafhopper vector of the rice waika virus, *Nephotettix malayanus* Ishihara et Kawase (Hemiptera: Cicadellidae). *Appl. Ent. Zool.* 12: 197-199.

Inoue, H. (1978). Strain S, a new strain of leafhopper-borne rice waika virus. *Plant Dis. Rep.* 62: 867-871.

Ishiie, T., Doi, Y., Yora, K., and Asuyama, H. (1967). Suppressive effects of antibiotics of tetracycline group on symptoms development of mulberry dwarf disease. *Ann. Phytopathol. Soc. Japan* 33: 267-275.

Ito, Y., Miyashita, K., and Sekiguchi, K. (1962). Studies on predators of rice crop insect pests, using the insecticidal check method. *Jap. J. Ecol.* 12: 1-11.

Jensen, D. D. (1956). Insect transmission of virus between tree and herbaceous plants. *Virology* 2: 249-260.

Jensen, D. D. (1957a). Differential transmission of peach yellow leafroll virus to peach and celery by the leafhopper, *Colladonus montanus. Phytopathology* 47: 575-578.

Jensen, D. D. (1957b). Transmission of peach yellow leafroll virus by *Fieberiella florii* (Stål) and a new vector, *Osbornellus borealis* DeL. & M. *J. Econ. Entomol.* 50: 668-672.

Jensen, D. D. (1959). A plant virus lethal to its insect vector. *Virology* 8: 164-175.

Jensen, D. D. (1968). Influence of high temperature on the pathogenicity and survival of western X-disease virus in leafhoppers. *Virology* 36: 662-667.

Jensen, D. D. (1969). Comparative transmission of western X-disease virus by *Colladonus montanus, C. geminatus,* and a new leafhopper vector, *Euscelidius variegatus. J. Econ. Entomol.* 62: 1147-1150.

Jensen, D. D. (1971a). Herbaceous host plants of western X-disease agent. *Phytopathology* 61: 1465-1470.

Jensen, D. D. (1971b). Vector fecundity reduced by western X-disease. *J. Invert. Path.* 17: 389-394.

Kaloostian, G. H. (1951). Transmission of western X little cherry virus from chokecherry to peach by *Colladonus germinatus* (Van D.). *Plant Dis. Rep.* 35: 374.

Kaloostian, G. H., and Pollard, H. N. (1962). Experimental control of phoney peach virus vectors with di-syston. *J. Econ. Entomol.* 55: 566-567.

Kaloostian, G. H., Pollard, H. N., and Turner, W. F. (1962). Leafhopper vectors of Pierce's disease virus in Georgia. *Plant Dis. Rep.* 46: 292.

Kaloostian, G. H., Oldfield, G. N., Pierce, H. D., Calavan, E. C., Granett, A. L., Rana, G. L., and Gumpf, D. J. (1975). Leafhopper-natural vector of citrus stubborn disease? *Calif. Agric.* 29(2): 14-15.

Kaur, P., Bindra, O. S., and Singh, S. (1971). Biology of *Circulifer opacipennis* (Lethierry), a leafhopper vector of sugar-beet curly-top virus. *Indian J. Agr. Sci.* 41: 1-10.

Kim, C. J. (1966). Witches' broom of jujube tree (*Zizyphus jujube* Mill, var. *inermis* Rehd.). *Proc. 11th Pacific Sc. Congr.*, Tokyo, pp. 188-194.

Kimura, I., and Shikata, E. (1968). Structural model of rice dwarf virus. *Proc. Japan Acad.* 44: 538-543.

Kimura, T., Maejima, I., and Nishi, Y. (1975). Transmission of the rice waika virus by *Nephotettix virescens* Distant. *Ann. Phytopath Soc. Japan* 41: 115.

Kiritani, K. (1972). Strategy in integrated control of rice pests. *Rev. Plant Prot. Res.* 5: 76-104.

Kiritani, K., Hokyo, N., Sasaba, T., and Nakasuji, F. (1970). Studies on population dynamics of the green rice leafhopper *Nephotettix cincticeps* Uhler. Regulatory mechanism of the population density. *Res. Popul. Ecol.* 12: 137-153.

Kiritani, K., Kawahara, S., Sasaba, T., and Nakasuji, F. (1971). An attempt of rice pest control by integration of pesticides and natural enemies. *Gensei* 22: 19-23.

Kiritani, K., Kawahara, S., Sasaba, T., and Nakasuji, F. (1972). Quantitative evaluation of predation by spiders on the green rice leafhopper, *Nephotettix cincticeps* Uhler, by a sight-count method. *Res. Popul. Ecol.* 13: 187-200.

Klein, M. (1970). Safflower phyllody-a mycoplasma disease of *Carthamus tinctorius* in Israel. *Plant Dis. Rep.* 54: 735-738.

Knowlton, G. F. (1932). The beet leafhopper in northern Utah. *Utah Agr. Exp. Stn. Tech. Bull.* No. 234, 64 pp.

Kobayashi, T. (1961). Effect of insecticidal applications to the rice stem borer on leaf-hopper populations. *Special Rep. on Predication of Pests, Ministry of Agric. Forest.* 13: 1-126.

Kunkel, L. O. (1924). Insect transmission of aster yellows. *Phytopathology* 14: 54.

Kunkel, L. O. (1932). Celery yellows of California not identical with the aster yellows of New York. *Contrib. Boyce Thompson Inst.* 4: 405-414.

Kunkel, L. O. (1933). Insect transmission of peach yellows. *Contrib. Boyce Thomp. Inst. Plant Res.* 5: 19-28.

Kunkel, L. O. (1946). Leafhopper transmission of corn stunt. *Proc. Nat. Acad. Sci. U.S.A.* 32: 246-247.

Kunkel, L. O. (1953). Aster yellows. *In* "Plant Diseases." U.S.D.A. 1953 Yearbook of Agriculture.

Kuno, E. (1968). Studies on the population dynamics of rice leafhopper in a paddy field. *Bull. Kyushu Nat. Agric. Exp. Stn.* 14: 131-246.

Kuno, E. (1973). Population ecology of rice leafhoppers in Japan. *Rev. Pl. Protect. Res.* 6: 1-16.

Kuwahara, M. (1974). Studies on the diel activity of the rhombic-marked leafhopper, *Hishimonus sellatus* Uhler. *Jap. J. Appl. Ent. Zool.* 18: 89-93.

Laboucheix, J., Van Offeren, A., et Desmidts, M. (1972). Mise en évidence du rôle vecteur d'*Orosius cellulosus* (Lindberg) (Homoptera, Cicadelloidea) dans la virescence florale du cotonnier en Haute Volta. *Coton Fibr. Trop.* 27: 393-394.

Landis, B. J., Powell, D. M., and Hagel, G. T. (1970). Attempt to suppress curly top and beet western yellows by control of the beat leafhopper and the green peach aphid with insecticide-treated sugarbeet seed. *J. Econ. Entomol.* 63: 493-496.

Lawson, F. R., Chamberlin, J. C., and York, G. T. (1951). Dissemination of the beet leaf-hopper in California. *U.S.D.A. Tech. Bull.* No. 1030, 59 pp.

Lee, P. E. (1967). Morphology of wheat striate mosaic virus and its localization in infected cells. *Virology* 33: 84-94.

Lim, G. S. (1972). Studies on penyakit merah disease of rice. III. Factors contributing to an epidemic in North Krian, Malaysia. *Malaysian Agr. J.* 48: 278-294.

Lindsten, K., Vacke, J., and Gerhardson, B. (1970). A preliminary report on three cereal virus diseases new to Sweden spread by *Macrosteles* and *Psammotettix* leafhoppers. *Stat. Växtskydanstr. medd.* 14: 285-297.

Ling, K. C. (1966). Nonpersistence of the tungro virus of rice in its leafhopper vector, *Nephotettix impicticeps. Phytopathology* 56: 1252-1256.

Ling, K. C. (1970). Ability of *Nephotettix apicalis* to transmit the rice tungro virus. *J. Econ. Entomol.* 63: 582-586.

Ling, K. C. (1975). Experimental epidemiology of rice tungro disease. I. Effect of some factors of vector (*Nephotettix virescens*) on disease incidence. *Phil. Phytopathol.* 11: 11-20.

Ling, K. C., and Carbonell, M. P. (1975). Movement of individual viruliferous *Nephotettix virescens* in cages and tungro infection of rice seedlings. *Phil. Phytopathol.* 11: 32-45.

Ling, K. C., and Tiongco, E. R. (1975). Effect of temperature on the transmission of rice tungro virus by *Nephotettix virescens. Phil. Phytopathol.* 11: 46-57.

Ling, K. C., and Tiongco, E. R. (1979). *In* "Leafhopper Vectors and Plant Disease Agents" (K. Maramorosch and K. F. Harris, eds.), pp. 349-366. Academic Press, New York.

Lo, T. C., and Han, Y. H. (1969). Studies on witches' broom of soyabean. *Bot. Bull. Acad. Sinica* 10: 10-22.

Lowe, S. K., Nyland, G., and Mircetich, S. M. (1976). The ultrastructure of the almond leaf scorch bacterium with special reference to topography of the cell wall. *Phytopathology* 66: 147-151.

Lukens, R. J., Miller, P. M., Walton, G. S., and Hitchcock, S. W. (1971). Incidence of X disease of peach and eradication of chokecherry. *Plant Dis. Rep.* 55: 645-647.

MacLeod, R., Black, L. M., and Moyer, F. H. (1966). The fine structure and intercellular localization of potato yellow dwarf virus. *Virology* 29: 540-542.

Magie, R. O., Smith, F. F., and Brierly, P. (1952). Occurrence of western aster yellows virus infection in gladiolus in eastern United States. *Plant Dis. Rep.* 36: 468-470.

Magyarosy, A. C., and Duffus, J. E. (1977). The occurrence of highly virulent strains of the beet curly top virus in California. *Plant Dis. Rep.* 61: 248-251.

Manns, T. F., and Manns, M. M. (1935). The dissemination of peach yellows and little peach. *Del. Agr. Exp. Stn. Bull.* 192: 40-44.

Maramorosch, K. (1952). Direct evidence for the multiplication of aster yellows virus in its insect vector. *Phytopathology* 42: 59-64.

Maramorosch, K. (1955). The occurrence of two distinct types of corn stunt in Mexico. *Plant Dis. Rep.* 39: 896-898.

Maramorosch, K. (1958). Cross protection between two strains of corn stunt virus in an insect vector. *Virology* 6: 448-459.

Maramorosch, K. (1963). Arthropod transmission of plant viruses. *Ann. Rev. Entomol.* 8: 369-414.

Maramorosch, K. (1974). Mycoplasmas and rickettsiae in relation to plant diseases. *Ann. Rev. Microbiol.* 28: 301-324.

Maramorosch, K. (1976). Plant mycoplasma diseases. *In* "Encyclopedia of Plant Physiology" (P. R. Williams and R. Haytefuss, eds.). Springer: pp. 150-171.

Maramorosch, K. (1979). Aster yellows spiroplasma ATCC 29747. Abstracts 79th Annu. Meeting Amer. Soc. Microbiol.: 85.

Maramorosch, K., and Harris, K. F. (eds.) (1979). "Leafhopper Vectors and Plant Disease Agents." Academic Press, New York. 654 pp.

Maramorosch, K., Hirumi, H., and Granados, R. R. (1970). Mycoplasma diseases of plants and insects. *Adv. Virus Res.* 16: 135-193.

Markham, P. G., and Townsend, R. (1974). Transmission of *Spiroplasma citri* to plants. *INSERM, Paris*, 11-17 Sept. 1974, 33: 201-206.

Markham, P. G., Townsend, R., Plaskitt, K., and Saglio, P. (1977). Transmission of corn stunt to dicotyledonous plants. *Plant Dis. Rep.* 61: 342-345.

Martin, M. W. (1970). Developing tomatoes resistant to curly top virus. *Euphytica* 19: 243-252.

Martinez-Lopez, G. (1977). New maize virus diseases in Colombia. *Proc. Maize Virus Dis. Colloq. Workshop*, 16-19 Aug. 1976. Ohio Agric. Res. and Develop. Ctr., Wooster, pp. 20-29.

Matsumoto, T., Lee, C. S., and Teng, W. S. (1969). Studies on sugarcane white leaf disease of Taiwan, with special reference to the transmission by a leafhopper, *Epitettix hiroglyphicus* Mats. *Ann. Phytopath. Soc. Japan* 35: 251-259.

Meade, A. B., and Peterson, A. G. (1964). Origin of populations of the six-spotted leafhopper, *Macrosteles fascifrons* in Anoka County, Minnesota. *J. Econ. Entomol.* 57: 885-888.

Menzies, J. D. (1944). Transmission studies with alfalfa witches' broom. *Phytopathology* 34: 936.

Mircetich, S. M., Lowe, S. K., Moller, W. J., and Nyland, G. (1976). Etiology of almond leaf scorch disease and transmission of the causal agent. *Phytopathology* 66: 17-24.

Mitsuhashi, J. (1979). Artificial rearing and aseptic rearing of leafhopper vectors: applications in virus and MLO research. *In* "Leafhopper Vectors and Plant Disease Agents" (K. Maramorosch and K. F. Harris, eds.), pp. 369-412. Academic Press, New York.

Miura, K. I., Kimura, I., and Suzuki, M. (1966). Double-stranded ribonucleic acid from rice dwarf virus. *Virology* 28: 571-597.

Miyata, T., and Saito, T. (1976). Mechanism of malathion resistance in the green rice leafhopper, *Nephotettix cincticeps* Uhler (Hemiptera: Deltocephalidae). *J. Pesticide Sci.* 1: 23-29.

Moreau, J. P., et Boulay, C. (1967). Mode de piqûre de trois cicadelles vectrices de virus, *Euscelis plebejus* Fall., *Macrosteles sexnotatus* Fall. et *Aphrodes bicinctus* Schrk. Étude histologique. *Ann. Épiph.* 18: 133-141.

Müller, H. J. (1979). Effects of photoperiod and temperature on leafhopper vectors. *In* "Leafhopper Vectors and Plant Disease Agents" (K. Maramorosch and K. F. Harris, eds.), pp. 29-94. Academic Press, New York.

Mumford, D. L., and Peay, W. E. (1970). Curly top epidemic in western Idaho. *J. Am. Soc. Sugar Beet Technol.* 16: 185-187.

Musil, M. (1960). Übertragung des Stolbur-, Kleeverzwergungs-und Kleeverlaubungsvirus durch die Zikade *Aphrodes bicinctus* (Schrk.). *Biologia* 15: 721-728.

Musil, M. (1962). Přénos viru parastolburu křisken *Euscelis plebejus* (Fallén). *Biologia* 17: 332-339.

Musil, M. (1967). The properties of yellows-type viruses recently isolated in Slovakia and Moravia. *Proc. 6th Conf. Czech. Plant Virologists*, Olomouc 1967, pp. 226-233.

Musil, M., and Valenta, V. (1958). Prenos stolburu a pribuzných virusov pomocou niektorých cikád (Transmission of the stolbur and related viruses by some leafhoppers). *Biologia* 13: 133-136.

Nagaich, B. B., and Giri, B. K. (1971). Marginal flavescence—a leafhopper transmitted disease of potatoes. *Indian Phytopath.* 24: 824-826.

Nagaich, B. B., and Sinha, R. C. (1974). Eastern wheat striate: a new viral disease. *Plant Dis. Rep.* 58: 968-970.

Naito, A. (1976). Studies on the feeding habits of some leafhoppers attacking the forage crops. II. A comparison of the feeding habits of the green rice leafhoppers in different developmental stages. *Jap. J. Appl. Entomol. Zool.* 20: 51-54.

Nakasuji, F. (1974). Epidemiological study of rice dwarf virus transmitted by the green rice leafhopper, *Nephotettix cincticeps. Japan Agri. Res. Quart.* 8: 84-91.

Nakasuji, F., and Kiritani, K. (1970). Ill-effects of rice dwarf virus upon its vector, *Nephotettix cincticeps* Uhler (Hemiptera: Deltocephalidae) and its significance for changes in relative abundance of infected individuals among vector populations. *Appl. Entomol. Zool.* 5: 1-12.

Nakasuji, F., and Kiritani, K. (1971). Inter-generational changes in relative abundance of insects infected with rice dwarf virus in populations of *Nephotettix cincticeps* Uhler (Hemiptera: Deltocephalidae). *Appl. Entomol. Zool.* 6: 75-83.

Nakasuji, F., and Kiritani, K. (1972). Descriptive models for the system of the natural spread of infection of rice dwarf virus (RDV) by the green rice leafhopper, *Nephotettix cincticeps* Uhler (Hemiptera: Deltocephalidae). *Res. Popul. Ecol.* 14: 18-35.

Nakasuji, F., and Kiritani, K. (1977). Epidemiology of rice dwarf virus in Japan. *Trop. Agri. Res. Ser.* 10: 93-98.

Narayanasamy, P. (1972). Influence of age of rice plants at the time of inoculation on the recovery of tungro virus by *Nephotettix impicticeps* (Ishihara). *Phytopath. Z.* 74: 109-114.

Nasu, S. (1963). Studies on some leafhoppers and planthoppers which transmit virus diseases of rice plant in Japan. *Bull. Kyushu Agric. Exp. Stn.* 18: 153-349.

Nasu, S. (1965). Electron microscopic studies on transovarial passage of rice dwarf virus.

Jap. J. Appl. Entomol. Zool. 9: 225-237.

Nasu, S., Sugiura, M., Wakimoto, T., and Iida, T. (1967). On the etiologic agent of rice yellow dwarf disease. *Ann. Phytopath. Soc. Japan* 35: 343-344.

Nault, L. R., and Bradfute, O. E. (1977). Reevaluation of leafhopper vectors of corn stunting pathogens. *Proc. Am. Phytopathol. Soc.* 4: 172.

Nault, L. R., and Bradfute, O. E. (1979). Corn stunt: involvement of a complex of leafhopper-borne pathogens. *In* "Leafhopper Vectors and Plant Disease Agents" (K. Maramorosch and K. F. Harris, eds.), pp. 561-586. Academic Press, New York.

Nault, L. R., Styer, W. E., Knoke, J. K., and Pitre, H. N. (1973). Semipersistent transmission of leafhopper-borne maize chlorotic dwarf virus. *J. Econ. Entomol.* 66: 1271-1273.

Nault, L. R., Gordon, D. T., Robertson, D. C., and Bradfute, O. E. (1976). Host range of maize chlorotic dwarf virus. *Plant Dis. Rep.* 60: 374-377.

Niederhauser, J. S., and Cervantes, J. (1950). Transmission of corn stunt in Mexico by a new insect vector, *Baldulus elimatus. Phytopathology* 40: 20-21.

Nielson, M. W. (1968). The leafhopper vectors of phytopathogenic viruses (Homoptera, Cicadellidae) taxonomy, biology, and virus transmission. *U.S.D.A. Tech. Bull.* No. 1382, 386 pp.

Nielson, M. W. (1979). Taxonomic relationships of leafhopper vectors of plant pathogens. *In* "Leafhopper Vectors and Plant Disease Agents" (K. Maramorosch and K. F. Harris, eds.), pp. 3-27. Academic Press, New York.

Nishi, Y., Kimura, T., and Maejima, I. (1975). Causal agent of "waika" disease of rice plants in Japan. *Ann. Phytopathol. Soc. Japan* 41: 223-227.

Okuda, S., and Nishimura, N. (1974). Witches' broom of *Cryptotaenia japonica* Hassk. *Ann. Phytopath. Soc. Japan* 40: 439-451.

Oldfield, G. N., Kaloostian, G. H., Pierce, H. D., Calavan, E. C., Granett, A. L., and Blue, R. L. (1976). Beet leafhopper transmits citrus stubborn disease. *Calif. Agric.* 30(6): 15.

Oman, P. W. (1949). The nearctic leafhoppers (Homoptera:Cicadellidae). A generic classification and check list. *Entomol. Soc. Wash. Mem.* No. 3, 253 pp.

Ou, S. H., Rivera, C. T., Navaratnam, S. J., and Goh, K. G. (1965). Virus nature of "Penyakit merah" disease of rice in Malaysia. *Plant Dis. Rep.* 49: 778-782.

Ouchi, Y., and Suenaga, H. (1963). On the transmissibility by the leafhopper *Nephotettix apicalis* of rice yellow dwarf virus. *Prof. Ass. Plant Protect. Kyushu* 9: 60-61.

Oya, S. (1978). Effects of photoperiod on the induction of diapause in the green rice leafhopper, *Nephotettix cincticeps* Uhler (Hemiptera: Cicadellidae). *Jap. J. appl. Ent. Zool.* 22: 108-114.

Pathak, M. D., Cheng, C. H., and Fortuno, M. E. (1969). Resistance to *Nephotettix impicticeps* and *Nilaparvata lugens* in varieties of rice. *Nature* 223: 502-504.

Pitre, H. N. (1967). Greenhouse studies of the host range of *Dalbulus maidis*, a vector of corn stunt virus. *J. Econ. Entomol.* 60: 417-421.

Pitre, H. N., and Boyd, F. J. (1970). A study of the role of weeds in corn fields in the epidemiology of corn stunt disease. *J. Econ. Entomol.* 63: 195-197.

Posnette, A. F., and Ellenberger, C. E. (1963). Further studies of green petal and other leafhopper-transmitted viruses infecting strawberry and clover. *Ann. Appl. Biol.* 51: 69-83.

Pring, D. R., Zeyen, R. J., and Banttari, E. E. (1973). Isolation and characterization of oat blue dwarf virus ribonucleic acid. *Phytopathology* 63: 393-396.

Purcell, A. H. (1974). Spatial patterns of Pierce's disease in the Napa Valley. *Am. J. Enol. Vitic.* 25: 162-167.

Purcell, A. H. (1975). Role of the blue-green sharpshooter, *Hordnia circellata*, in the epidemiology of Pierce's disease of grapevines, *Environ. Entomol.* 4: 745-752.

Purcell, A. H. (1976). Seasonal changes in host plant preference of the blue-green sharpshooter *Hordinia circellata. Pan-Pacific Entomol.* 52: 33-37.

Purcell, A. H. (1978). Lack of transtadial passage by Pierce's disease vector. *Phytopathology News* 12: 217-218.

Purcell, A. H. (1979). Leafhopper vectors of xylem-borne plant pathogens. *In* "Leafhopper Vectors and Plant Disease Agents" (K. Maramorosch and K. F. Harris, eds.). pp. 603-625. Academic Press, New York.

Purcell, A. H., and Finlay, A. (1979). Evidence for noncirculative transmission of Pierce's disease bacterium by sharpshooter leafhoppers. *Phytopathology* 69: 393-395.

Raine, J. (1967). Leafhopper transmission of witches' broom and clover phyllody viruses from British Columbia to clover, alfalfa and potato. *Can J. Bot.* 45: 441-445.

Rangaswami, R. S., and Griffith, A. L. (1941). Demonstration of *Jassus indicus* (Walk.) as a vector of the spike disease of sandal (*Santalum album* Linn.). *Indian Forester* 67: 387-394.

Ritenour, G., Hills, F. J., and Lang, W. H. (1970). Effect of planting date and vector control on the suppression of curly top and yellows in sugarbeet. *J. Am. Soc. Sugar Beet Technol.* 16: 78-84.

Rivera, C. T., and Ou, S. H. (1965). Leafhopper transmission of "tungro" disease of rice. *Plant Dis. Rep.* 49: 127-131.

Rivera, C. T., Ou, S. H., and Pathak, M. D. (1963). Transmission studies of the orange-leaf disease of rice. *Plant Dis. Rep.* 47: 1045-1048.

Rose, D. J. W. (1962). Insect vectors of maize streak. *Zool. Soc. S. Afr. News Bull.* 3(3): 11.

Rose, D. J. W. (1972a). Times and sizes of dispersal flights by *Cicadulina* species (Homoptera:Cicadellidae), vectors of maize streak disease. *J. Anim. Ecol.* 41: 495-506.

Rose, D. J. W. (1972b). Dispersal and quality in populations of *Cicadulina* species (Cicadellidae). *J. Anim. Ecol.* 41: 589-609.

Rose, D. J. W. (1973). Distances flown by Cicadulina spp. (Hem., Cicadellidae) in relation to distribution of maize streak disease in Rhodesia. *Bull. Ent. Res.* 62: 497-505.

Rose, D. J. W. (1974). The epidemiology of maize streak disease in relation to population densities of *Cicadulina* spp. *Ann. Appl. Biol.* 76: 199-207.

Rose, D. J. W. (1978). Epidemiology of maize streak disease. *Ann. Rev. Entomol.* 23: 259-282.

Rosenberger, D. A., and Jones, A. L. (1978). Leafhopper vectors of the peach X-disease and its seasonal transmission from chokecherry. *Phytopathology* 68: 782-790.

Saglio, P., L'Hospital, M., Laflèche, D., Dupont, G., Bové, J. M., Tully, J. G., and Freundt, E. A. (1973). *Spiroplasma citri* gen. and sp. n.: A mycoplasma-like organism associated with "stubborn" disease of citrus. *Int. J. Syst. Bacteriol.* 23: 191-204.

Saito, Y., Nabheerong, N., Hino, T., Hibino, H., Srimarut, S., and Wungkobkiat, S. (1971). Purification of rice yellow orange leaf virus and observation on ultrathin sections of the disease. *Ann. Phytopathol. Soc. Japan* 37: 195.

Saito, Y., Roechan, M., Tantera, D. M., and Iwaki, M. (1975). Small bacilliform particles associated with penyakit habang (tungro-like) disease of rice in Indonesia. *Phytopathology* 65: 793-796.

Saito, Y., Iwaki, M., and Usugi, T. (1976). Association of two types of particles with tungro-group diseases of rice. *Ann. Phytopathol. Soc. Japan* 42: 375.

Sakai, S. (1937). On the transmission of a mosaic-like disease of mulberry by *Eutettix disciguttus* Walk. *Nagano-Ken. Ser. Exp. Stn. Rep.* 39: 1-14.

Sasaba, T. (1974). Computer simulation studies on the life system of the green rice leafhopper, *Nephotettix cincticeps* Uhler. *Rev. Plant Protec. Res.* 7: 81-98.

Sasaba, T., and Kiritani, K. (1975). A system model and computer simulation of the green rice leafhopper populations in control programmes. *Res. Popul. Ecol.* 16: 231-244.

Satomi, H., Hirao, J., and Kimura, T. (1975). Transmission of the rice waika virus by a new leafhopper vector, *Nephotettix nigropictus* (Stål) (Homoptera:Cicadellidae). *Proc. Assoc. Pl. Prot. Kyushu* 21: 60-63.

Sauer, H. F. G. (1946). A cigarrinha *Agallia albidula* Uhl. (Hom., Cicadel.) vectora de uma doenca de virus do tomateiro. *Biologico* 12: 176-178.

Savulescu, A., and Ploaie, P. (1962). Studies on the incidence of the "stolbur" virus disease in relation to the vectors, host plants and ecological conditions. *In* "Plant Virology," *Proc. 5th Conf. Czech. Plant Virologists*, Prague, 1962, pp. 195-202.

Schultz, H. N., and Dean, L. L. (1947). Inheritance of curly top disease reaction in the bean, *Phaseolus vulgaris. J. Am. Soc. Agron.* 39: 47-51.

Schvester, D., Carle, P., et Moutous, G. (1961). Sur la transmission de la flavescence dorée des vignes par une cicadelle. *C. R. Acad. Agric. Fr.* 47: 1021-1024.

Scott, G. E., Rosenkranz, E. E., and Nelson, L. R. (1977). Yield loss of corn due to corn stunt disease complex. *Agron. J.* 69-: 92-94.

Severin, H. H. P. (1929). Yellows disease of celery, lettuce and other plants, transmitted by *Cicadula sexnotata* (Fall.). *Hilgardia* 3: 543-583.

Severin, H. H. P. (1930). Life history of the beet leafhopper, *Eutettix tenellus* (Baker), in California. *Calif. Univ. Entomol. Pub.* No. 5: 37-88.

Severin, H. H. P. (1933). Field observations on the beet leafhopper, *Eutettix tenellus*, in California. *Hilgardia* 7: 281-360.

Severin, H. H. P. (1934a). Experiments with the aster-yellows virus from several states. *Hilgardia* 8: 305-325.

Severin, H. H. P. (1934b). Transmission of California aster and celery-yellows virus by three species of leafhoppers. *Hilgardia* 8: 339-361.

Severin, H. H. P. (1939). Factors affecting curly top infectivity of beet leafhopper, *Eutettix tenellus. Hilgardia* 12: 497-530.

Severin, H. H. P. (1945). Evidence of nonspecific transmission of California aster-yellows virus by leafhoppers. *Hilgardia* 17: 21-59.

Severin, H. H. P. (1946). Transmission of California aster-yellows virus by the first reported leafhopper vector in Gyponinae. *Hilgardia* 17: 141-150.

Severin, H. H. P. (1947). Newly discovered leafhopper vectors of California aster-yellows virus. *Hilgardia* 17: 511-524.

Severin, H. H. P. (1948). Transmission of California aster-yellows virus by leafhoppers in *Thamnotettix* group. *Hilgardia* 18: 203-216.

Severin, H. H. P. (1949). Transmission of the virus of Pierce's disease of grapevines by leafhoppers. *Hilgardia* 19: 190-206.

Severin, H. H. P. (1950a). Spittle-insect vectors of Pierce's disease virus. II. Life history and virus transmission. *Hilgardia* 19: 357-382.

Severin, H. H. P. (1950b). *Texananus incurvatus*. II. Transmission of California aster-yellows virus. *Hilgardia* 19: 544-545.

Severin, H. H. P., and Frazier, N. W. (1945). California aster-yellows on vegetable and seed crops. *Hilgardia* 16: 573-586.

Shaw, H. B. (1910). The curly top of beets. *U.S. Dept. Agr. Bur. Plant Indus. Bull.* No. 181, 46 pp.

Shiga Agricultural Experiment Station. (1899). Results of experiments with insect pests. Rep. 2, 1-26.

Shikata, E. (1979). Cytopathological changes in leafhopper vectors of plant viruses. *In* "Leafhopper Vectors and Plant Disease Agents" (K. Maramorosch and K. F. Harris, eds.), pp. 309-325. Academic Press, New York.

Shikata, E., and Chen, M. (1969). Electron microscopy of rice transitory yellowing virus. *J. Virology* 3: 261-264.

Shinkai, A. (1958). Transovarial passage of rice dwarf virus in *Inazuma dorsalis* (Motsch.). *Ann. Phytopathol. Soc. Japan* 23: 26.

Shinkai, A. (1962). Studies on insect transmission of rice virus diseases in Japan. *Nat. Inst. Agr. Sci. Bull.* C 14: 1-112.

Shinkai, A. (1964). Studies on insect transmission of sweet potato witches' broom disease in the Ryukyu Islands. *Spec. Res. Rep., Agri. Sec., Econ. Dept., Govt. Ryukyu Is.*, 44 pp.

Singh, K. G. (1969). Penyakit merah disease, a virus infection of rice in Malaysia. *In* "The

Virus Diseases of the Rice Plant" (*Proc. Symp. IRRI, 1967*), pp. 75-78. Johns Hopkins Press, Baltimore, Maryland.

Sinha, R. C. (1965). Recovery of potato yellow dwarf virus from hemolymph and internal organs of an insect vector. *Virology* 27: 118-119.

Sinha, R. C. (1970). *Elymana virescens*, a newly described vector of wheat striate mosaic virus. *Can. Plant Dis. Surv.* 50: 118-120.

Sinha, R. C., and Chiykowski, L. N. (1967). Multiplication of wheat striate mosaic virus in its leafhopper vector *Endria inimica*. *Virology* 32: 402-405.

Sinha, R. C., and Chiykowski, L. N. (1968). Distribution of clover phyllody virus in the leafhopper *Macrosteles fascifrons* (Stål). *Acta Virol.* 12: 546-550.

Sinha, R. C., and Chiykowski, L. N. (1969). Synthesis, distribution and some multiplication sites of wheat striate mosaic virus in a leafhopper vector. *Virology* 38: 679-684.

Slykhuis, J. T. (1953). Striate mosaic, a new disease of wheat in South Dakota. *Phytopathology* 43: 537-540.

Slykhuis, J. T. (1961). The causes and distribution of mosaic diseases of wheat in Canada in 1961. *Can. Plant Dis. Surv.* 41: 330-343.

Slykhuis, J. T. (1962). Wheat striate mosaic, a virus disease to watch on the Prairies. *Can. Plant Dis. Surv.* 42: 135-142.

Slykhuis, J. T. (1963). Vector and host relations of North American wheat striate mosaic virus. *Can. J. Bot.* 41: 1171-1185.

Sogawa, K. (1973). Feeding of the rice plant- and leafhoppers. *Rev. Plant Protec. Res.* 6: 31-43.

Staples, R., Jansen, W. P., and Andersen, L. W. (1970). Biology and relationship of the leafhopper *Aceratagallia calcaris* to yellow vein disease of sugarbeets. *J. Econ. Entomol.* 63: 460-463.

Stoddard, E. M. (1947). The X-disease of peach and its chemotherapy. *Conn. Agr. Exp. Stn. Bull.* No. 506, 19 pp.

Stoner, W. N. (1953). Leafhopper transmission of a degeneration of grape in Florida and its relation to Pierce's disease. *Phytopathology* 43: 611-615.

Storey, H. H. (1924). The transmission of a new plant virus disease by insects. *Nature* 114: 245.

Storey, H. H. (1936). Virus diseases of East African plants. V. Streak disease of maize. *East Afr. Agric. J.* 1: 471-475.

Storey, H. H. (1937). A new virus of maize transmitted by *Cicadulina* spp. *Ann. Appl. Biol.* 24: 87-94.

Sukhov, K. S., and Vovk, A. M. (1945). On the identity between yellows of koksaghyz and yellows of aster and its possible relation to big bud in tomato. *Compt. rend. Acad. Sc. USSR* 48: 365-368.

Sukhov, K. S., and Vovk, A. M. (1947). New virus disease of the tomato leaf crinkles and its leafhopper vector, *Agallia venosa* Fall. *Akad. Nauk S.S.S.R. Dok.* 56: 433-435.

Swenson, K. G. (1971). Environmental biology of the leafhopper *Scaphytopius delongi*. *Ann. Entomol. Soc. Am.* 64: 809-812.

Takata, K. (1895). Results of experiments with dwarf disease of rice plant. *J. Japan Agr. Soc.* 171: 1-4.

Thomas, K. M., and Krishnaswami, C. S. (1939). Little leaf-a transmissible disease of brinjal. *Proc. Indian Acad. Sci.*, Sect. B, 10: 201-212.

Thomas, P. E. (1969). Thirty-eight new hosts of curly top virus. *Plant Dis. Rep.* 53: 548-549.

Thomas, P. E. (1972). Mode of expression of host preference by *Circulifer tenellus*, the vector of curly top virus. *J. Econ. Entomol.* 65: 119-123.

Thomas, P. E., and Boll, R. K. (1977). Effect of host preference on transmission of curly top virus to tomato by the beet leafhopper. *Phytopathology* 67: 903-905.

Thomas, P. E., and Martin, M. W. (1971). Vector preference, a factor of resistance to curly top virus in certain tomato cultivars. *Phytopathology* 61: 1257-1260.

Thornberry, H. H. (1954). Preliminary report on insect transmission of eastern peach X-disease virus in Illinois. *Plant Dis. Rep.* 38: 412-413.

Thornberry, H. H. (1966). Plant pests of importance to North American agriculture. *In* "Index of Plant Virus Diseases." U.S.D.A. Agri. Handbook No. 307, 446 pp.

Ting, W. P., and Paramsothy, S. (1970). Studies on penyakit merah disease of rice. I. Virus-vector interaction. *Malaysian Agr. J.* 47: 290-298.

Tomlinson, W. E., Jr., Marucci, P. E., and Doehlert, C. A. (1950). Leafhopper transmission of blueberry stunt disease. *J. Econ. Entomol.* 43: 658-662.

Turner, W. F. (1949). Insect vectors of phony peach disease. *Science* 109: 87-88.

Turner, W. F., and Pollard, H. N. (1955). Additional leafhopper vectors of phony peach. *J. Econ. Entomol.* 48: 771-772.

Turner, W. F., and Pollard, H. N. (1959). Insect transmission of phony peach disease. *U.S.D.A. Tech. Bull.* No. 1193, 27 pp.

Vacke, J. (1961). Wheat dwarf virus disease. *Biol. Plant* 3: 228-233.

Vacke, J. (1973). Wheat pale green dwarf found in Czechoslovakia. *In* "Plant Virology," *Proc. 7th Confr. Czech. Plant Virol.*, 1971, pp. 523-528.

Valenta, V. (1958). Problem stolbura i srodnik žutica u Čehoslovackoj. (The problem of stolbur and related yellows type virus diseases in Czechoslovakia.) *Agron. Glasnik* 8: 87-102.

Valenta, V., Musil, M., and Mišiga, S. (1961). Investigations on European yellows-type viruses. I. The stolbur virus. *Phytopath. Z.* 42: 1-38.

Vasudeva, R. S., and Sahambi, H. S. (1955). Phyllody in sesamum (*Sesamum orientale* L.). *Indian Phytopath.* 8: 124-129.

Wallace, J. M., and Murphy, A. M. (1938). Studies on the epidemiology of curly top in southern Idaho, with special reference to sugar beets and weed hosts of the vector *Eutettix tenellus. U.S.D.A. Tech. Bull.* No. 624, 47 pp.

Wathanakul, L., and Weerapat, P. (1969). Virus diseases of rice in Thailand. *In* "The Virus Diseases of the Rice Plant" (*Proc. Symp. IRRI, 1967*), pp. 79-85. Johns Hopkins Press, Baltimore, Maryland.

Westdal, P. H. (1968). Host range studies of oat blue dwarf virus. *Can. J. Bot.* 46: 1431-1435.

Westdal, P. H. (1969). Reaction of some oat varieties to aster yellows. *Can. J. Plant Sci.* 49: 630-631.

Westdal, P. H., and Richardson, H. P. (1966). The painted leafhopper, *Endria inimica* (Say), a vector of wheat striate mosaic in Manitoba. *Can. Entomol.* 98: 922-931.

Westdal, P. H., and Richardson, H. P. (1969). The susceptibility of cereals and wild oats to an isolate of the aster yellows pathogen. *Can. J. Bot.* 47: 755-760.

Westdal, P. H., Barrett, C. F., and Richardson, H. P. (1961). The six-spotted leafhopper, *Macrosteles fascifrons* (Stål) and aster yellows in Manitoba. *Can. J. Plant Sci.* 41: 320-331.

Whitcomb, R. F., and Davis, R. E. (1970). Mycoplasma and phytarboviruses as plant pathogens persistently transmitted by insects. *Ann. Rev. Entomol.* 15: 405-464.

Whitcomb, R. F., and Jensen, D. D. (1968). Proliferative symptoms in leafhoppers infected with western X-disease virus. *Virology* 35: 174-177.

Whitcomb, R. F., Jensen, D. D., and Richardson, J. (1966). The infection of leafhoppers by western X-disease virus. II Fluctuation of virus concentration in the hemolymph after injection. *Virology* 28: 454-458.

Whitcomb, R. F., Jensen, D. D., and Richardson, J. (1967). The infection of leafhoppers by western X-disease virus. III. Salivary, neural, and adipose histopathology. *Virology* 31: 539-549.

Whitcomb, R. F., Jensen, D. D., and Richardson, J. (1968a). The infection of leafhoppers by western X-disease virus. IV. Pathology in the alimentary tract. *Virology* 34: 69-78.

Whitcomb, R. F., Jensen, D. D., and Richardson, J. (1968b). The infection of leafhoppers by western X-disease virus. VI. Cytopathological interrelationships. *J. Invert. Pathol.* 12: 202-221.

Wilcox, R. B. (1951). Tests of cranberry varieties and seedlings for resistance to the leafhopper vector of false-blossom disease. *Phytopathology* 41: 722-735.

Wilcox, R. B., and Beckwith, C. S. (1933). A factor in the varietal resistance of cranberries to the false blossom disease. *J. Agr. Res.* 47: 583-590.

Williamson, D. L., and Whitcomb, R. F. (1975). Plant mycoplasmas: a cultivable spiroplasma causes corn stunt disease. *Science* 188: 1018-1020.

Windsor, I. M., and Black, L. M. (1973). Evidence that clover club leaf is caused by a rickettsia-like organism. *Phytopathology* 63: 1139-1148.

Wolfe, H. R. (1955). Relation of leafhopper nymphs to the western X-disease virus. *J. Econ. Entomol.* 48: 588-590.

Wolfe, H. R., Anthon, E. W., and Jones, L. S. (1950). Transmission of western X-disease of peaches by the leafhopper *Colladonus geminatus* (Van D.). *Phytopathology* 40: 971.

Wolfe, H. R., Anthon, E. W., Kaloostian, G. H., and Jones, L. S. (1951). Leafhopper transmission of western X-disease. *J. Econ. Entomol.* 44: 616-619.

Wright, N. S. (1954). The witches' broom virus disease of potatoes. *Amer. Pot. J.* 31: 159-164.

Wright, N. S. (1957). Potato witches' broom in North America. *Proc. 3rd Conf. Potato Virus Diseases*, Lisse-Wageningen, pp. 239-245.

Yokoyama, S., Sakai, H., Hidari, M., and Inoue, T. (1974). Studies on the "Waisei" disease of rice plant. 3. Transmission of the causal agent by *Nephotettix cincticeps* Uhler. *Proc. Assoc. Pl. Prot. Kyushu* 20: 140.

Zazhurilo, V. K., and Sitnikova, G. M. (1939). Mosaic of winer wheat. *Comp. Rend. Acad. Sci. U.S.S.R.* 25: 798-801.

Zeyen, R. J., and Banttari, E. E. (1972). History and ultrastructure of oat blue dwarf virus infected oats. *Can J. Bot.* 50: 2511-2519.

Chapter 5

EPIDEMIOLOGY OF HELPER-DEPENDENT PERSISTENT APHID TRANSMITTED VIRUS COMPLEXES

B. W. Falk

Department of Plant Pathology,
University of California,
Riverside, California

J. E. Duffus

Science and Education Administration,
U.S. Department of Agriculture,
Salinas, California

5.1 INTRODUCTION

Mixed infections of plant viruses in nature are probably more common than infections by single viruses. The complex virus diseases affecting sugarbeets, potatoes, legumes, grasses, and the fruit crops are examples. In the Salinas Valley of California, as many as nine distinct virus entities have been isolated from individual spinach plants (Duffus, 1979). However, most of the work with plant viruses has been with single entities, as pure as possible, isolated from the naturally occurring complexes. From an epidemiological standpoint, the pure cultures as studied in the laboratory rarely occur alone in nature.

Several types of interactions between related and unrelated viruses may affect the epidemiology of these viruses. For instance, the lack of the interference phenomenon apparently greatly increases the severity of the beet curly top disease in resistant sugarbeets in western United States. It is also common for the infection of a plant by more than one virus to result in more serious effects (disease) on the host than infection by either virus alone (Zink and Duffus, 1972; Gill and Comeau, 1977; Watson and Serjeant, 1964). It has also become evident in recent years that mixed infections have effects on the properties of the progeny virions in these infections. In fact, in many cases, the mixed infection may be necessary for the survival of some viruses such as helper-dependent satellite viruses (Schneider, 1977). Recent articles have reviewed some of these subjects (Dodds and Hamilton, 1976; Rochow, 1972; 1977).

Under natural conditions most plant viruses in mixed infections are probably transmitted independently. However, more and more evidence is accumulating, as the field aspects of these complex diseases are studied, that indicates the dependent transmission phenomenon, first described by Clinch et al. (1936), is more common than previously thought. Transmission of these viruses by insect vectors is dependent on the presence of another virus or virus product. This is found in nonpersistent, semipersistent, and persistent virus vector relationships (Freitag, 1969; Rochow, 1977).

The dependent transmission of nonpersistent viruses has been recently reviewed (Pirone, 1977). Basically, it is found to be due to a helper component produced in the virus infected plant that is necessary for the virus to become transmissible by aphids. Some virus isolates lack the ability to produce the helper component and thus are not transmissible by aphids independently. They can become aphid transmissible when they are in a mixed infection with a virus that does produce the helper component—the viruses generally must be in the same taxonomic group (Pirone, 1977). They may also become transmissible by allowing aphids to feed or probe plants containing helper component prior to feeding or probing the plant infected with the non-helper component producing virus. Thus, aphids can acquire the helper component and the virus simultaneously, or they may acquire the helper component prior to acquiring the virus.

Dependent transmission of semipersistent viruses has one recognized example, parsnip yellow fleck virus (PYFV) and the helper virus, anthriscus yellows (AYV). The helper mechanism is thought to result from the helper virion itself (Elnagar and Murant, 1976a, 1976b). The helper and dependent viruses could be acquired simultaneously from a mixed infection, or the helper virus could be acquired first and then the dependent virus acquired by aphids.

The dependent transmission of persistent viruses differs from nonpersistent and semipersistent systems in that both viruses must be acquired simultaneously (helper and dependent viruses) (Rochow, 1972). The best understood examples of persistent virus dependent transmission are those of barley yellow dwarf virus (BYDV) (Rochow, 1970; 1972; 1977) and pea enation mosaic virus (PEMV) (Harris, 1977, 1979). One BYDV isolate, MAV, is specifically transmitted by *Macrosiphum avenae*. Another isolate, RPV, is specifically transmitted by *Rhopalisiphum padi*. As a result of the mixed infection of these two isolates, some MAV is transmissible by *R. padi* and some is still transmissible by *M. avenae*. It is believed that in the infected plant, some of the nucleic acid of the MAV isolate is coated by the coat protein of the RPV isolate by the heterologous encapsidation or genomic masking mechanism, and thus MAV is transmitted by *R. padi*. Serological studies have demonstrated that the *R. padi* transmissible MAV can be neutralized by antiserum to RPV (Rochow, 1970). Rochow and Gill (1978) also have found other BYDV isolates that behave similarly. A similar system of heterologous encapsidation and reciprocity between recognition sites on viral coat protein and aphid salivary gland membranes appears to control PEMV aphid-transmissibility and vector specificity (Harris *et al.*, 1975; Harris, 1977, 1979; Clarke and Bath, 1977; Hull, 1977; Adam *et al.*, 1979).

Another type of persistent virus dependent transmission involves pairs of viruses that differ in physical properties, tissue specificity, insect transmission, and are probably of remote virus groups. The helper dependent virus complexes make up this group. Examples thus far described have a mechanically transmitted component that is aphid transmissible only when in association with a helper virus. The helper viruses are commonly found alone in nature, and except for one possible exception, are luteoviruses.

In this discussion we will consider some real and potential epidemiological characteristics that result from the mixed infections of some aphid transmitted viruses, specifically the helper dependent virus complexes. Although aspects of this topic have been recently reviewed (Rochow, 1972; 1977) new information on the nature of some of these viruses and their mechanisms of transmission warrants a new look from an epidemiological standpoint.

5.2 EXAMPLES OF PERSISTENT HELPER-DEPENDENT TRANSMISSION

The persistent aphid transmitted virus complexes are a group of specific mixed infections that cause some very important diseases in many parts of the world.

The members of this group are listed in Table I, along with two potential members that exhibit many typical properties of the complexes.

Smith (1946) first observed that two viruses were involved in the tobacco rosette disease when he found that mechanical transmission from rosetted plants yielded symptoms different than those obtained by aphid transmission from the same plants. The sap transmissible virus was very easily and efficiently mechanically transmitted, but it was not aphid transmissible after mechanical transmission. Together, both viruses are persistent in their relationship with the aphid vector, *Myzus persicae*, and occasionally one or the other of the viruses may be inoculated alone by aphids viruliferous for both, but generally the aphid transmitted both viruses.

Stubbs (1948) reported a very severe disease of carrots from Australia that was called carrot motley dwarf. Carrot motley dwarf has been a very serious disease in Australia and is still important in many areas of England and in California. In the early 1940's, it restricted carrot growing in Australia so much that carrots were abandoned in some parts of Australia. Watson *et al.* (1964) reported losses of yield in carrots to be as much as 40-60% in England as a result of motley dwarf, especially when aphids moved into plots in early spring. The virus(es) was transmitted in a persistent manner by *Cavariella aegopodii*. Work on carrot motley dwarf disease in England showed that with English motley dwarf, two viruses are involved with the disease (Watson and Serjeant, 1964). They showed that the mechanically transmissible virus, carrot mottle virus (CMotV), was dependent upon the presence of a helper virus, carrot red leaf virus (CRLV), for aphid transmission. CMotV is easily mechanically transmitted, but CRLV is not.

TABLE I. Persistent Helper-Dependent Complexes

Complex Name	Helper Virus(es)	Dependent Virus	Vector
Tobacco rosette	Tobacco vein distorting	Tobacco mottle	*Myzus persicae*
Carrot motley dwarf	Carrot red leaf	Carrot mottle	*Cavariella aegopodii*
Tobacco yellow vein	Tobacco yellow vein assistor	Tobacco yellow vein	*Myzus persicae*
Groundnut rosette	Groundnut rosette assistor	Groundnut rosette	*Aphis craccivora*
Lettuce speckles	Beet western yellows	Lettuce speckles mottle	*Myzus persicae*
Bean yellow vein-banding	Pea enation mosaic virus Bean leaf roll	Bean yellow vein-banding	*Acyrthosiphon pisum*
Possible Members			
Late mosaic of cotton	Cotton anthocyanosis	Brazilian tobacco streak	*Aphis gossypii*
Celery yellow spot	Unknown	Celery-yellow spot	*Rhopalosiphum conii*

Two additional complexes were reported from Africa, groundnut rosette (Hull and Adams, 1968) and tobacco yellow vein diseases (Adams and Hull, 1972). Both have mechanically transmissible components that are dependent upon helper viruses for aphid transmission.

Another complex recently reported from England (Cockbain, 1978) is induced by the bean yellow vein-banding virus, which is mechanically transmissible and helper dependent for aphid transmission, and a helper virus, pea enation mosaic (PEMV). PEMV is a persistent aphid borne virus but is not a typical luteovirus. Another virus, pea (bean) leaf roll that is a typical luteovirus, can also serve as a helper in this complex. In pot tests, field beans infected with bean yellow vein-banding virus (BYVBV) and pea enation mosaic, yielded 77% fewer pods and 80% less seed than plots infected with PEMV alone (Cockbain, 1978).

In California, we have found a similar complex. It induces a disease in head lettuce, sugar beets, and spinach, and is termed the speckles disease because of the characteristic symptoms induced in the outer leaves of early spring lettuce (Falk *et al.*, 1979a). It can be severe in early spring crops, but generally disappears with the coming of summer. The speckles disease is caused by an association of beet western yellows virus (BWYV), and lettuce speckles mottle virus (LSMV). Both viruses are transmitted together by *Myzus persicae*, the green peach aphid. BWYV is the helper virus and LSMV the dependent virus. BWYV is a typical luteovirus, non-mechanically transmissible, and has a host range inclusive of many plant species (Duffus, 1973). LSMV is, so far, easily mechanically transmitted to a few species and is not aphid transmissible unless in a mixed infection with BWYV.

5.3   DEPENDENT TRANSMISSION MECHANISMS

Much work has been done studying the mechanisms of the dependent transmission in the motley dwarf, tobacco yellow vein, groundnut rosette, and speckles complexes. In all of the complexes, the dependent virus must be associated with the helper virus for it to be aphid transmitted.

We investigated various combinations of LSMV and BWYV and determined the aphid and mechanical transmission of each virus in these different combinations. Groups of *M. persicae* were allowed acquisition access to either LSMV or BWYV infected plants followed by acquisition on plants infected with the other virus. Aphids were then tested for the ability to transmit LSMV and BWYV. LSMV was never transmitted alone, but BWYV was transmitted in nearly 100% of the cases (Falk *et al.*, 1979a). From control plants infected with both viruses, LSMV and BWYV were both aphid transmitted. Thus, sequential acquisition access to both viruses failed to make LSMV aphid transmissible. This is a significant difference from the helper dependent mechanisms of nonpersistent and semipersistent aphid transmitted viruses. The genomic masking of BYDV is similar to the persistent dependent complexes in that both viruses must be simultaneously acquired from the same plant for the dependent transmission to occur (Rochow, 1965; 1970).

Clarified preparations of both viruses were tested for aphid and mechanical transmissibility. Aphid transmissions were done by allowing aphids to feed on the preparations of either virus alone or both viruses through parafilm membranes, and then allowing the aphids inoculation access to indicator plants. Mixtures of LSMV and BWYV prepared from plants infected separately with either of the viruses and then mixed yielded only aphid transmission of BWYV, but LSMV was recoverable from these preparations by mechanical inoculation. Thus, LSMV was present but non-aphid transmissible. However, combinations of both viruses prepared from plants infected simultaneously with both viruses yielded aphid transmissible LSMV and BWYV, as well as mechanically transmissible LSMV. Thus, the physical presence of the helper virus was not enough to make LSMV aphid transmissible in the preparations prepared from separately infected plants. Similar results have been obtained by Elnagar and Murant (1978a; 1978b) for CMotV and CRLV. They created various mixtures of CMotV and CRLV in aphids by injection and feeding experiments. CMotV was aphid transmissible only when aphids fed on doubly infected plants or were injected with preparations from aphids viruliferous for both viruses, but the helper, CRLV was aphid transmissible in all cases.

The mixed infection then changes a major epidemiological property of the dependent virus by giving it the ability to be dispersed by an aphid vector. The dependent viruses from mixed infections gain transmission properties nearly identical to those of the helper virus. Smith (1946) showed that with the tobacco rosette complex, both TMotV and TVDV are persistent in their relationship with the aphid, although TVDV has a persistent relationship regardless of the presence of TMotV. Similarly, for speckles and carrot motley dwarf, both helpers are persistent regardless if the dependent virus is present or not, and both are found in nature alone (Watson et al., 1964; Duffus, 1960a). Studies on the transmission of the dependent virus from mixed infections have shown that aphids require minutes to hours for efficient acquisition and minutes to hours for efficient inoculation (Elnagar and Murant, 1978a; Falk et al., 1979a), which is typical of the helpers and other persistent viruses. Also, the dependent virus is retained by the vector for a period of days (Falk et al., 1979a; Elnagar and Murant, 1978a; Smith, 1946), again demonstrative of a persistent relationship with the aphid.

In addition to acquiring the transmission characters of the helper virus from the mixed infection, the dependent virus acquires some of the physical properties of the helper virus. Most of the dependent viruses (in single infections) have been shown to possess susceptibility to short periods of aging in vitro (Smith, 1946; Falk et al., 1979a; Watson and Serjeant, 1964; Murant et al., 1969) as determined by mechanical inoculation. LSMV was infective at 24 hrs, but not at 48 hrs when mechanically inoculated from single infected plants. However, from the mixed infection with BWYV, LSMV was mechanically transmissible for as long as 96 hrs (Falk *et al.*, 1979a). BWYV has a longevity *in vitro* of 8 days as determined by membrane feeding (Duffus, 1969), so again LSMV acquires a stability more like BWYV, the helper from the mixed infection. Elnagar and Murant (1978a) have also shown that aphid transmitted CMotV from aphids viruliferous for CMotV and CRLV is ether resistant, while mechanically transmissible CMotV from singly infected plants is susceptible to ether.

The dependent transmission exhibited between strains of BYDV has also been demonstrated to be due to genomic masking (transcapsidation) (Rochow, 1970; 1972). In fact, most of the evidence for the mechanism of persistent virus transmission by aphids suggests specific coat proteins are involved (Rochow, 1970; Clarke and Bath, 1977; Harris, 1977, 1979; Hull, 1977). Genomic masking also has been suggested to be the mechanism responsible for the dependent transmission of the virus complexes (Rochow, 1972; 1977; Murrant, 1974; Elnagar and Murant 1978b; Adams and Hull, 1972).

Studies on the serological properties of LSMV and BWYV from single and doubly infected plants have given evidence suggesting why the dependent viruses exhibit different properties in the mixed infection as compared to single infections. A number of serological procedures have been used with luteoviruses but one of the most useful has been infectivity neutralization that permits the evaluation of antigen-antibody reactions by bioassay with aphids (Gold and Duffus, 1967; Duffus and Rochow, 1978). This method has the advantage that antibodies in the serum to normal plant constituents do not affect its usefulness, even with crude virus preparations. Antiserum to healthy shepherd's purse (*Capsella bursa-pastoris*), and to BWYV showed no difference in effects on mechanical transmission of LSMV from single or double infected plants although both considerably reduced the mechanical transmission. On the other hand, BWYV antiserum neutalized the aphid transmissibility of LSMV from mixed infections while antiserum to healthy shepherd's purse had no effect. This demonstrates that LSMV is serologically related to BWYV when in the mixed infection and probably gains the BWYV coat protein as a result of genomic masking.

There is a major difference between the genomic masking (transcapsidation) for the BYDV isolates and that of the complexes. The genomic masking for the BYDV isolates (MAV and RPV) occurs between related luteoviruses (Rochow, 1965; 1970; Brakke and Rochow, 1974; Scalla and Rochow, 1977). However, the genomic masking among BWYV and LSMV (and probably the other helper-dependent complexes) occurs between two viruses that are drastically different in many properties.

5.4 THE HELPER VIRUS

The property of a given virus to act as a helper for the aphid transmission of the dependent viruses of the complexes has been investigated in a few cases and is not specifically limited to the helper associated with the dependent virus from the field. Murant and Goold (1971) tested the semipersistent helper virus, anthriscus yellows virus (AYV), for the ability to help the aphid transmission of CMotV. AYV was unable to act as a helper for CMotV. Adams and Hull (1972), however, demonstrated that groundnut rosette assistor virus (GRAV) could help TYVV from the tobacco yellow vein complex. By finding a common host of GRAV and TYVV, and one upon which the aphid would feed, *Aphis craccivora* transmitted GRAV and TYVV to groundnuts. TYVV then was transmitted by *A. craccivora*, the vector of groundnut rosette. However, TYVV is normally transmitted by *M. persicae* when in the tobacco yellow vein complex. The change in vector for the dependent virus according to the helper again suggests genomic masking is involved and that coat protein determines vector specificity. Cockbain (1978) showed that although pea enation mosaic has been associated with all isolates of bean yellow vein-banding virus collected from the field, pea (bean) leaf roll can also serve as a helper virus. In our experiments, LSMV also was not specific as to the helper virus that determines its aphid transmission. LSMV was inoculated to plants which were inoculated later with the BWYV isolate from the speckles complex, another isolate of BWYV from radish, or TuYV, a serologically related luteovirus transmitted by *M. persicae* (Falk *et al.*, 1979a). Aphid recoveries showed that all of these luteoviruses were able to effect transmission of LSMV.

The luteoviruses are extremely common and affect most of the world's important food crops. Approximately 30 members or possible members of this group have been reported (Rochow and Duffus, 1979). The damage induced by these viruses results in vast economic losses (Duffus, 1977a). Interestingly, the dependent transmission complexes thus far described induce more damage or have the potential to induce more damage than the luteovirus helper viruses alone.

5.5 PHYSICAL PROPERTIES OF THE VIRUSES

5.5.1 Helper Virus Properties

Luteoviruses are not mechanically transmissible, are phloem limited, and induce yellowing type symptoms on susceptible hosts (Rochow and Duffus, 1979). Two isolates of BYDV, the type member of the luteoviruses, have been characterized (Brakke and Rochow, 1974; Scalla and Rochow, 1977). Both are small isometric viruses with one single species of ss-RNA of approximately 2.0×10^6 daltons. They also have a major species of coat protein of about 24,000 daltons. BWYV has been shown to be a small isometric virus like BYDV (Esau and Hoefert, 1972a; 1972b). We have examined two isolates of BWYV. One has a major species of ss-RNA of approximately 2.0×10^6 daltons like BYDV, but preliminary investigations of the other isolate show two species of ss-RNA of about 2.0 and 1.0×10^6 daltons. Two coat proteins have been distinguished from BWYV, one of 24,000 daltons and another of 61,000 daltons. Another persistent aphid transmitted virus, pea enation mosaic virus (PEMV), which serves as a helper for bean yellow vein-banding virus (Cockbain, 1978), also has two coat proteins. Although not considered a member of the luteovirus group, it has similar particle morphology and a persistent relationship with its aphid vector (Hull, 1977). It induces a mosaic pattern on susceptible hosts and is both mechanically and aphid transmissible.

5.5.2   Dependent Virus Properties

The properties of the dependent viruses are vastly different from the properties of the helper luteoviruses. All are mechanically transmissible and induce parenchymatic symptoms on susceptible hosts. TVDV (Smith, 1946), CMotV (Murant *et al.*, 1969), and LSMV have been shown to have very short longevity *in vitro*. In addition, LSMV and CMotV are both sensitive to low concentrations of RNase, and conditions that favor free RNA unstable viruses such as phenol extraction and addition of zinc or bentonite to the extraction media in some cases have increased the infectivity of these viruses in tissue extracts (Falk *et al.*, 1979b; Murant *et al.*, 1969; Watson *et al.*, 1964). Murant *et al.* (1969; 1973) have suggested CMotV to be a 52 nm membranous spherical particle, unique among plant viruses. We have found similar particles in LSMV infected tissues. However, our data suggest that the spherical particles found in LSMV-infected plants may not be virus particles, but that LSMV may be an unstable-defective virus lacking a functional coat protein (Falk *et al.*, 1979b). Extracts of LSMV infected tissues were 3 to 4 times more infectious when extracted with phenol versus buffer. This and the sensitivity to short aging *in vitro* and low amounts of RNase are classical properties exhibited by unstable plant viruses lacking a functional coat protein (Siegel *et al.*, 1962; Cadman, 1962). Three species of ds-RNA and one species of ss-RNA were associated with LSMV infection, both from whole infected plants and from partially clarified preparations. The ds-RNAs had molecular weights of 1.0, 1.1, and 2.7×10^6 daltons, and the ss-RNA 1.4×10^6 daltons. Infectivity was associated with the ss-RNA fraction (Falk *et al.*, 1979b).

5.5.3   Dependent Virus Properties as a Result of Genomic Masking

As a result of dependent transmission (genomic masking), the dependent viruses gain some interesting properties. The data showing that luteoviruses other than those isolated directly with the field complexes may effect aphid transmission of some of the dependent viruses has already been presented. In several instances, the host range of the dependent virus appears to be different depending whether it is aphid transmitted or mechanically transmitted. Most of the luteoviruses have host ranges limited to a single plant family, with the major exception being BWYV and a few closely related viruses which can infect many species (Duffus, 1973). CRLV is limited to the Umbelliferae, BYDV to the Gramineae, and pea leaf roll virus to the Fabaceae (Watson *et al.*, 1964; Rochow and Duffus, 1979). The dependent viruses studied thus far have host ranges inclusive of a few species in a few families (Murant *et al.*, 1969; Falk *et al.*, 1979a; Adams and Hull, 1972). CMotV infects species in the Chenopodiaceae, Solanaceae, and Fabaceae, and TYVV infects species in the Solanaceae and Fabaceae. When the host range of LSMV was examined by both mechanical and aphid transmission, the hosts

TABLE II. Host Range of Lettuce Speckles Mottle Virus

	Mode of Inoculation	
Test Plant	Mechanical[a]	Aphid[b]
Amaranthaceae		
Amaranthus hybridus L.	$-^c$	+
Boraginaceae		
Amsinckia douglasiana A. DC.	−	+
Chenopodiaceae		
Chenopodium quinoa Willd.	+	−
C. capitatum L.	+	+
C. murale L.	−	+
Beta vulgaris L.	−	+
Spinacea oleracea L.	−	+
Compositae		
Lactuca sativa L.	−	+
Cruciferae		
Brassica napus L.	−	+
Capsella bursa-pastoris (L.) Medic.	−	+
Geraniaceae		
Erodium moschatum	−	+
Solanaceae		
Nicotiana clevelandii Gray.	+	+
N. glutinosa L.	−	+
N. glutinosa X *clevelandii*	+	+
Physalis floridana Rydb.	−	+
Hyoscyamus niger L.	+	+

[a]Mechanical inoculations were performed by rubbing test plants with freshly ground LSMV-infected plant tissue using a Q-tip. A test plants were checked for LSMV at 3 to 6 weeks post-inoculation, whether they developed symptoms or not, by mechanical inoculation back to *N. clevelandii*.

[b]Aphid transmissions were performed by allowing aphids 24-hr acquisition-access feeding periods on detached LSMV-infected leaves in a petri dish. Aphids were then transferred to the test hosts for a 48-hr inoculation period. After 3 to 6 weeks, recovery was performed back to *N. clevelandii* by aphid and mechanical transmission from test hosts.

[c]Test plants were nonsusceptible (−) or susceptible (+) to inoculation with LSMV.

that LSMV could infect were much greater when it was aphid transmitted (Table II). In addition, when LSMV was aphid transmitted it even could infect some hosts that are not hosts for BWYV, the helper virus, such as *Chenopodium murale* and *Nicotiana glutinosa*. When this occurs, LSMV is no longer aphid transmissible, and can only be detected from these infections by mechanical inoculation. This also has been found with TYVV and CMotV (Adams and Hull, 1972; Watson *et al.*, 1964). Such a condition may or may not be advantageous to the dependent virus. It gains in host range, but then loses the immediate possibility of further being aphid transmitted as a member of that complex.

5.6   REAL AND POTENTIAL EPIDEMIOLOGICAL CONSIDERATIONS

Helper dependent transmission of viruses by aphids can be very significant in the development of plant virus diseases. All of the virus disease complexes mentioned result from specific mixed infections of two viruses, the helper and the dependent virus. The helper virus alone may or may not cause serious disease, but generally when both helper and dependent virus infect the same plant, the disease is more severe. For carrot motley dwarf, Watson and Serjeant (1964) have shown that infection by CRLV does not stunt carrots nearly as much as infection with both viruses. The same is true for the speckles disease in lettuce. BWYV does decrease weight of crisphead lettuce in greenhouse experiments (Ryder and Duffus, 1966), but infection by BWYV and LSMV severely stunts plants especially when infected young. Aphid transmission and the resulting disease caused by the dependent virus is probably a result of simultaneous replication and genomic masking.

5.6.1   Importance of a Varied Approach to Virus Epidemiology

Many studies regarding the incidence of virus diseases in crops utilize only mechanical transmission or serology as testing procedures. While these methods are rapid, and in some instances, very accurate for individual viruses, they may give misleading information as to the importance of these entities in the field aspects of disease. The lack of mechanical transmission from plants showing virus-like symptoms does not mean that a virus is not infecting the plant. Luteoviruses are a good example of non-mechanically transmissible viruses which are widespread in nature. In many cases, the infection of a plant by one of these viruses is overlooked due to the investigator not using vectors, or at least the correct vector to recover virus from these plants. The properties of the virus complexes now indicate even more reasons why a varied approach to the detection of viruses from field infected plants should be used.

The host range of the dependent virus is sometimes greater when they are aphid transmitted. These viruses may only be able to infect a given host when they are aphid transmitted and mechanical transmission attempts would therefore be negative. Indicator species possibly could be mechanically infected in the greenhouse from field carrots or lettuce infected with the motley dwarf or speckles complexes, respectively, but the dependent virus could not be inoculated back into these hosts, and Koch's postulates could not be fulfilled.

Symptoms induced by the dependent viruses are parenchymatic symptoms generally associated with mechanically transmissible viruses. Conversely, the symptoms of many luteoviruses are not parenchymatic, but overall yellowing or reddening occurs and these are often mistaken for nutrient deficiencies (Duffus, 1960b; 1977; Costa, 1956). Mechanical transmission from these plants could detect the dependent virus, but leave the luteovirus behind. Therefore, one could isolate a virus from infected plants, but a true understanding of the disease would not be achieved. With the tobacco yellow vein disease, the symptoms are identical if plants are infected with both viruses, or only with TYVV. In the latter case, however, TYVV is non-aphid transmissible.

Some plant species are resistant to infection by the helper virus, but susceptible to the dependent virus. In such a case, aphid transmission would fail to detect the dependent virus, and mechanical transmission would be necessary.

Serological properties of BWYV and LSMV so far are similar from the mixed infection. Tests have failed to show that the two aphid transmissible viruses from the mixed infection of BWYV and LSMV are serologically distinct as infectivity of both may be neutralized from the mixed infection by the antiserum of the helper (BWYV). Thus, serological tests could mislead the investigator and suggest that only one virus is infecting the plant.

5.6.2 Host Range Differences

Despite the fact that many animal viruses exhibit host range influences as a result of coat protein, evidence for the role of the coat protein in plant virus infection and host range is meager. In the case of alfalfa mosaic and related viruses, coat protein or coat protein messenger RNA (RNA4), is necessary along with RNAs 1, 2, and 3 for infectivity, but coat protein appears to not exhibit any host range characteristics (Jaspars, 1974). Dodds and Hamilton (1972; 1974) have shown for TMV and barley stripe mosaic virus (BSMV), in mixed infections, TMV is genomically masked by BSMV coat protein. TMV does not efficiently replicate in barley alone, but does so in the mixed infection with BSMV. However, the fact that both viruses must be simultaneously infecting barley does not suggest a role of the coat protein of the BSMV genomically masked TMV to replicate in barley, but probably the mixed infection allows TMV to replicate in barley. The late mosaic disease of cotton in Brazil is somewhat similar. Brazilian tobacco streak virus (BTSV) can infect cotton by mechanical inoculation, and yields local lesions (Costa, 1969). All field cotton plants showing late mosaic symptoms, however, contain cotton anthocyanosis virus (CAV) in addition to systemic BTSV. This somewhat also resembles the aphid transmitted complexes except that the vector of BTSV is unknown.

In the case of the dependent viruses like LSMV from the helper dependent complexes, the host range is probably expanded as a result of the aphid transmissibility gained from the original mixed infection. In the cases of transmission of the dependent virus to non-hosts of the helper virus, simultaneous replication cannot then be responsible for the expanded host range. Data is accumulating for the involvement of a specific cost protein of the virus particle in determining vector specificity. If this is applied to at least the speckles complex, then aphid transmissible LSMV (genomically masked LSMV) has a greater host range, indirectly as a result of its coat protein.

There is no evidence for mechanical transmission of BWYV or any of the other luteoviruses from single infections. There is evidence, however, of mechanical transmission (10-15%) of BWYV from speckles complex infected plants (Falk *et al.*, 1979a). Although the reasons for this were not investigated, multiplication or enhanced movement of BWYV through parenchyma tissue with the presence of the dependent virus was suspected.

5.6.3 Distribution

Persistently transmitted viruses have a great advantage epidemiologically over nonpersistent or semipersistent viruses (Duffus, 1973). A high percentage of aphids leaving infected breeding hosts, for instance, may carry the virus. Swenson (1968) has indicated that relatively few aphids that develop on diseased plants (with nonpersistent type viruses) are infective when they fly from the plant. Persistent viruses are retained by the vectors for periods up to the life of the vector. The virus is not lost with the molt at ecdysis, thus all stages of the insect may acquire and transmit the virus. With these transmission characteristics, virus spread is much more general and widespread than the marginal distribution of nonpersistent or semipersistent transmitted viruses (Duffus, 1973). Through the ability to use the coat protein of the helper virus, BWYV, LSMV gains significantly epidemiologically by exhibiting a persistent relationship with the aphid.

Aphids that are viruliferous for BWYV and LSMV can initiate three types of infections: LSMV + BWYV, BWYV alone, or LSMV alone. In the latter case, infection yields non-aphid transmissible LSMV. This situation occurs in the greenhouse when aphids viruliferous for the speckles complex are allowed to inoculate *N. glutinosa* or *C. capitatum*. LSMV can infect these hosts, but BWYV cannot. In nature, LSMV or the other dependent viruses could still be a serious disease threat in such a case. Parsnip yellow fleck virus (PYFV), which is helper dependent upon anthriscus yellows virus (AYV) for semipersistent aphid transmission, is the most common cause of virus-like symptoms in parsnip in Britain (Murant and Goold, 1968), yet parsnip is resistant to the helper virus, AYV. Therefore, only primary spread of the virus PYFV, can occur from an outside source of primary inoculum and not from parsnip to parsnip. It is possible that the dependent viruses, such as LSMV could behave the same way for nonhosts of BWYV, and then LSMV would not be detectable by serological methods or by aphid recoveries from such infected plants.

There are a number of diseases with similar epidemiology patterns, transmitted by various vector groups that could perhaps also be explained in this manner. Two whitefly-transmitted diseases of soybeans in Brazil, crinkle mosaic and dwarf mosaic, are caused by viruses transmitted only by vectors that acquire the virus from natural weed reservoirs. Virtually no virus spread within soybean plantings occurs in spite of high insect density in the crop (Costa, 1975). The celery yellow spot virus, transmitted persistently by the honeysuckle aphid [*Rhopalosiphium conii* (Dvd.)] naturally infected up to 40% of the celery plants in some fields, yet could not be transmitted from celery (Freitag and Severin, 1945).

5.6.4 More Than One Helper Virus

The dependent viruses from the complexes have been demonstrated to be able to use more than one helper virus for their aphid transmission. We found that LSMV was able to use an isolate of BWYV from radish and turnip yellows virus, a serologically related luteovirus also transmitted by *M. persicae*. Artifical mixes were created by mechanically inoculating *N. clevelandii* with LSMV and one week later inoculating with BWYV or TUYV. Recoveries then were made using *M. persicae* to *N. clevelandii*. Adams and Hull (1972) found that groundnut rosette assistor virus (GRAV) could act as a helper for tobacco yellow vein virus. In this case, TYVV, which is normally transmitted by *M. persicae* when in mixed infections with the helper virus tobacco yellow vein assistor virus, was transmitted by *A. craccivora* when in the mixed infection with GRAV. So far, the limitations for studying the role of other luteoviruses as helpers has been finding common hosts for the dependent virus and the helper virus.

5.7 SUMMARY AND CONCLUSIONS

The increased host range and ability to use various helper viruses suggests that LSMV and other dependent viruses could have very complicated epidemiological characters. Such theoretical possibilities were discussed by Rochow (1977) in a recent review concerning dependent transmission. In many ways, the different properties exhibited as a result of genomic masking are not unlike the behavior exhibited by some animal viruses. Phenotypic mixing and genomic masking are very common among many animal viruses. Often, host ranges of animal viruses are very much influenced by the coat protein of the virion (Vogt, 1967; Levy, 1977; Dodds and Hamilton, 1976). Levy has stated that "the inability to detect a virus by serologic or host range studies is not conclusive evidence that the virus is not present." The same can be said for the dependent viruses like LSMV from the helper virus complexes.

It has been demonstrated that LSMV may infect hosts that are not hosts of the helper, BWYV, and the same is true from some of the other dependent viruses. In this case, aphid transmissibility would cease. However, if another helper virus (luteovirus) were to infect this plant, it might then be able to act as a helper virus and the dependent virus could again be aphid transmissible, possibly even by a different aphid species.

Dodds and Hamilton (1976) stated that mixed infections and resulting structural interactions may be considered survival mechanisms for some defective viruses. The dependent viruses from the aphid transmitted complexes almost surely depend upon the mixed infections for their survival and dispersal by the aphid vector. The properties exhibited by the helper dependent viruses are different from many other well characterized plant viruses. The properties suggest that these viruses may represent a group of viruses that survive in nature by genomic masking and aphid transmission. The reported host ranges of the dependent viruses are similar, yet each has a few diagnostic species suggesting that there are differences among them. More work on the relationships of these viruses to each other and their ability to use other viruses as helpers is necessary.

Speculation on the origin and evolution of viruses have considered a number of possibilities, as descendants of primitive forms, degenerate microorganisms, and development from normal cell constituents (Matthews, 1970). In probability, all of these mechanisms and probably other mechanisms are operational in all directions. Thus, LSMV may lose components and become defective, but gain other components from helper viruses and thereby be more able to survive epidemiologically.

The comparable transmission properties and particle morphology of pea enation mosaic virus (PEMV) (Hull, 1977; Adam et al., 1979) to the luteoviruses, but symptomatology and mechanical transmission like mosaic viruses, suggests a similarity to the mixed infection diseases such as lettuce speckles. This does not suggest that PEMV is a complex disease like lettuce speckles, but that with so many similarities, they may have a similar origin. Both are transmitted in a persistent manner by aphids, both can be mechanically transmitted, and induce a mottle in susceptible hosts. Selection of isolates following a series of mechanical inoculations will eliminate aphid transmissions of both diseases.

The main difference structurally between the helper dependent complex (speckles) and PEMV is the ability to separate from the speckles complex a component which is aphid transmitted, but not mechanically transmitted. Could it be that such an entity exists with PEMV or has this association progressed beyond this stage?

One theory for the origin of multicomponent viruses is that they arose from two or more coinfecting viruses (Reijnders. 1978). Reijnders (1978) also presented a hypothesis that vector transmission by insects, fungi, etc., may help explain why plants have multicomponent viruses, whereas none have yet been found to infect animals. Small RNA viruses of animals show a lack of vector transmission, but are spread through air, water, food, etc. (Reijnders, 1978). A specific vector (insect) may be better able to hold together the necessary components of a multicomponent virus for infection than a more random spread by air or water. The association and coinfection of the helper and dependent viruses also would fit this requirement.

Mixed infections of plant viruses are common in nature. This has been stressed by Rochow and others. New properties that are a result of the mixed infections are just being discovered. The possible importance of these in plant disease and plant virus epidemiology needs to be investigated. So far, the best evidence for the involvement of genomic masking in dependent transmission of the virus complexes is for the speckles complex. It is possible that there could be other mechanisms for persistent virus dependent transmission as has been stated by Rochow (1977), and this needs to be considered. In any case, properties such as expanded host range and the use of more than one helper virus for the dependent viruses creates new possibilities to be considered in epidemiological investigations of aphid transmitted viruses. The properties exhibited by the virus complexes also reinforces the importance for using vectors in the study of plant viruses, for serological and mechanical transmission techniques alone would not have uncovered many of the properties of the complexes.

5.8 REFERENCES

Adam, G., Sander, E., and Shepherd, R. J. (1979). Structural differences between pea enation mosaic virus strains affecting transmissibility by *Acyrthosiphon pisum* (Harris). *Virology* 92: 1-14.

Adams, A. N., and Hull, R. (1972). Tobacco yellow vein, a virus dependent on assistor viruses for its transmission by aphids. *Ann. Appl. Biol.* 71: 135-140.

Brakke, M. K., and Rochow, W. F. (1974). Ribonucleic acid of barley yellow dwarf virus. *Virology* 61: 240-248.

Cadman, C. H. (1962). Evidence for association of tobacco rattle virus nucleic acid with a cell component. *Nature* 193: 49-52.

Clarke, R. G., and Bath, J. E. (1977). Serological properties of aphid-transmissible and non aphid-nontransmissible pea enation mosaic virus isolates. *Phytopathology* 67: 1035-1040.

Clinch, P. E. M., Loughnane, J. B., and Murphy, P. A. (1936). A study of the aucuba or yellow mosaics of the potato. *Roy. Soc. Sci.* (Dublin) *Proc.* 21: 431: 448.

Cockbain, A. J. (1978). Bean yellow vein-banding virus. *Rep. Rothamsted Exp. Stn.*, 1977: 221-222.

Costa, A. S. (1956). Anthocyanosis, a virus disease of cotton in Brazil. *Phytopathol. Z.* 28: 167-186.

Costa, A. S. (1969). Conditioning of the plant by one virus necessary for systemic invasion of another. *Phytopathol. Z.* 65: 219-230.

Costa, A. S. (1975). Increase in the population density of *Bemisia tabaci*, a threat of widespread infection of legume crops in Brazil. *In* "Tropical Diseases of Legumes" (J. Bird and K. Maramorosch, eds.), pp. 27-49. Academic Press, New York.

Dodds, J. A., and Hamilton, R. I. (1972). The influence of barley stripe mosaic virus on the replication of tobacco mosaic virus in *Hordeum vulgare* L. *Virology* 50: 404-411.

Dodds, J. A., and Hamilton, R. I. (1974). Masking of the RNA genome of tobacco mosaic virus by the protein of barley stripe mosaic virus in doubly infected barley. *Virology* 59: 418-427.

Dodds, J. A., and Hamilton, R. I. (1976). Structural interactions between viruses as a consequence of mixed infections. *Adv. Virus Res.* 20: 33-86.

Duffus, J. E. (1960a). Radish yellows, a disease of radish, sugar beet, and other crops. *Phytopathology* 50: 389-394.

Duffus, J. E. (1960b). Two viruses that induce symptoms typical of "June yellows" in lettuce. *Plant Dis. Reptr.* 44: 406-408.

Duffus, J. E. (1969). Membrane feeding used in determining the properties of beet western yellows virus. *Phytopathology* 59: 1668-1669.

Duffus, J. E. (1973). The yellowing virus diseases of beet. *Adv. Virus Res.* 18: 347-386.

Duffus, J. E. (1977). Aphids, viruses, and the yellow plague. *In* "Aphids as Virus Vectors" (K. E. Harris and K. Maramorosch, eds.), pp. 361-383. Academic Press, New York.

Duffus, J. E. (1979). Unpublished data.

Duffus, J. E., and Rochow, W. F. (1978). Neutralization of beet western yellows virus by antisera against barley yellow dwarf virus. *Phytopathology* 68: 45-49.

Elnagar, S., and Murant, A. F. (1976a). Relations of the semi-persistent viruses, parsnip yellow fleck and anthriscus yellows, with their vector, *Cavariella aegopodii*. *Ann. Appl. Biol.* 84: 153-167.

Elnagar, S., and Murant, A. F. (1976b). The role of the helper virus, anthriscus yellows, in the transmission of parsnip yellow fleck virus by the aphid, *Cavariella aegopodii*. *Ann. Appl. Biol.* 84: 169-181.

Elnagar, S., and Murant, A. F. (1978a). Relations of carrot red leaf and carrot mottle viruses with their aphid vector, *Cavariella aegopodii*. *Ann. Appl. Biol.* 89: 237-244.

Elnagar, S., and Murant, A. F. (1978b). Aphid-injection experiments with carrot mottle virus, and its helper virus, carrot red leaf. *Ann. Appl. Biol.* 89: 245-250.

Esau, K., and Hoefert, L. L. (1972a). Ultrastructure of sugarbeet leaves infected with beet western yellows virus. *J. Ultrastructure Res.* 40: 556-571.

Esau, K., and Hoefert, L. L. (1972b). Development of infection with beet western yellows virus in sugarbeet. *Virology* 48: 724-738.

Falk, B. W., Duffus, J. E., and Morris, T. J. (1979a). Transmission, host range, and serological properties of the viruses causing lettuce speckles disease. *Phytopathology* 69: 612-617.

Falk, B. W., Morris, T. J., and Duffus, J. E. (1979b). Unstable infectivity and sedimentable ds-RNA associated with lettuce speckles mottle virus. *Virology* 96: 239-248.

Freitag, J. H. (1969). Interactions of plant viruses and virus strains in their insect vectors. *In* "Viruses, Vectors, and Vegetation" (K. Maramorosch, eds.), pp. 303-326. Wiley-Interscience, New York.

Freitag, J. H., and Severin, H. H. P. (1945). Transmission of celery-yellowspot virus by the honeysuckle aphid, *Rhopalosiphum conii* (Dvd.). *Hilgardia* 16: 375-384.

Gill, C. C., and Comeau, A. (1977). Synergism in cereals between corn leaf aphid-specific and aphid-nonspecific isolates of barely yellow dwarf virus. *Phytopathology* 67: 1388-1392.

Gold, A. H., and Duffus, J. E. (1967). Infectivity neutralization–a serological method as applied to persistent viruses of beets. *Virology* 31: 308-313.

Harris, K. F. (1977). An ingestion-egestion hypothesis of noncirculative virus transmission. *In* "Aphids as Virus Vectors" (K. F. Harris and K. Maramorosch, eds.), pp. 165-220. Academic Press, New York.

Harris, K. F. (1979). Leafhoppers and aphids as biological vectors: Vector-virus relationships. *In* "Leafhopper Vectors and Plant Disease Agents (K. Maramorosch and K. F. Harris, eds.), pp. 217-308. Academic Press, New York.

Harris, K. F., Bath, J. E., Thottappilly, G., and Hooper, G. R. (1975). Fate of pea enation mosaic virus in PEMV-injected pea aphids. *Virology* 65: 148-162.

Hull, R. (1977). Particle differences related to aphid-transmissibility of a plant virus. *J. Gen. Virol.* 34: 183-187.

Hull, R., and Adams, A. N. (1968). Groundnut rosette and its assistor virus. *Ann. Appl. Biol.* 62: 139-145.

Jaspars, E. M. J. (1974). Plant viruses with a multipartite genome. *Adv. Virus Res.* 19: 37-149.

Levy, J. A. (1977). Murine xenotropic type C viruses. II. Phenotypic mixing with mouse and rat ecotropic type C viruses. *Virology* 77: 797-810.

Matthews, R. E. F. (1970). Plant Virology. Academic Press, New York.

Murant, A. F. (1974). Carrot mottle virus. *CMI/AAB Descriptions of Plant Viruses* No. 137.

Murant, A. F., and Goold, R. A. (1968). Purification, properties and transmission of parsnip yellow fleck, a semi-persistent, aphid-borne virus. *Ann. Appl. Biol.* 62: 123-137.

Murant, A. F., and Goold, R. A. (1971). Specificity of dependence on assistor viruses for aphid transmission. *Rep. Scot. Hort. Res. Inst.*, 1970: 53.

Murant, A. F., Goold, R. A., Roberts, I. M., and Cathro, J. (1969). Carrot mottle–a persistent, aphid-borne virus with unusual properties and particles. *J. Gen. Virol.* 4: 329-341.

Murant, A. F., Roberts, I. M., and Goold, R. A. (1973). Cytopathological changes and extractable infectivity in *Nicotiana clevelandii* leaves infected with carrot mottle virus. *J. Gen. Virol.* 31: 269-283.

Pirone, T. P. (1977). Accessory factors in nonpersistent virus transmission. *In* "Aphids as Virus Vectors," (K. F. Harris and K. Maramorosch, eds.), pp. 221-235. Academic Press, New York.

Reijnders, L. (1978). The origin of multicomponent small ribonucleoprotein viruses. *Adv. Virus Res.* 23: 79-102.

Rochow, W. F. (1965). Apparent loss of vector specificity following double infection by two strains of barley yellow dwarf virus. *Phytopathology* 55: 62-68.

Rochow, W. F. (1970). Barley yellow dwarf virus: phenotypic mixing and vector specificity. *Science* 167-: 875-878.

Rochow, W. F. (1972). The role of mixed infections in the transmission of plant viruses by aphids. *Annu. Rev. Phytopath.* 10: 101-124.

Rochow, W. F. (1973). Selective virus transmission by *Rhopalosiphon padi* exposed sequentially to two barley yellow dwarf viruses. *Phytopathology* 63: 1317-1322.

Rochow, W. F. (1977). Dependent virus transmission from mixed infections. *In* "Aphids as Virus Vectors" (K. F. Harris and K. Maramorosch, eds.), pp. 253-273. Academic Press, New York.

Rochow, W. F., and Duffus, J. E. (1979). Luteoviruses and yellows diseases. *In* "Comparative Diagnosis of Viral Diseases." Academic Press, New York (in press).

Rochow, W. F., and Gill, C. C. (1978). Dependent virus transmission by *Rhopalosiphun padi* from mixed infections of various isolates of barley yellow dwarf virus. *Phytopathology* 68: 451-456.

Ryder, E. J., and Duffus, J. E. (1966). Effects of beet western yellows and lettuce mosaic viruses on lettuce seed production, flowering time, and other characters in the greenhouse. *Phytopathology* 56: 842-844.

Scalla, R., and Rochow, W. F. (1977). Protein component of two isolates of barley yellow dwarf virus. *Virology* 78: 576-580.

Schneider, I. R. (1977). Defective plant viruses. Chapter 13 *in* Virology in Agriculture I. Beltsville Symposia in Agricultural Research. Landmark Studies, New York.

Siegel, A., Zaitlin, M., and Sehgal, O. P. (1962). The isolation of defective tobacco mosaic virus strains. *Proc. Natl. Acad. Sci.* 48: 1845-1851.

Smith, K. M. (1946). Tobacco rosette: a complex virus disease. *Parasitology* 37: 21-24.

Stubbs, L. L. (1948). A new virus disease of carrots: its transmission, host range and control. *Aust. J. Sci. Res.* 1: 303-332.

Swenson, K. C. (1968). Role of aphids in the ecology of plant viruses. *Ann. Rev. Phytopathol.* 6: 351-374.

Vogt, P. K. (1967). Phenotypic mixing in the avian tumor virus group. *Virology* 32: 708-717.

Watson, M., and Serjeant, E. P. (1964). The effect of motley dwarf virus on yield of carrots and its transmission in the field by *Cavariella aegopodii* Scop. *Ann. Appl. Biol.* 53: 77-93.

Watson, M., Serjeant, E. P., and Lennon, E. A. (1964). Carrot motley dwarf and parsnip mottle viruses. *Ann. Appl. Biol.* 54: 153-166.

Zink, F. W., and Duffus, J. E. (1972). Association of beet western yellows and lettuce mosaic viruses with internal rib necrosis of lettuce. *Phytopathology* 62: 1141-1144.

Chapter 6

ECOLOGY AND CONTROL OF SOYBEAN MOSAIC VIRUS

Michael E. Irwin

Office of Agricultural Entomology,
Department of Plant Pathology, and International Soybean Program (INTSOY),
University of Illinois and Illinois Natural History Survey,
Urbana, Illinois

Robert M. Goodman

Department of Plant Pathology and International Soybean Program (INTSOY),
University of Illinois,
Urbana, Illinois

6.1 INTRODUCTION

In less than fifty years, soybeans (*Glycine max* (L.) Merr.) have become a major agricultural crop in the Western Hemisphere. In spite of their role as a staple in the Orient for centuries, soybeans were, until after the turn of the 20th century, a botanical curiosity in the eyes of most Western agriculturalists. Soybean germplasm was introduced from Manchuria to Europe and North America in the 18th and 19th centuries. Along with the early accessions (the United States Department of Agriculture soybean germplasm collection was begun in 1898) came the seedborne pathogens of the species. One of these no doubt was soybean mosaic virus (SMV) (Fig. 1). The disease caused by SMV was noted in the Western scientific literature as early as 1916 by Clinton, soon after the beginning of serious scientific work in America to develop improved cultivars. While the introduction of SMV into the U.S. is not documented, most of what is known about the virus and its epidemiology make it reasonable to conclude that SMV was

FIG. 1. Symptoms of soybean mosaic virus infection on field-grown soybean plant. (Photograph courtesy of M. Kogan)

brought with the crop from the soybean center of origin in north central China.

Soybean mosaic virus belongs to the large and important potato virus Y (poty-virus) group (Gibbs and Harrison, 1978). The virus particles are flexuous and rod-shaped (Bos, 1972). The single-stranded RNA genome of the virus is contained within a capsid having helical symmetry and made up of a single type of coat protein subunit. Molecular weights of the RNA and coat protein are 3.25×10^6 (Hill and Benner, 1979a) and 28,300 (Hill and Benner, 1979b), respectively.

The physical and chemical properties of SMV are typical of potyviruses. The virus remains infective in plant sap for 2 to 4 days, has a dilution end point titer of about 10^{-3}, and a thermal inactivation point of less than $60°C$ (Bos, 1972). The virus is a moderately good antigen; rabbit antiserum titers of 1:2048 to 1:4096 have been reported (Ross, 1967).

SMV has a relatively narrow host range that, with the exceptions of two species in the genus *Chenopodium*, is confined to the family Leguminosae (Galvez, 1963). In this respect it is one of the less promiscuous members of the potyvirus group; the group also contains viruses with quite wide and extensive host ranges. SMV is transmitted in the field and in the laboratory by aphids (Homoptera: Aphididae) in a nonpersistent manner and is also seed transmitted in soybeans.

It is seed transmissibility that accounts for the worldwide distribution of SMV. Soybean germplasm collections, such as those maintained by USDA laboratories at Urbana, Illinois and Stoneville, Mississippi, are repositories of virus as well as soybean germplasm (Cho and Goodman, 1979). International exchange of germplasm resources, necessary for continued improvement of crop species, carries a risk of virus spread. Economic consequences of this situation have not been serious in the major soybean growing areas of the world as of the late 1970's. SMV is not considered a serious problem in soybean production in the United States or in Brazil. the two countries where world soybean production is now concentrated.

That the virus can do damage, however, is beyond question. Published and unpublished reports of serious epidemics and yield loss caused by SMV in such diverse areas as Korea (Cho *et al.*, 1977), Indonesia (Whigham, 1980), Ecuador (Irwin, 1980b), the Soviet Union (Muraveva, 1973; Ovchinnikova, 1973), and Japan (Koshimizu and Iizuka, 1963) document the potential importance of the disease. The disease is widespread in the southern soybean growing areas of the United States also, and in mixed infection with bean pod mottle virus can cause significant economic loss (Lee and Ross, 1972; Gray, 1980). Recent reports indicate that SMV is widespread and of concern in mainland China as well (Wang, 1980).

In studies on the pathogenic variation of seedborne SMV isolates recovered from the USDA germplasm collections, Cho and Goodman (1979) found that none of the soybean cultivars considered resistant to SMV were resistant to all isolates of the virus. Severity of symptoms induced in susceptible cultivars and the virulence they displayed against putatively SMV-resistant cultivars varied considerably among SMV isolates. Each source of SMV resistance currently in use provides useful protection to only a few of the SMV strains found in the germplasm collections. These results, together with the universal susceptibility of soybean cultivars commonly grown on large acreages in the midwestern U.S. and Brasilian soybean growing regions, suggest the obvious possibility that SMV could become a major problem in soybean production, and raise the question why has it not already become so.

The purpose of this chapter is to review the status of our knowledge on the ecology and control of SMV. We draw extensively on our own recent studies which are part of an internationally-oriented research program to generate information that can be used to prevent spread of and losses due to soybean viruses before they become a problem in areas where soybeans are being newly introduced and established as a crop. Much of what we have learned, together with important work by other investigators extending back to the early years of the 20th century, can be applied to help us understand why SMV, which is always present wherever soybeans are grown, is not more of a problem than it has heretofore been in areas where soybeans have been grown for many years. We may even be able to predict the impact of changing agricultural practices on SMV spread and its economic consequences in future years. Following a review of present knowledge about seed transmission and aphid transmission of the virus, we discuss the integrated use of this knowledge for management of the pathogen and its spread by vectors.

6.2 SEED TRANSMISSION

Transmission of viruses through seeds of higher plants plays an important role in the spread and survival of a number of agriculturally important plant viruses. Often these are viruses of annual herbaceous hosts that have no perennial hosts or overseasoning infective vectors or they are viruses that are otherwise transmitted in nature by sedentary animal vectors, such as nematodes. In the latter case, it is through the seeds of perennial wild hosts that the viruses are most effectively dispersed. Seed transmission thus plays different roles in the ecology of different viruses.

In the case of SMV, the virus has a limited host range and is transmitted by aphids in a nonpersistent manner. Evidence presented later in this chapter shows that seed transmission is the primary source of inoculum for field spread of SMV and therefore is a major means of survival from season to season. Because soybeans are not vegetatively propagated and the vectors do not remain infective for long periods, seed transmission is also the route by which the virus has been and continues to be spread from one geographical region to another. Seed transmission plays a pivotal role in the epidemiology of SMV. The virus is not reported to be seed transmitted in any of its other host plants and to our knowledge does not often occur naturally in species other than *G. max* and its close relatives.

Soybean seeds produced by SMV-infected plants exhibit a characteristic symptom termed mottling (Kennedy and Cooper, 1967). Mottling is a discoloration of the seed coat of yellow-seeded soybeans caused by the bleeding of hilum pigment into other regions of the seed coat (Wilcox and Laviolette, 1968). While a symptom of SMV infection of the plant producing the seeds, mottling has long been recognized to be influenced also by environmental and genetic factors (Owen, 1927; Dimmock, 1936; Koshimizu and Iizuka, 1963; Cooper, 1966; Ross, 1970). Seed coat mottling is not a symptom of seed transmission (Koshimizu and Iizuka, 1963; Ross, 1970; Goodman *et al.*, 1979), although Ross (1968) reported that, among seeds from infected plants, the seed transmission incidence was twice as high for mottled seeds as for nonmottled seeds. We (Goodman *et al.*, 1979) and others (Porto and Hagedorn, 1975; Kennedy and Cooper, 1967) have found soybean lines that show seed transmission but no mottling, others that show mottling but no or very low incidence of seed transmission, and still others in which the incidence and severity of mottling varied significantly from one year to the next. Seed coat mottling, when it occurs, is an indication that the plant producing the seeds was probably infected, but nonmottled seeds from such plants may transmit and mottled seeds may not.

6.2.1 Factors Influencing the Incidence of Seed Transmission

The incidence of seed transmission of SMV in soybeans is affected by two principle factors, viz., the time of infection in relation to plant development and the plant genotype. The effect of time of inoculation on seed transmission incidence was discounted by the early results of Kendrick and Gardner (1924). Later reports (Koshimizu and Iizuka, 1963; Iizuka, 1973) indicated that inoculation after initiation of flowering resulted in no seed transmission. Bowers and Goodman (1979a) reported a reduction but not elimination of seed transmission from inoculations during and after flowering (Table I).

There are several difficulties in experiments on the time of inoculation in relation to seed transmission that may account for the conflicting results reported by various workers. Highly determinant plants which flower over a relatively short period might give no seed transmission from inoculations after flowering begins while indeterminant plants, on which pods are filling at the bottom of the plant while late flowers appear at the top, may produce transmitting seeds if inoculation occurred early enough to result in systemic infection before fertilization of the late flowers. In recent experiments designed to test one aspect of this hypothesis, we found evidence that a small percentage of immature seeds on a highly determinant variety became SMV-infected when inoculation of the mother

TABLE I.   The Effects of Time of Inoculation with Soybean
Mosaic Virus upon Incidence of Seed Coat Mottling and
Seed Transmission in Williams Soybeans[a]

Time of inoculation (wks after planting)	Developmental stage[b]	Seed coat mottling[c] (%)	Seed Transmission[d] (%)
3		69.0	18.0
4		65.8	19.0
5	R1	80.5	11.0
6	R2	72.3	2.0
7	R4	80.0	4.0
8		69.8	4.0
9	R6	77.0	3.0
10		78.5	4.0

[a] Data adapted from Bowers and Goodman (1979) by permission of the American Phytopathological Society.

[b] According to the descriptions in Fehr et al. (1971).

[c] Sample of 400 seeds.

[d] Sample of 100 seeds. The correlation coefficient (r) between week of inoculation and incidence of seed transmission was -0.826.

plant was delayed until after flowering had ceased (Bowers and Goodman, 1980). Another uncertainty is the role, if any, of pollen transmission. Soybeans that produce fertile male and female gametes are largely self-pollinated, but the extent of fertilization of male-sterile plants under field conditions (Singh, 1980) indicates that outcrossing is at least a theoretical possibility for field spread of SMV if, as reported by Iizuka (1973), the virus is pollen transmissible.

There are no published reports on the effect of SMV infection on timing of pollen production or vigor of pollen produced. Soybean plants infected with tobacco ringspot virus produced fewer, less germinable pollen grains, and shorter stigmas and pollen germ tubes (Yang and Hamilton, 1974).

The investigators who first reported seed transmissibility of SMV also noted that the incidence of transmission varied from cultivar to cultivar (Kendrick and Gardner, 1924); transmission frequency ranged from nil to over 60%. Casual observations of the same sort have been reported since (Porto and Hagedorn, 1975; Kennedy and Cooper, 1967; Koshimizu and Iizuka, 1963). In an extensive recent study of this phenomenon, we (Goodman et al., 1979; Goodman and Oard, 1980) screened nearly 900 soybean germplasm accessions from temperate (maturity groups 0, II, and III) and tropical (maturity groups VIII, IX, and X) maturity groups. In the first round of tests, 200 seeds per accession from SMV infected plants were planted and resulting seedlings indexed for SMV. Lines producing no infected seedlings in this test were planted the second year and inoculated with SMV; 1000 of the seeds of each line harvested were tested. From this second round of tests, 19 temperate lines and 4 tropical lines were found to be free of seed transmission (Goodman et al., 1979). In a third round of tests with tropical lines, the apparent nontransmitters from the second round were found to transmit SMV through seeds at an incidence of from 0.05 to 0.2% (Goodman and Oard, 1980). Since the incidence of seed transmission of SMV in improved cultivars is typically 5 to 20% when plants are inoculated before the onset of flowering, use of germplasm with lower rates of seed transmission offers the possibility of reducing primary inoculum levels by a factor of 10 or more.

It is noteworthy that among the tropical accessions with low incidence of seed transmission are two named varieties with superior agronomic qualities, Improved Pelican and UFV-1 (Goodman and Oard, 1980). Among the temperate accessions with low incidence of transmission are two early soybean introductions, Mukden and Manchu 2204, which appear in the pedigrees of many named cultivars released in the United States; this may account for the generally lower incidence of transmission of such cultivars when compared with the incidence of transmission in the germplasm as a whole (Goodman et al., 1979). The possible use of reduced seed transmission in control of SMV will be addressed in a later section of this chapter.

Work is currently under way in our group to determine the possible influence of virus strain on SMV transmission through seeds. Different strains of certain viruses are seed transmitted in their hosts at different rates (Anderson, 1957; McKinney and Greely, 1965; Frosheiser, 1974; Ghanekar and Schwenck, 1974), but our preliminary results suggest that what differences exist may be slight in the case of SMV in soybeans (Bowers and Goodman, 1979b). The related question of whether soybean accessions with low incidence of seed transmission of one strain of SMV behave similarily when infected with other strains of the virus is also being studied by us and our colleagues.

6.2.2 Mechanism of Seed Transmission

Embryo infection appears to be a prerequisite for the transmission of most seed transmitted plant viruses (Bennett, 1969; Shepherd, 1972), although there are exceptions to this generalization (Taylor et al., 1961; McDonald and Hamilton, 1972). Immature embryos from SMV infected soybean plants contain infective SMV (Koshimizu and Iizuka, 1963; Iizuka, 1973, Porto and Hagedorn, 1975; Bowers and Goodman, 1979a). The incidence of infective SMV in harvest-mature embryos is the same as the incidence of seed transmission, suggesting that for SMV, embryo infection is a prerequisite for transmission.

Bowers and Goodman (1979a) investigated the development of seeds on SMV infected plants of two cultivars, one with relatively high incidence of seed transmission, and another in which seed transmission has yet to be demonstrated. For both cultivars, infective SMV was present in immature seeds up to and including the reproductive stage R4 (Fehr et al., 1971). Thereafter, recovery of infective virus from seeds dropped as maturity progressed until harvest maturity when no infective virus was recovered from seeds of the nontransmitting cultivar. In the transmitting cultivar, the incidence of infective virus in mature embryos was the same as the incidence of seed transmission. These results ruled out the hypothesis that the incidence of seed transmission is a function of the incidence of flower or immature seed infection. The results also appear to rule out the concept that immature embryos are resistant or immune to SMV infection.

The picture that emerges from work reported so far is that in a susceptible, SMV-infected plant, flowers become infected and lead to the formation of virus-infected embryos. Whether the microgametophyte as well as the megagametophyte can serve as the route of infection is not yet known, although there is one report of pollen transmission (Iizuka, 1973). The same course of events occurs in transmitting and nontransmitting genotypes. During the course of embryo development, embryos of the nontransmitting genotype in some way lose infection. A similar process of virus inactivation or some other therapeutic mechanism occurs in transmitting genotypes, except that a proportion of the seeds, when they reach harvest maturity, still contains infective virus and, when germinated, produce virus infected seedlings.

The role of infective virus particles in seed transmission has not been demonstrated, and there is no *a priori* reason to assume that infection requires the presence in cells of intact virus particles. Some other form of the virus, such as a replication intermediate, perhaps associated with cell membranes, could just as well be the form in which virus infection is maintained in embryo cells. Sufficient intact virus is produced, however, to make its detection possible by infectivity tests.

During immature stages, infective virus can also be recovered from cotyledons and testas (Koshimizu and Iizuka, 1963; Iizuka, 1973; Porto and Hagedorn, 1975; Bowers and Goodman, 1979a). At maturity, cotyledon infection in one series of tests was the same as embryo infection, but the incidence of infective virus in testas fell off drastically between stages R7 and R8 (Bowers and Goodman, 1979a). Viral antigen is readily detectable in such testas, but the virus appears to have been inactivated.

The mechanism by which SMV is inactivated in maturing embryos of seed nontransmitting or seed transmitting genotypes is not known. Differences among soybean lines and cultivars in the incidence of seed transmission of the same isolate of the virus and the apparent susceptibility of immature embryos of both genotypes to infection would appear to argue against desiccation or other generalized physical inactivation processes. Processes by which virus or viral intermediates undergo chemical degradation in cells are poorly understood. The possibility of inhibitors or specific chemicals that inactivate the virus and/or prevent systemic spread of the virus once cell division and growth resume in the germinating embryo cannot be ruled out. A fascinating area of plant-virus interrelationship awaits elucidation.

6.2.3 Seed Transmission and Seed Quality

Mention has already been made of the mottling of soybean seed coats as a result of infection by SMV. While not strictly related to SMV seed transmission, mottling is an important aspect of the impact of SMV on soybean seed quality, as the discoloration of the seed coat reduces the value of soybeans in commercial trade. Discolored seeds of whatever cause lead to a reduced price paid per bushel at the elevator.

SMV infection also reduces the size of seeds produced by infected plants. The effect, which may be due to the diversion of cell energy and the reduction of leaf area and photosynthetic capacity associated with symptoms, is probably the result of reduced ability of the plant to fill out the endosperm in developing seeds. Seed size, measured as the mass of 1000 seeds, was reduced 10 to 20 percent in experiments conducted at the University of Illinois (Milbrath and McLaughlin, 1980; Irwin and Goodman, 1980a).

SMV-infected seedling arising from seeds infected by many isolates of SMV give the appearance of being less vigorous than noninfected siblings. In the field, one often finds that less vigorous infected seedlings from a seed lot with a low incidence of seed transmission lose out in the competition for canopy space and soon are overtaken and shaded out by nearby, more vigorous, healthy plants. Thus, in a season or location where early-season spread of the virus does not occur, the consequences of an occasional SMV infected plant in a seed lot may be negligible. One of the interesting epidemiological questions that has not yet been answered is at what level of incidence does seed transmission of SMV become significant. Because soybeans, like other legumes, can compensate when neighboring plants are removed during vegetative growth, severely infected seedlings that are eliminated during the vegetative development of the crop probably could be tolerated at quite high incidence with no loss of yield as long as SMV spread to healthy plants does not occur. Seedlings that are virus infected but are relatively vigorous may result in greater losses, both because they extend the time during which they may serve as sources of inoculum for spread by aphids and because they take up space in the canopy with less potential for seed production due to virus infection. This question is one that will have to be answered by a series of carefully controlled field studies with a variety of SMV strains and soybean cultivars and under a variety of environments.

There have been conflicting reports in the literature about the impact of SMV infection on germinability of soybean seeds. Gardner and Kendrick (1921) noted that SMV infection of soybeans did not appear to reduce the rate of germination of seeds from infected plants, while later reports suggested that seeds from SMV infected plants had reduced germinability (Dunleavy et al., 1970). Ross (1977) noted the association of seed infection by *Phomopsis sojae*, the pod and stem blight pathogen, with seeds from SMV infected plants, an observation that suggested that it was *P. sojae* rather than the virus that reduced germination. This hypothesis was experimentally verified in work done by our group (Hepperly *et al.*, 1979). Some, but not all, soybean cultivars investigated showed a predisposition to *P. sojae* infection as a consequence of SMV infection and some cultivars ordinarily resistant to infection by *P. sojae* became susceptible to the fungus when the plants were SMV infected. The ability of seedborne *P. sojae* to reduce germinability of soybean seeds is well established (Wallen and Seaman, 1963; Hepperly and Sinclair, 1978), and results from controlled inoculation experiments showed that it was the seedborne fungus rather than virus infection that caused loss of germinability in seed lots from SMV infected plants (Hepperly *et al.*, 1979). There is no evidence linking embryo infection by SMV with reduced germination.

6.3 APHID TRANSMISSION

Soybean mosaic virus is transmitted from plant to plant within the field by members of the Family Aphididae (Insecta: Homoptera) (Heinze and Kohler, 1941; Conover, 1948; Nariana and Pingaley, 1960; Koshimizu and Iizuka, 1963; DeVasconcelos, 1966; Usman *et al.*, 1973; Abney *et al.*, 1976; Benigno and Boonarkka, 1979). Laboratory tests confirm the nonpersistent manner by which SMV is transmitted; using clones of apterous *Myzus persicae* (Sulzer) maintained in the laboratory, we demonstrated optimum transmission from acquisition probes of 30-60 sec; rates of transmission were lower with acquisition access of 15 min or longer and with acquisition access of 15 sec or less (Schultz *et al.*, 1980).

In Asia and parts of Africa, three species of aphids colonize soybean fields: *Aphis glycines* Matsumura, *Aulacorthum solani* (Kaltenbach), and *Aphis craccivora* (Arunin, 1978; Khamala, 1978; Litsinger *et al.*, 1978). All three species can be pests as well as vectors of SMV. Aphids seldom colonize soybean fields in the rest of the world. Two species are known on rare occasions to form small, temporary colonies on seedling soybeans in the Western Hemisphere: *A. craccivora* in the high jungles of Peru and in a late-planted experimental plot in central Illinois and *Aphis gossypii* Glover in the coastal areas of Ecuador (Irwin, 1980b). All of our transmission studies have been conducted in central Illinois where transient alate aphids are wholly responsible for the field spread of SMV. The timing and rate of SMV spread in a soybean field is, to a large extent, a function of the activity of transient aphids that land on soybean leaves, probe, move to new plants and probe again.

6.3.1 Monitoring Landing Rates of Aphids in Soybean Fields

To study the spread of plant viruses in relation to aphid activity, a monitoring device that accurately measures the numbers and species alighting on plant foliage is essential. Several types have been used in aphid-borne virus epidemiology work throughout the past century. Three are in common use today: the yellow pan trap; the vertical cylindrical sticky trap, and the Johnson-Taylor suction trap (Irwin, 1980a). Yellow pan traps are differentially attractive to different species of aphids (Eastop, 1955; Roach and Agee, 1972; Taylor and Palmer, 1972). Vertical cylindrical traps give a relative measure of aerial density with respect to species composition, but the numbers collected on the trap are very much a function of wind velocity (Taylor and Palmer, 1972) and may not represent accurately the

FIG. 2. The horizontal ermine lime (HEL) trap. An ermine lime colored ceramic tile 15.3 cm square was covered with tanglefoot (not shown) or placed in a slightly larger Plexiglass tray (shown) containing a 50% aqueous solution of ethylene glycol. The trap is mounted on a ring clamp and secured to a metal rod. As the season progresses, the trap is raised or lowered to keep the trap at the level of the soybean canopy. Reproduced by permission by Springer-Verlag, New York, Inc., from Irwin (1980a).

numbers or species composition of aphids alighting on plants. The Johnson-Taylor suction trap gives near absolute counts of aerial density and, while very useful for aphid migration studies, provides data that may have little to do with how many of which species land on plants.

For virus-vector monitoring, what seems desirable is to know how many of which species of aphids are landing in the canopy per unit of area over a given time interval. To be effective, the traps should be neither more nor less attractive or repellent to any species of aphid relative to the crop under study.

We have designed and tested a sampling device, the horizontal ermine lime (HEL) trap (Fig. 2), which gives data closer to absolute landing rates of aphids on a soybean canopy than obtained with other traps tested (Irwin, 1980a). The device consists of a square (15.2 X 15.2 cm) ceramic tile of rugose texture, and of a color closely resembling that of soybean leaves based on data from reflectance spectrophotometry. The trap is mounted in such a way that it can be maintained at canopy level throughout the growing season. In the first three years of our work (1976-1978), the traps were coated with Tanglefoot; in 1979 we placed the ermine lime tiles in square Plexiglass trays 15.7 cm on a side and 3.5 cm deep (inside dimensions) filled with 50% solution of ethylene glycol in water. We found the use of liquid-covered traps cut down on labor costs and made it easier to identify the aphids caught.

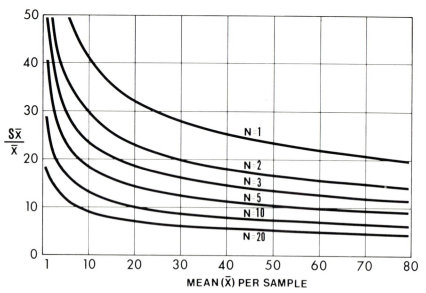

FIG. 3. Relationship between the number of horizontal ermine lime (HEL) traps (N), aphids per trap per week ($\bar{x}$), and accuracy of the estimate of numbers per HEL trap ($\frac{S\bar{x}}{\bar{x}}$). Graph based on regression of three years' catch of aphid alatae; mean and variance plotted on a log-log scale. Regression equation $s^2 = 0.8355\bar{x}^{1.3054}$, where s^2 = variance, $\bar{x}$ = mean; n = 46; r = 0.87. Reproduced by permission of Springer-Verlag, New York, Inc., from Irwin (1980a).

From data obtained by HEL traps during three soybean growing seasons, 1976-1978, we calculated the number of traps needed per field depending on the mean number of alate aphids collected per trap and the desired level of reliability. For virus-vector studies, a measure of the landing rate within 20% of the mean is probably adequate; from our studies a conservative estimate of aphid captures per week is 30 alatae. For such a situation, two traps per plot are needed (Fig. 3).

6.3.2　Aphid Species That Transmit SMV

Twenty-four aphid species have been reported to transmit SMV. To this list another seven have been added as a result of our laboratory and field experiments (Table II). Results of our laboratory tests (Schultz *et al.*, 1980) indicate that for *M. persicae*, adult apterae and nymphs transmit with about the same efficiency as do alatae.

While laboratory tests of various aspects of transmission are important, what seems of even more importance is field transmission. What species from a mixture of wild clones are capable of transmitting SMV? What are their apparent field transmission efficiencies? We conducted studies addressing these questions during the 1976 and 1978 growing seasons. Alate aphids were collected from vertical nets 1.4 m high by 4.6 m long erected downwind of a soybean field (30 X 30 m) in which all plants were sap inoculated with the Illinois severe isolate of SMV (SMV-G2 strain of Cho and Goodman, 1979). Aphids landing on the windward side of the net were individually placed on healthy soybean seedlings (growth stage V1), caged with a cylindrical Mylar tube, and maintained in the shade until the seedlings were taken to the greenhouse. At least 1 hr after the beginning of inoculation access, the aphid caged on each seedling was removed and placed in ethanol for identification, and the seedling maintained in an aphid-free greenhouse for two weeks, after which the seedlings were indexed for SMV (Halbert *et al.*, 1980a).

More than 60 species of aphids were captured during the two seasons. The results indicated, however, that five species of aphids were responsible for about 93% of all transmissions: *A. craccivora, Macrosiphum euphorbiae* (Thomas), *M. persicae, Rhopalosiphum maidis* (Fitch), and *Rhopalosiphum padi* (L.). Among the five species, *R. maidis* and *R. padi* were relatively less efficient transmitters while *A. craccivora, M. euphorbiae,* and *M. persicae* were more efficient (Table III). These five species comprised nearly 64% of all aphid specimens collected on the vertical nets. The overall transmission efficiency of all aphids tested was 4.2% (Table III). When numbers of specimens of each of the five most important vectors were taken into account, we found that *R. maidis*, a relatively inefficient transmitter, had the greatest impact on virus transmissions, mainly because large numbers of that species were collected. *R. padi* had a low impact because of its low numbers and its relatively low efficiency in transmitting SMV (Table III).

TABLE II. Aphids Known to Transmit Soybean Mosaic Virus

Aphid Species	Name Reported	Source
Acyrthosiphon pisum (Harris)	*Acyrthosiphon pisum* Harris	DeVasconcelos 1966
	Macrosiphum pisi Kalt.	Conover 1948
Aphis armoraciae Cowen	*Aphis armoraciae* Cowen	Halbert *et al.*, 1980a
Aphis citricola Van der Goot	*Aphis citricola* Van der Goot	Schultz *et al.*, 1980a
Aphis craccivora Koch	*Aphis craccivora* Koch	Benigno and Boonarkka 1979
	Aphis craccivora Koch	DeVasconcelos 1966
	Aphis craccivora Koch	Nariana and Pingaley 1960
	Aphis craccivora Koch	Usman *et al.*, 1973
Aphis fabae Scopoli	*Aphis fabae* Scopoli	DeVasconcelos 1966
	?Doralis fabae (Scopoli)	Heinze and Kohler 1941
Aphis glycines Matsumura	*Aphis glycines* Matsumura	Benigno and Boonarkka 1979
	Aphis glycines Matsumura	Koshimizu and Iizuka 1963
Aphis gossypii Glover	*Aphis gossypii* Glover	DeVasconcelos 1966
	Doralis frangulae (Koch)	Heinze and Kohler 1941
	Aphis gossypii Glover	Nariani and Pingaley 1960
	Aphis gossypii Glover	Usman *et al.*, 1973
Aphis laburni Kaltenbach	*Aphis laburni* Kaltenbach	Koshimizu and Iizuka 1963
Aphis nasturtii Kaltenbach	*Doralis rhamni* (Boyer)	Heinze and Kohler 1941
	Aphis nasturtii Kaltenbach	DeVasconcelos 1966
Aphis nerii Boyer de Fonscolombe	*Aphis nerii* Boyer de Fonscolombe	Usman *et al.*, 1973
Aphis rumicis (L.)	*?Doralis fabae* (Scop)	Heinze and Kohler 1941
Aulacorthum (*Neomyzus*) *circumflextum* (Buckton)		
	Neomyzus circumflexus Buckt.	DeVasconcelos 1966
	Neomyzus circumflexus (Buckton)	Heinze and Kohler 1941
Aulacorthum solani (Kaltenbach)	*Macrosiphum solani*	Koshimizu and Iizuka 1963
	Aulacorthum solani Theob	DeVasconcelos 1966
	Aulacorthum pseudosolani (Theob.)	Heinze and Kohler 1941
Capitophorus elaeagni (del Guer.)	*Capitophorus elaeagni* (del Guer.)	Halbert *et al.*, 1980a
Hysteroneura setariae (Thomas)	*Hysteroneura setariae* Thomas	Benigno and Boonarkka 1979

Species	Synonym/determination	Reference
Lipaphis erysimi (Kaltenbach)	*Lipaphis erysimi* Kalt.	Nariani and Pingaley 1960
Macrosiphum euphorbiae (Thomas)	*Macrosiphum euphorbiae* Thomas	DeVasconcelos 1966
	Macrosiphon solanifolii (Ashm.)	Heinze and Kohler 1941
	Macrosiphum euphorbiae (Thomas)	Abney *et al.*, 1976
Macrosiphum rosae L.	*Macrosiphum rosae* L.	DeVasconcelos 1966
Megoura viciae Buckton	*Megoura viciae* Buckton	DeVasconcelos 1966
Melanaphis sacchari (David) forma *indosacchari*	*Melanopsis indosacchari* David	Benigno and Boonarkka 1979
Myzus ornatus Laing	*Myzus ornatus* Laing	DeVasconcelos 1966
	Myzus ornatus Laing	Heinze and Kohler 1941
Myzus (Nectarosiphon) persicae (Sulzer)	*Myzus persicae* Sulz.	DeVasconcelos 1966
	Myzus persicae Sulzer	Conover 1948
	Myzodes persicae (Sulz.)	Heinze and Kohler 1941
	Myzus persicae Sulz.	Benigno and Boonarkka 1979
	Myzus persicae	Koshimizu and Iizuka 1963
	Myzus persicae Sulz.	Nariani and Pingaley 1960
	Myzus persicae Sulzer	Usman *et al.*, 1973
	Myzus persicae (Sulzer)	Abney *et al.*, 1976
Rhopalosiphum insertum (Walker)	*?Rhopalosiphum prunifolii*	Koshimizu and Iizuka 1963
Rhopalosiphum maidis (Fitch)	*Rhopalosiphum maidis* (Fitch)	Abney *et al.*, 1976
Rhopalosiphum padi (L.)	*?Rhopalosiphum prunifolii*	Koshimizu and Iizuka 1963
Schizaphis graminum (Rondani)	*Schizaphis graminum* (Rondani)	Schultz *et al.*, 1980a
Therioaphis trifolii (Monell)	*Therioaphis trifolii* (Monell)	Halbert *et al.*, 1980a
Uroleucon ambrosiae (Thomas)	*Dactynotus ambrosiae* (Thomas)	Abney *et al.*, 1976
Uroleucon ? nigrotibium Olive	*Uroleucon ? nigrotibium* Olive	Halbert *et al.*, 1980a
Uroleucon ? nigrotuberculatum Olive	*Uroleucon ? nigrotuberculatum* Olive	Halbert *et al.*, 1980a
Uroleucon sonchi (L.)	*Dactynotus sonchi* L.	DeVasconcelos 1966

195

TABLE III. Abundance, Transmission Efficiency, and Relative Transmission Efficiency of Live-Trapped Aphids. Urbana, Illinois 1976-1978. Modified from Halbert *et al.*, 1980a.

Aphid species	Number of specimens tested	Percentage of catch contributed by species	Percentage transmission per species	Percentage transmission contributed by species
A. craccivora	109	6.4	13.8	20.8
M. euphorbiae	145	8.5	8.3	16.7
M. persicae	148	8.7	12.2	25.0
R. maidis	605	35.4	3.1	26.4
R. padi	81	4.7	3.7	4.2
Subtotal	1088	63.7	6.2	93.1
Other species (55+)	621	36.3	0.8	6.9
Total	1709	100.0	4.2	100.0

Therefore, percentage transmission contributed by a given species is a function of the number of specimens of that species collected and tested and the percentage of specimens of that species that transmitted. Our results also indicated that species of the Pemphaginae and *Lipaphis erysimi* (Kaltenbach) were unimportant in transmitting SMV in central Illinois soybean fields. The study failed to account for species of aphids that did not land on the vertical net but might have landed in the field and the possible importance of the effect of crop growth stages on ability of viruliferous aphids to acquire virus from or infect plants on which they probed or fed.

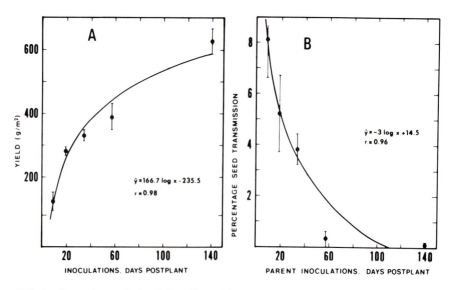

FIG. 4. Regression analysis of the effect of inoculation time on (A) yields and (B) seed transmission from field-grown Williams soybean plants inoculated with the Illinois severe isolate of soybean mosaic virus. Error bars indicate standard errors of the mean.

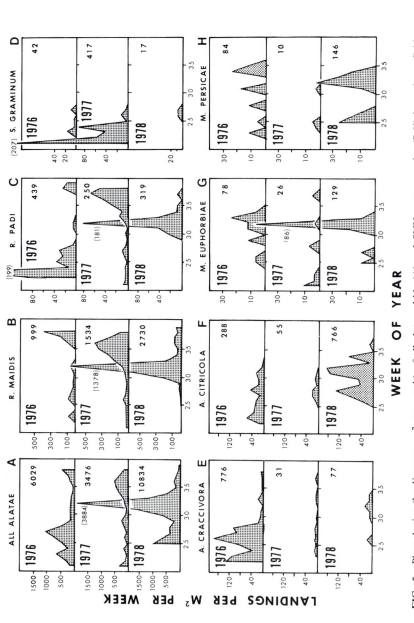

FIG. 5. Phenologies (landings per m² per week) of all alate aphids (A) and SMV vector species (B-H) in soybean fields at Urbana, Illinois, in 1976, 1977, and 1978 as monitored by trapping with horizontal ermine lime traps. Note the different scales used (ordinate axis) for the various species. Numbers in upper right corner of each graph are cumulative totals for each species in season indicated by date in upper left corner.

At least one species of aphid could be an important vector of SMV but went undetected by field methods employed in the study. Horizontal ermine lime traps demonstrated that *Aphis citricola* Van der Goot commonly landed on soybean foliage during the 1976, 1978, and 1979 seasons (Fig. 5F). However, very few were captured on the vertical nets. *A. citricola* readily transmits SMV under laboratory conditions (Schultz *et al.*, 1980) and in 1979 was the only aphid species caught in significant numbers at a time when field data indicated rapid spread of SMV (Cho and Goodman, 1980). Another potentially important vector is *Schizaphis graminum* (Rondani). *S. graminum*, which transmitted SMV in laboratory tests, was collected in low numbers in years (1976 and 1978) when the vertical net experiments were done, but in 1977 large numbers of this species were caught early in the season (Fig. 5D).

6.3.3 Flight Phenologies of Important Vector Species

The potential consequences of SMV infection (seed transmission incidence and yield loss) are greatest when infection occurs early in the season (Fig. 4). Thus, it is important to consider the timing of flights, as well as the transmission efficiency, of important vector species.

The phenology and numbers of aphids visiting soybean fields varied considerably among species and from year to year (Fig. 5). For instance, the landing frequency of *R. padi* in 1976 was higher in the spring, but in 1977 and 1978 it was higher in the fall (Fig. 5C). Landing rates of *A. craccivora* were high in the spring of 1976 but low throughout the 1977 and 1978 seasons (Fig. 5E). Certain patterns can be detected, however. Landing rates for *R. maidis* were always higher than for any other species, and the rate was always highest in the summer and fall (Fig. 5B). Landing rates of *M. persicae* and *M. euphorbiae* were generally relatively low and were concentrated in the middle of the season (Fig. 5G, H). Only *A. citricola*, *A. craccivora*, *S. graminum*, and occasionally *R. padi*, had landing rates concentrated in the spring and early summer in central Illinois (Fig. 5C, D, E, F).

Because different species of aphids alight with different frequencies at different times during the growing season, a timing factor must be included in attempts to assess possible economic loss and reduced seed quality due to resulting SMV infection.

Based on data showing the catch on HEL traps approximates the total number of alate aphid landings per unit area over a given time interval, we have calculated the number of alate aphids landing in a soybean canopy over a season. The number of landings of all aphid species per m^2 over the season vary considerably from year to year in central Illinois: 1976 = 6,029; 1977 = 3,476; 1978 = 10,834 (Fig. 5A; Irwin and Goodman, 1980b). During the 1978 growing season, the density of aphid landings on the soybean canopy averaged one per cm^2. For a crop that is not colonized by aphids, the numbers of aphids alighting in the crop are staggering.

6.3.4 Plant, Virus, Vector Interactions

Aphid vectors of SMV act as messengers of the virus, acquiring it from one host plant and giving it to another. In this section we explore some of the important factors that govern from whence and how quickly SMV can be acquired from one host and given to another.

One of the more important factors we studied was the latent period of the virus in a soybean plant. In greenhouse trials, when an aphid inoculated a plant on day 1, the earliest another aphid could recover the virus was on day 6. In some cases, it took an additional 2 or 3 days before the virus could be recovered, probably due largely to abiotic factors such as temperature. Field tests of this phenomenon gave similar results (Schultz et al., 1980). The movement of the virus in a field is largely governed by the number of infective plants present. The short lag period between inoculation and the time when a plant can serve as a source for transmission suggests that secondary and tertiary spread may be occurring. As mentioned previously, this could be especially important if it occurs early in the growing season.

We have also tested the ability of aphids to acquire SMV from various parts of infected plants. The results from experiments done in the laboratory indicate that almost all parts of an infected plant can serve as sources for SMV acquisition and subsequent transmission by aphids. The plant parts tested included the bud complex, flowers, folded (young) trifoliolate leaves, recently expanded trifoliolate leaves, mature leaves, senescing leaves, and green pods (Schultz et al., 1980). Apterous M. persicae acquired SMV from plants of different ages (from 1 to 9 weeks after infection in tests where each plant was infected for two weeks before acquisition testing. Furthermore, alatae of M. persicae acquired virus from plants of different ages infected for different amounts of time and transmitted to healthy seedlings with equal efficiency (Schultz et al., 1980). Neither age of plant nor plant part greatly influence the process of transmission; an infected plant remains a source plant for much of the season and many of its parts are infectious. We have recently discovered, however, that M. persicae did not transmit SMV following acquisition probes on soybean plants inoculated at R5 even though the inoculated plants indexed positively for SMV (Schultz et al., 1980).

Soybean plants infected during early growth stages by severe isolates of SMV are stunted and often covered over by healthy, more vigorous neighbors. Therefore, the reaction of the plant to individual virus strains is important to an understanding of the relative accessibility of the virus source to the vector and the infective vector to healthy versus already-infected plants.

6.3.5 Effect of Time of Inoculation on Yield Loss

How can one predict the impact of virus spread on soybean yield and seed quality? In the previous section of this chapter we pointed out that aphid species differ in their potential to transmit SMV. They differ because individuals of different species are dissimilar in their efficiency of transmission. They also differ in alighting phenology. We have developed a preliminary model of virus spread in relation to timing of inoculation by vectors to illustrate the importance of time of inoculation on soybean yield.

Results from yield loss studies with the Illinois severe isolate of SMV on Williams soybean indicated that during early vegetative development of the crop infection caused more serious losses and higher rates of transmission through the seed than infection after flowering begins (Fig. 4; Irwin and Goodman, 1980a; Bowers and Goodman, 1979a). These data allowed us to calculate a formula for yield of Williams soybeans infected by the Illinois severe isolate of SMV in 1977 (Fig. 4A):

$$\hat{y} = 166.7 \log x - 235.5 \tag{1}$$

where $\hat{y}$ = yield in g/m^2, and log x = natural logarithm of the number of days postplant that a given percentage of plants were infected.

If yield loss (YL) is the difference between yield when virus infection occurs at harvest maturity, 140 days postplant (i.e., no yield loss) (y_0) and yield if the crop is infected at x days postplant $(\hat{y})$, then:

$$YL = y_0 - \hat{y}$$

Substituting from equation (1),

$$YL = (166.7 \log 140 - 235.5) - (166.7 \log x - 235.5)$$

$$YL = 166.7 \log \left(\frac{140}{x}\right) \tag{2}$$

where yield loss (YL) is a function of infection days postplant (x).

Let us assume that a flight of aphids of a given vector species alights on a soybean canopy and results in 20% of the plants becoming infected with SMV 20 days postplant (case 1) and 70 days postplant (case 2). The difference in yield loss between case 1 and case 2 will represent a delay in infection of 50 days.

case 1:

$$YL = (80\%) \left(166.7 \log \left(\frac{140}{140}\right)\right) + (20\%) \left(166.7 \log \left(\frac{140}{20}\right)\right)$$

$$= 0.2 \, (166.7 \log 7)$$

$$= 64.8 g/m^2$$

case 2:

$$YL = (80\%) \left(166.7 \log \left(\frac{140}{140}\right)\right) + (20\%) \left(166.7 \log \left(\frac{140}{70}\right)\right)$$

$$= 0.2 \, (166.7 \log 2)$$

$$= 23.3 g/m^2$$

Therefore, by delaying a flight of aphids for 50 days, one cuts yield loss due to SMV by more than half. This simple example illustrates the importance of timing on yield loss. A similar analysis can be used for seed transmission (Fig. 4B).

6.3.6 Field Spread

We monitored the movement of SMV from spreader to adjacent reader rows in an experiment designed to measure the potential for virus spread during various parts of the season (Schultz *et al.*, 1980). The experimental design allowed four separate readings of virus spread under conditions where the age of the plants was not a factor. The first plot was planted on May 31. At the V1 stage, two center rows were sap inoculated, and transmission from spreader to adjacent reader rows was allowed for 3 weeks (June 22-July 13). All plants were then indexed, percentage infected recorded, and the plot was plowed under and replanted. Meanwhile, another plot had been planted on June 15, and infection from its spreader to reader rows was allowed for 3 weeks (July 13-August 3). This sequence was repeated twice again, for a total of four 3-week trials. Landings of alate aphids were recorded in HEL traps throughout the experiment.

The results showed that the amount of movement of SMV from spreader to reader rows differed through the season, and was positively correlated with aphid landing rates (Fig. 6). Within a 3-week period of July, the disease incidence reached 31%, and this correlated well with a flight of *A. citricola*. The greatest percentage infection in reader rows was in mid-August when aphid populations were very high. Within a 9-day period beginning on August 6, the disease incidence reached 59%, and at the end of 21 days, the incidence was 94%. During this period, the most abundant species of aphids were *R. maidis*, a known vector of SMV, and *Myzocallis asclepiadis* (Monell), not previously tested for its ability to transmit SMV. As a result of this experiment, we have identified *M. asclepiadis* as a candidate for laboratory experiments to determine its potential as a vector of SMV. The results of this experiment showed that aphid vectors are capable of transmitting SMV in young soybean plantings throughout the growing season.

We also monitored the movement of SMV from a point source into a surrounding, initially virus-free, field of Williams soybeans during the 1976 growing season (Irwin and Goodman, 1980b). Two plots, each 92 X 92 m (0.83ha) were established within a large field. All plants in the center 0.5% of the area of the downwind plot were sap inoculated with the Illinois severe isolate of SMV

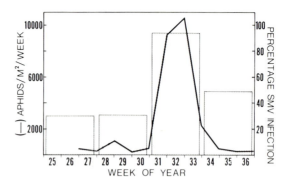

FIG. 6. Landing rates (aphids per m² per week) of alate aphids and spread of soybean mosaic virus (SMV) from inoculated rows to adjacent initially-healthy soybean plants during the 1978 growing season at Urbana, Illinois. Each shaded bar on the graph represents the final percentage of SMV infection for the 3-week period shown. Aphids were collected on horizontal ermine lime traps daily; the data reported are weekly totals. The inoculated source and initially-healthy reader plants for each 3-week trial were approximately the same age to eliminate plant age as a factor in assessing the potential of SMV acquisition or spread over the course of the growing season.

(SMV-G2 strain of Cho and Goodman, 1979). A grid of 240 intersecting points on 5.6 m centers was marked with flags in each plot, and leaf samples were gathered at random from plants at these points 6 times during the growing season, once before inoculation of the virus source at the center of the field and 5 times afterwards. Ten HEL traps were placed at regular intervals along in a line running through the center of both plots to monitor aphid landing rates during the season.

The spread outward from the source was greater downwind than upwind (Figs. 7 and 8). The number of infections declined exponentially with distance from the source, whether up- or downwind (Fig. 8). The infection farthest from the source at the end of the season was ca. 45 m along the vector of the prevailing wind. Thus, wind is an important factor in the directional flow of SMV infections in a field. Twenty-five percent of all spread occurred within 2 m of the source and 95% of all infections were within 17 m of the source. Spread in the check plot was almost nil, indicating no influx of virus from outside the plot.

The cumulative number of sampling sites that indexed SMV-positive was not exponential, probably because of the paucity of vectors during the latter third of the 1976 season (Fig. 9). The amount of secondary and tertiary spread was thus minimal in 1976, while in an experiment done in 1978, considerable secondary and tertiary SMV spread occurred (Schultz et al., 1980). At the end of the 1976 experiment, indexing confirmed 19 infected sites out of 240 tested, or ca. 8% in the inoculated plot.

Leaves on Williams soybeans infected by a severe isolate of SMV remain green for several days after leaves on healthy plants turn brown and fall off (Schultz et al., 1980). This phenomenon was useful in determining the number of infected

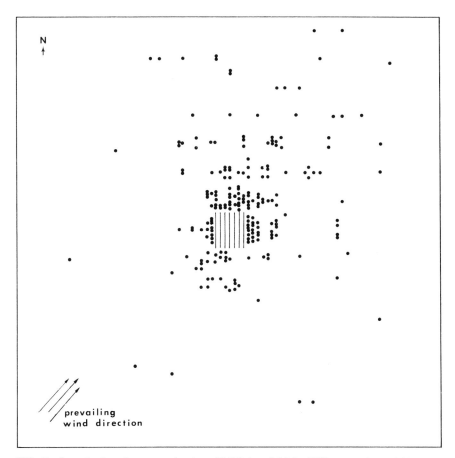

FIG. 7. Spread of soybean mosaic virus (SMV) in a 0.83 ha Williams soybean field near Tolono, Illinois in 1976. The seven parallel lines at the center of the field diagram represent the 0.5% of the plant population inoculated with SMV at the primary leaf stage. The dots represent the location of SMV-infected plants at the end of the growing season. The prevailing wind direction was as indicated.

plants in the inoculated and control plots of the 1976 experiment. Shortly after the leaves of healthy soybeans fell from the plants all plants retaining green leaves in the two plots were carefully inspected for foliar symptoms of SMV. Only two plants in the control (noninoculated) plot were found to be SMV infected, whereas 209 plants in the inoculated plot were infected. The positions of the infected plants in the inoculated plot are diagrammed in Fig. 7.

Movement of SMV within a field thus depends upon several factors, including primary inoculum level (soybean seedlings infected from seed); timing, numbers and species composition of transient alate aphids; and wind and other climatic conditions during the time of these flights. The pattern of spread from a source decreased exponentially as the distance from the source increased, with no infec-

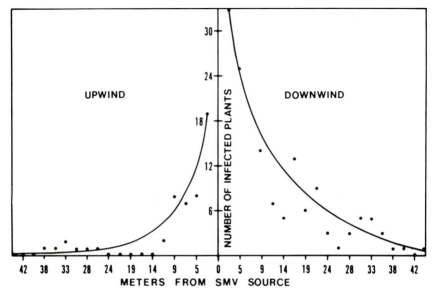

FIG. 8. Spread of soybean mosaic virus as a function of wind direction in a 0.83 ha Williams soybean field near Tolono, Illinois, in 1976 (see Fig. 7).

tion occurring more than 50 m from the source. Other factors not yet investigated may be involved, including vector biotype, plant cultivar, and virus strains.

6.4 DISEASE AND VECTOR MANAGEMENT

Some of the factors that govern or influence soybean mosaic virus spread in the field are understood to a degree that should allow at least preliminary formulation of control tactics. Some of these will be of quite general utility, both in terms of soybean crops grown in different ecological situations and in terms of their application in control strategies of this and other nonpersistently transmitted, seedborne viruses. Other approaches will, however, be more specific in their applicability, either because they depend on the environmental situation in which the crop is grown or because they rest on principles that reflect ways in which soybeans are a unique crop. That the following discussion refers almost exclusively to soybeans should not be taken as implying that we are ignoring the broader implications nor that we are unaware of the many similarities that exist between soybeans and other crops. If soybean mosaic has a claim to being unique among the plant virus diseases investigated in great epidemiological detail, it is that the system combines in an unusual way a virus that is seedborne in its principle host and has an otherwise restricted host range with a vector that transmits in a nonpersistent manner and is entirely transient, or noncolonizing, within the crop.

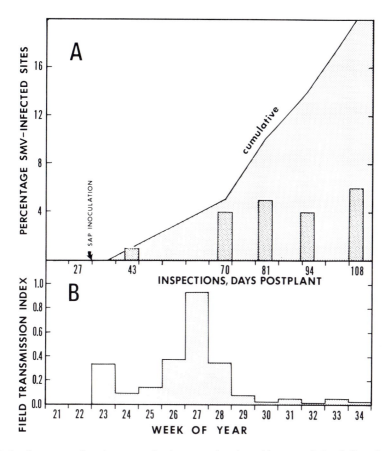

FIG. 9. Summary of soybean mosaic virus spread and weekly transmission indices for a 0.83 ha Williams soybean field near Tolono, Illinois, in 1976. (A) Samples of soybean foliage were collected at 240 equally spaced sites on the inspection dates shown (27, 43, 70, 81, 94 and 108 days after planting) and indexed for presence of SMV. Bars represent the number of sites at which SMV infected plants were found on each inspection date. The shaded curve represents the cumulative total of sites at which infected plants were found. Sap inoculation of 0.5% of the plant population in the field (see Fig. 7) was 31 days after planting. (B) Aphids were collected on horizontal ermine lime traps and the weekly totals of 5 vector species and their relative transmission efficiencies (see Table 3) were used to calculate a field transmission index which represents the relative likelihood of virus transmission occurring in each week during the growing season.

6.4.1 Resistance to Infection

We have stated elsewhere in this article that all of the soybean varieties in wide commercial use in the United States are susceptible to SMV. Notwithstanding the apparent unimportance of SMV as a pathogen of soybeans in the U.S., considerable work by a few investigators has been done to incorporate resistance to one or more isolates of SMV into improved cultivars. The principle genetic sources of resistance to SMV have been the variety Tokyo, which occurs in the pedigrees of the cultivars Davis and Ogden (Weiss, 1953; Koshimizu and Iizuka, 1963; Caviness and Walters, 1966; Hymowitz *et al.*, 1977), and more recently the germplasm accession PI 96983 (Ross, 1969, 1977).

In the early U.S. work establishing the resistance of these cultivars to SMV, little attention was paid to the possibility that variation in virulence of the virus might exist and that as a consequence cultivars resistant to one isolate of the virus might be susceptible to isolates from other localities differing in virulence.

In the last decade, investigators, particularly Ross (1969, 1975) and Han and Murayama (1970), began to recognize differences among SMV isolates as regards their differential ability to infect or cause markedly different symptoms in various soybean cultivars. In an extensive series of tests with 100 SMV isolates from the USDA soybean germplasm collection, Cho and Goodman (1979) recognized seven virulence classes or strains, based on the ability of isolates to infect and cause either mosaic or necrotic symptoms in six putatively SMV-resistant soybean cultivars (Table IV). Among the strains were several that could infect various of the resistant cultivars, three that had not before been reported from the U.S., and one that could infect all cultivars and lines of soybeans known to have resistance to SMV. These results show clearly the necessity, in breeding for SMV resistance in soybeans, to use more than one isolate of the virus for evaluation of segregating populations from crosses. They also show that the SMV resistant cultivars used as differentials had different genetic systems conferring virus resistance, in that just as the strains reacted differently on any one cultivar, so the cultivars reacted differently to any one strain.

None of the cultivars with resistance to SMV are widely grown in the U.S. today. Improved cultivars with SMV resistance against a wider range of SMV strains

TABLE IV. Reactions of Soybean Cultivars to Seven Soybean Mosaic Virus Strains Obtained from the USDA Soybean Germplasm Collections[a]

Soybean Cultivars	Symptoms caused by SMV strains[b]						
	SMV-G7	SMV-G6	SMV-G5	SMV-G4	SMV-G3	SMV-G2	SMV-G1
Clark	-/M	-/M	-/M	-/M	-/M	-/M	-/M
Rampage	-/M	-/M	-/M	-/M	-/M	-/M	-/M
Davis	-/M	-/M	-/M	-,N/M,N	-/-	-/-	-/-
York	-/M	-/M	-/M	-,N/M,N	-/-	-/-	-/-
Marshall	N/N	N/N	-/-	N/N	N/N	N/N	-/-
Ogden	N/N	-/-	-/-	N/N	N/N	-/-	-/-
Kwanggyo	N/N	N/N	N/N	-/-	-/-	-/-	-/-
Buffalo	N/N	-/-	-/-	-/-	-/-	-/-	-/-

[a] A list of accessions from which SMV isolates were obtained is available on request from the authors. From Cho and Goodman (1979) by permission of the American Phytopathological Society.

[b] Reactions on the inoculated primary leaves/reactions on the noninoculated trifoliolate leaves.

-: Symptomless, no virus detected by Top Crop indexing M: Mosaic symptoms N: Necrosis are in the developmental stages at several locations in the United States. These are based on the resistance found in PI 96983 (Kiihl, 1976), which Cho and Goodman (1979) showed resistant to 5 of the 7 strains they identified, and in the tropical cultivar Buffalo, a tropical soybean released in Rhodesia and found to be resistant to 6 of the 7 strains (Paschal and Goodman, 1978; Cho and Goodman, 1979).

SMV resistance has been found by most investigators to involve dominance at one or more loci. Koshimizu and Iizuka (1963) found resistant F_1 plants and a 3:1 ratio (resistant:susceptible) in the F_2 of one cross, susceptible F_1 plants and a 7:9 ratio in the F_2 of a second cross, and lethal necrosis in the inoculated F_1 plant from a third cross. Kiihl (1976) and Kiihl and Hartwig (1979) investigated the inheritance of SMV resistance from PI 96983 and Ogden. They proposed a multiple allelic series RSV, rsv^1 and rsv in which RSV is dominant to rsv^1 and rsv, and rsv^1 is dominant to rsv. PI 96983 carries RSV and is resistant to both virus isolates used. Ogden carries rsv^1 and is resistant to one but susceptible to the other isolate. Plants susceptible to both isolates carry rsv.

Work in our laboratory has centered on the mode of inheritance of SMV resistance from tropical soybean accessions (Paschal and Goodman, 1978), particularily from the maturity group VIII cultivar Buffalo (Bowers et al., 1980). A diallele series of crosses was made among Buffalo, Hardee LS (a late-maturing selection), Jupiter, PI 324068, and PI 341242. An additional cross was Buffalo x PI 96983. F_1, F_2, and F_3 plants are being evaluated for SMV reactions with six isolates of SMV representing 5 of the 7 strains designated by Cho and Goodman (1979). Preliminary results indicate a complex genetic situation involving at least two and possibly three loci, each with a dominant allele for resistance. One locus gives resistance to all SMV strains used while a second confers resistance to two of the five strains.

An exception to the usual observation of resistance being conditioned by dominant alleles is the recent report by Kwon and Oh (1979) that resistance to a virulent SMV isolate causing necrotic symptoms in the SMV-resistant Korean cultivar Kwanggyo is conditioned by a single recessive allele.

6.4.2 Resistance to Seed Transmission

Another approach to using resistance in the control of SMV is to develop soybean cultivars with little or no seed transmission. Because the primary source of virus inoculum for plant to plant spread in the field is infected seedlings arising from infected seeds, elimination or even reduction of seed transmission could have a major impact on virus spread. In view of previous reports cited earlier in this chapter that the incidence of SMV transmission through seeds varied from cultivar to cultivar, we undertook a major screening of temperate and tropical germplasm accessions to identify lines with low incidence of virus transmission from seeds produced on SMV infected plants (Goodman et al., 1979). The results, as yet incomplete, indicated that among the tropical lines used, 6 had low rates of seed transmission (less than 0.25%). The lowest rate encountered was in PI 86736 (maturity group VIII) with 0.05% transmission (Goodman and Oard, 1980). Tests with large numbers of seeds from temperate lines are incomplete, but at least 19 of the temperate lines gave no apparent seed transmission when 1000 seeds were planted and the emerging seedlings evaluated (Goodman *et al.*, 1979). The cultivar Merit was reported by Kennedy and Cooper (1967) as not transmitting SMV through seeds; we too have used Merit in numerous tests and have never found seed transmission although we have never attempted tests with thousands of seeds.

From these results we concluded that soybean breeding programs in which reduced seed transmission of SMV is an objective have ample material from which to chose. The next step in the process of developing improved cultivars with low incidence of seed transmission will be to determine whether through crossing one can combine this characteristic with other desirable traits. Some evidence for optimism in this regard comes from the observation that improved cultivars having in their pedigrees one or both of the early soybean accessions Mukden and Manchu 2204, both of which are on the list of lines with low incidence of transmission (Goodman *et al.*, 1979), have in general lower incidence of seed transmission than all lines taken together. At least two tropically adapted cultivars, Improved Pelican and UFV-1, had low rates of seed transmission (Goodman and Oard, 1980). Also remaining to be tested is whether soybean lines with low incidence of transmission of one strain of the virus will also prove to transmit other strains at low rates. Again, there are several indirect lines of evidence for expecting the resistance to hold up in comparisons with various virus strains, and experiments are currently in progress to test this hypothesis.

6.4.3 Resistance to Aphids

Much is known about resistance to colonization of various species of aphids by various plant hosts. Unfortunately, little is known about the influence of plant genotype and phenotype on landing/probing responses of various aphid species. Gibson and Plumb (1977) recently reviewed this topic. In general, it has been thought that aphids land indiscriminantly in plant canopies and discriminate only after probing. From our studies, this, in fact, appears to be true for some species that land in soybean fields (e.g., *R. padi, T. trifolii* [Monell]), but not for others (e.g., *R. maidis, A. citricola, M. asclepiadis, Capitophorus hippophaes* (Walker), and *M. euphorbiae*; (Irwin *et al.*, 1980). Wavelength reflectance can attract (e.g., yellow spectrum) or repel (e.g., ultraviolet spectrum) several aphid species (Kring, 1954; Smith *et al.*, 1964; Smith and Webb, 1969; Irwin and Halbert, 1980). We have recently found that soybeans with leaves of slightly different wavelength reflectance can influence the species composition and relative abundance of aphids caught in horizontal, transparent, sticky traps (see below). Thus by manipulating plant genetics, the species composition of the aphid fauna visiting soybean fields may be influenced, but whether control of SMV spread can be achieved by such manipulations has not yet been tested. In the following sections, we will discuss some methods of altering plant phenotype to modify aphid responses.

6.4.4   Altering Intensity and Pattern of SMV Spread

Several potential tools exist for altering the intensity and pattern of spread of a nonpersistent virus in a cropping system, some of the newer of which include various chemicals that act in one way or another to prevent infection, or at least to slow down its rate. Insecticides applied to a crop, however, are usually ineffective in reducing spread of nonpersistently-transmitted viruses because the chemicals kill the vectors too slowly (Broadbent, 1969). McLean (1957) found that if *M. persicae* were transferred successively at 20 minute intervals to plants treated with one of several highly toxic compounds, some specimens were still able to transmit a nonpersistent virus after the fifth transfer. Although not tested with SMV and its aphid vectors, it seems likely that insecticides would prove ineffective, especially since the vectors do not colonize the crop in the Western Hemisphere.

The spread of many aphid-borne nonpersistent viruses has been impeded and resulting percentage field infection reduced when crops were sprayed with oils or emulsified oils (Toba *et al.*, 1977; see review by Vanderveken, 1977). This was shown to apply to soybean mosaic virus by Lu (1970) who reported that vegetable oil emulsions sprayed at 5 day intervals retarded the spread of SMV by 27% over unsprayed controls.

Alarm pheromones are another potentially important group of chemicals for retarding SMV spread. Extracts of *M. persicae*, presumably containing alarm pheromones, deterred settling and subsequent larviposition by apterous adults of *M. persicae* (Griffiths *et al.*, 1978).

It is unclear whether responses of various aphid species to alarm pheromones might include reduced probing. Nault and Montgomery (1977) discovered that nonfeeding adult *M. persicae* were the most responsive morphs to a synthetic aphid alarm pheromone, *trans*-β-farnesene. Transient alatae might be relatively sensitive to a slow-release alarm pheromone, and this might possibly reduce total probes in a field, thereby retarding SMV spread. However, since aphids of the species *M. persicae* and *L. erysimi* seem more prone to dispersal under the influence of an alarm pheromone (Phelan *et al.*, 1976), one wonders whether this might increase the distance a nonpersistent virus, like SMV, is likely to be carried and what effect that might have on the spread pattern over a large area. Furthermore, alarm pheromones were found to have different effects on settling behavior of two species of aphids (Yang and Zettler, 1975). *A. craccivora* exposed to alarm pheromones settled on cowpea seedlings in about half the time required by aphids not exposed to the pheromone, while *M. persicae* exposed to pheromone wandered more than twice as long before probing compared to the time required by individuals not exposed to pheromones.

Antiviral chemicals (Gupta, 1977) may prove to be useful in reducing spread of nonpersistent viruses in field crops, but their efficacy for this purpose has not been demonstrated.

6.4.4.1 Influencing Landing Frequency of Aphids. Reflective aluminum foils and mulches have significantly reduced the landing rates of alate aphids in various crops, including melons in southern California (Boardman, 1977), in lettuce in New York (Nawrocka *et al.*, 1975), and potatoes in Maine (Shands and Simpson, 1972). In several cases, the decreased number of aphids was correlated with reduced virus spread and increased yields. While not experimentally explored to date as a means of altering SMV spread, reflective mulches probably hold little practical promise because of their relatively high cost in relation to the per acre value of crops like soybeans.

Some aphids respond to the relative amount of ground cover, a behavioral response that may provide a way of influencing alighting frequencies of aphids on crops. Some species of aphids are reported to occur more frequently in sparsely planted fields while other species are apparently unaffected by planting density (A'Brook, 1964, 1968; Iark and Smith, 1976). Results from our experiments in 1977 and 1978 demonstrated that different species of aphids were collected in different abundances on HEL traps placed in soybean fields differing in ground cover densities (Halbert and Irwin, 1980). Some species, e.g., *A. citricola* and *M. asclepiadis*, landed in significantly greater numbers on traps over denser cover. Other species, e.g., *Capitophorus elaeagni* (del Guer.), landed more frequently on traps over sparser cover. Yet other species, e.g., *L. erysimi, R. maidis, S. graminum*, landed with no statistically demonstrable preference for either cover regime. (All other aphid species were collected in insufficient numbers for statistical analysis.) Data from 1978 showed trends similar to those of 1977; thus, behavioral responses to ground cover seem consistent from one year to the next. Moreover, the influence ground cover has on landing responses of aphids seems species specific (A'Brook, 1968; Halbert and Irwin, 1980).

Canopy color can also influence aphid landing rates in soybean fields. In replicated 16 X 16 m blocks of solid darker or lighter green soybean foliage (obtained by roguing unwanted phenotypes from a segregating population based on Clark soybean background: 1 part dark green, 2 parts light green) we found that traps maintained from the time of canopy closure until about a week prior to leaf senescence at the end of the season collected 3.5 X and nearly 3 X as many *R. maidis* and *M. euphorbiae*, respectively, when placed well within the darker green than when similarly placed within the lighter green canopy (Table V; Irwin *et al.*, 1980). This result was observed in both the 1977 and 1978 seasons. All other species collected in sufficient quantity for statistical analysis (*M. asclepiadis*, *A. craccivora*, *T. trifolii*) showed no preference for one canopy color over the other (Table V). The influence of canopy color, like that of ground cover, proved to depend on aphid species.

6.4.4.2   Influencing Landing Pattern of Aphids.   As was shown in the preceding section, aphid landing frequencies can be modified. Another hypothetical means of changing field spread of SMV is to alter landing patterns of alate aphids. Our data (Irwin and Halbert, 1980) demonstrated that most species of aphids move with the wind under field conditions (see previous section). By modifying wind currents, therefore, it was thought possible to alter the horizontal distribution of alighting aphids. We tested this hypothesis in several experiments, all involving the use of sunflowers as a barrier crop.

Two plots of 20 concentric, circular rows of soybean 0.75 m apart, one of which had a circular row of sunflower between soybean rows 4 and 5 from the center, were planted to test the influence of sunflower barriers on the horizontal and vertical distribution of aphids. Three HEL traps were placed within the sunflower barrier and three equidistant between the sunflower barrier and the outside circular row of soybeans. The traps were placed in similar positions in the plot lacking a barrier. Many more aphids were collected outside than inside the barrier in the barrier plot, while no differences were observed in the corresponding traps in the plot lacking a barrier. Moreover, fewer aphids were trapped inside the barrier than were trapped in a similar position in the plot lacking a barrier. The total number of aphids trapped in each plot was similar. *A. citricola* was trapped more frequently on inside traps in both plots, indicating a possible nega-

TABLE V.   Mean Numbers of Alate Aphids Collected for Season on Horizontal Transparent Pan Traps at Canopy Level Over Dark Green and Light Green (Yellowish) Foliage of a Segregating Population on Clark Cultivar Background. Urbana, IL, 1978.

Species	Dark green	Light green
R. maidis	48.7	13.8**[1]
M. euphorbiae	2.3	0.8*
M. asclepiadis	19.1	19.7
T. trifolii	1.3	1.8
A. craccivora	0.7	0.6
All Alatae	167.0	62.0**

[1] Numbers in rows followed by one or two stars are significantly different at the 1% level (**) or 5% level (*) according to the Waller-Duncan test.

tive border attraction, *M. asclepiadis* was collected more frequently on the peripheral traps of both plots, a possible positive border attraction, and *R. maidis*
occurred much less frequently on traps inside the barrier than anywhere else.
Therefore, plot size and/or configuration probably had some influence on the
alighting pattern of some species of aphids while, for *R. maidis*, barriers had a
marked effect on landing distribution. Neither plot configuration (or size) nor
barrier seemed to influence landing patterns of *C. hippophaes* (Halbert *et al.*,
1980b).

Transparent glass or Plexiglass, vertically-oriented, sticky plates were placed
at sunflower canopy level and ca. 30 cm above and below this level in the plots
described above. Aphid catches were greatest on the highest plates, and least on
the lowest plates. There were more aphids at each of the three levels in the barrier plot than on plates placed at similar levels in the nonbarrier plot. These data
suggest that within the first meter or so above the canopy, the aphid biomass
may be more dense above canopy level than at the canopy level. Alternatively
the greater apparent number of impacting aphids on the higher traps may be because air velocity was greater (Taylor, 1960). Whatever the mechanism of the
barrier effect, comparisons of the plots with and without barriers suggest that by
placing a sunflower barrier in a soybean field, the vertical distribution of alate
aphids is elevated at the barrier.

Edges of soybean fields have a higher incidence of aphid alightings per unit of
area than do non-edge areas. Aphid catches were higher overall in HEL traps near
the edges of a 0.83 ha field than in similar traps away from the edges (Irwin and
Goodman, 1980b). This further supports the idea that the pattern and sizes of
fields may influence the distribution of aphid alightings in an area.

Sunflower barriers were planted along all sides of a heavily infected soybean
field (30 × 30 m) to test their effect on aphid abundance. Vertical nets (1.4 m
high × 4.6 m long) were erected on the windward and leeward sides of the barrier on the downwind side of the field. Vertical nets were placed in similar positions along the downwind side of the field but where no barrier existed. The
average number of aphids collected per hour per net position was calculated for
the combined 1977 and 1978 seasons. Significantly more aphids were collected
on the windward net than on the one leeward of the barrier; conversely, significantly more aphids were collected on the leeward net than on the one windward
of the same positions but where no barrier existed. This suggests that sunflowers
probably acted as a barrier by diverting aphids over and around them. A possible
explanation of why more aphids were trapped downwind of no barrier is that
the barriers may have tended to funnel the aphids between them, concentrating
more aphids behind than at the line of the sunflower barrier (Halbert *et al.*,
1980b).

6.4.4.3 Relationship Between Aphid Landing Rates and Virus Spread.   The
correlation between alate aphid landing rate and the spread of SMV has proven
positive and high in several of our experiments. Some evidence for this has al-

ready been presented (Fig. 6). Further evidence for this relationship comes from results of experiments in which landing rates of aphid vectors and SMV spread were examined in plots of soybeans with different color foliage.

Several species of aphids landed more frequently (Table V) and virus spread was higher in the darker than in the lighter green canopy (Irwin *et al.*, 1980). In laboratory studies, foliage color had no demonstrable effect on quantitative transmissibility of SMV, either to or from a given color, by alatae of *M. persicae* (Irwin *et al.*, 1980). Thus the field results indicate that the extent of virus spread was probably a direct consequence of vector abundance.

6.5 CONCLUSIONS

Our present knowledge of the ecology of soybean mosaic virus in relation to its hosts and vectors is sufficiently detailed to allow at least provisional conclusions as to why the virus has not been a major economic problem in the principle soybean growing areas of the United States. Although the varieties grown throughout the region in question are susceptible and the aphid vectors are present, our results show that the inoculum level in commercial certified seed lots, at least in Illinois, is low (Redborg *et al.*, 1980). This is in turn probably explained in part by our data showing that the period during the season when most virus spread occurs is too late to have a significant impact on the level of seed transmission in the next generation and little if any effect on yield. The results of numerous different kinds of experiments summarized in foregoing sections of this chapter indicate that SMV can and will spread in midwestern soybean fields if sufficient primary inoculum is provided.

Certain agricultural practices have the potential for increasing the seedlot incidence of SMV and thus increasing losses to this potentially important virus. For example, soybeans planted in late June or early July as a second crop after winter wheat is harvested flower during or after the time when SMV field spread is more likely (Fig. 6). As a result, the incidence of seedborne SMV could, over the course of several seasons, build up to high levels in the seedlots harvested from such fields. We therefore recommend that seedlots from late-planted soybeans not be used for planting.

In the absence of inoculum sources other than seedborne virus for SMV spread to the soybean crop, it would seem likely that a high level of disease control might be achieved by means that keep the incidence of seed transmission at a low level. One approach to this goal would be the use of cultivars that produce few if any seeds infected with the virus. It may be possible in time to develop improved cultivars that possess a useful degree of resistance to SMV seed transmission. Our results indicate that certain soybean lines and a few cultivars have relatively low incidence of SMV seed transmission, but it is not yet established whether this characteristic will be heritable in a useful way, nor do we know whether this resistance will be effective against a wide range of strains or in situations that might arise where highly efficient spread occurs early in the growing season. The percentage of seed transmission that is epidemiologically significant has not been determined.

The timing of visits by aphid vectors carrying SMV is an important factor in keeping the level of SMV seed transmission low, because the incidence of seed transmission is substantially lower when infection occurs after the beginning of flowering (Table I). Clearly, the tendency of aphid vectors not to be abundant early in the cropping season in the U.S. midwest contributes to reduced seed transmission (Irwin and Goodman, 1978), but in other environments, early season aphid flights may be common. Knowledge of the aphid species important in field spread of the virus is thus an important first step in developing a rational approach to SMV control in soybeans in a particular environment.

We obtained data that led to identification of the important SMV vectors in our area by a combination of infectivity testing of live-trapped aphids given access to SMV-infected soybeans in the field and season-long monitoring of aphids visiting soybeans using HEL traps. Data from the live-trapping studies gave a measure of the relative efficiencies of transmission by the species trapped while data from HEL traps provided a continuous record of the aphids visiting soybeans and their relative abundancies. While somewhat labor-intensive, these methods require only the simplest of laboratory and other facilities and so should be as useful at remote field stations as at major agricultural research centers. The major difficulty with such studies is the need to accurately identify large numbers of aphid specimens.

In the situation studied, data showed that the flights of aphids in general or SMV vectors in particular were generally light in the early part of the growing season and that the seasonal pattern of flights varied markedly from year to year (Fig. 5). The usual absence of significant aphid flights in the first month of the soybean season probably contributes significantly to the relative unimportance of SMV as a disease in midwestern soybeans. A major vector of SMV in this area is *R. maidis*, the corn leaf aphid, which builds up high populations in corn, the other major crop in the U.S. midwest, and does not usually appear in large numbers in soybean fields until large numbers of alatae are produced, presumably in response to population pressure and the approaching vegetative maturity of the corn. This interpretation implies that significant changes in cropping in other areas, or the variety of minor crops found in locations where multiple cropping is the norm would have a major effect on the species composition and phenologies of SMV vectors. For example, if winter small grains rather than corn were the other major crop in the soybean-growing areas of the U.S. midwest, one might expect early season flights of *S. graminum* to present a serious threat to soybean production because of the possibility of large numbers of aphids visiting soybeans and possibly transmitting SMV to young plants with the consequent yield loss and seed transmission impact suggested by the data of Fig. 4.

In certain situations, use of genetic resistance or immunity to SMV infection may be an important weapon in the arsenal against losses due to the virus. However, much work remains to be done to provide and validate the durability of SMV resistant cultivars. Numerous putatively resistant cultivars have been shown to be susceptible to more virulent SMV strains (Cho and Goodman, 1979), and in at least one case (Cho *et al.*, 1977) the deployment of resistance over a wide area was followed by spread of an SMV strain that caused a more severe disease in the "resistant" cultivar than that caused by less virulent strains in susceptible cultivars.

The data summarized in this chapter and the principles we have deduced from these data are being used by our group as the basis for a computer model which we hope will be useful in analyzing SMV epidemics, in predicting the impact of changing agricultural practices on SMV spread, and allowing analysis of economic injury levels in relation to aphid vector populations and the amount of seed-transmitted virus in the seedlot which constitutes the primary inoculum. The model, which is based on general parameters describing plant growth and yield components, should be flexible enough to accept parameters based on soybeans grown in different environments so that it will be useful also over a wide range of environmental and agroclimatic conditions.

6.6 ACKNOWLEDGMENTS

Research in our laboratories and writing of this chapter were supported by a competitive grant from the Illinois Agricultural Experiment Station and by research contracts cm/ta-c-73-19 and TA/C-1294 from the U.S. Agency for International Development (USAID) to the International Soybean Program (INTSOY). The views and interpretations are those of the authors and should not be attributed to USAID or to any individual acting on its behalf.

We are grateful to Dr. Marvin Rode and Illinois Foundation Seeds, Inc. for cooperation in conducting field studies and to the Illinois Crop Improvement Association for a grant used to develop indexing procedures for SMV in seed lots. We thank Elizabeth A. Allison, Glenn R. Bowers, Jr., E. K. Cho, Susan E. Halbert, Gail E. Kampmeier, William G. Ruesink, Gerald A. Schultz, and Teresa L. Shock for their contributions to various parts of these studies. The technical assistance of Christy Barnes, Walton Fehr, LuAnn DiPietro, Robert DiPietro, Lauran Luschen, Maureen Murray, Michael Orfanedes, Mary Phelan, Michal Rubin, Carol Schuh, Tom Shaw, Patricia Sipple, Betty Warfield, and James Westervelt is gratefully acknowledged, as is the assistance of Nora Simkus in bibliographic research and Roxanne Cross and Sandra McGary for typing.

6.7 REFERENCES

Abney, T. S., Sillings, J. O., Richards, T. L., and Broersma, D. B. (1976). Aphids and other insects as vectors of soybean mosaic virus. *J. Econ. Entomol.* 69: 254-256.

A'Brook, J. (1964). The effect of planting date and spacing on the incidence of groundnut rosette disease of the vector, *Aphis craccivora* Koch, at Mokwa, northern Nigeria. *Ann. Appl. Biol.* 54: 199-208.

A'Brook, J. (1968). The effect of plant spacing on the numbers of aphids trapped over the groundnut crop. *Ann. Appl. Biol.* 61: 289-294.

Anderson, C. W. (1957). Seed transmission of three viruses in cowpea. *Phytopathology* 47: 515.

Arunin, A. (1978). Pests of soybean and their control in Thailand. *In* "Pests of Grain Legumes: Ecology and Control" (S. R. Singh, H. F. van Emden, and T. Ajibola Taylor, eds.), pp. 43-46. Academic Press, London. 454 pp.

Benigno, D. R. A., and Boonarkka, S. (1979). Aphid vectors of soybean mosaic virus in the Philippines. *World Soybean Research Conference II Abstracts.* p. 21.

Bennett, C. W. (1969). Seed transmission of plant viruses. *Adv. Virus Res.* 14: 221-261.

Boardman, R. M. (1977). UC scientists 'foil' aphids with aluminum. *Univ. of Calif. Coop. Extension Service News*, Oct. 7, 1977. 3 pp.

Bos, L. (1972). Soybean mosaic virus. *CMI/AAB Descriptions of Plant Viruses*, No. 93. Commonwealth Mycological Institute, Kew, England.

Bowers, G. R., Jr., and Goodman, R. M. (1979a). Soybean mosaic virus: Infection of soybean seed parts and seed transmission. *Phytopathology* 69: 569-572.

Bowers, G. R., Jr., and Goodman, R. M. (1979b). Effect of soybean mosaic virus strains on several soybean seed characters. IX International Congress of Plant Protection, Abstract 602. *Phytopathology* 69: in press.

Bowers, G. R., Jr., and Goodman, R. M. (1980). Unpublished data.

Bowers, G. R., Jr., Paschal, E. H., II, and Goodman, R. M. (1980). Unpublished data.

Broadbent, L. (1969). Disease control through vector control. *In* "Viruses, Vectors, and Vegetation" (K. Maramorosch, ed.), pp. 593-630. Wiley-Interscience Publ., New York. 666 pp.

Caviness, E. E., and Walters, H. J. (1966). Registration of Davis soybeans. *Crop Sci.* 6:503.

Cho, E. K., and Goodman, R. M. (1979). Strains of soybean mosaic virus: Classification based on virulence in resistant soybean cultivars. *Phytopathology* 69: 467-470.

Cho, E. K., and Goodman, R. M. (1980). Unpublished data.

Cho, E. K., Chung, B. J., and Lee, S. H. (1977). Studies on identification and classification of soybean virus diseases in Korea. II. Etiology of a necrotic disease of *Glycine max*. *Plant Dis. Reptr.* 61: 313-317.

Clinton, G. P. (1916). Report of the botanist for 1915, soybeans. *Ann. Rept. Conn. Agric. Expt. Sta.*, 1915. p. 446.

Conover, R. A. (1948). Studies of two viruses causing mosaic diseases of soybean. *Phytopathology* 38: 724-735.

Cooper, R. L. (1966). A major gene for resistance to seed coat mottling in soybean. *Crop Sci.* 6: 290-292.

DeVasconcelos, F. A. T. (1966). Contribuião para o estudo do virus do mosaico da soja. *Anais do Instituto Superior de Agronomia* 26: 181-221.

Dimmock, F. (1936). Seed mottling in soybeans. *Sci. Agr.* 17: 42-49.

Dunleavy, J. M., Quiniones, S. S., and Krass, C. J. (1970). Poor seed quality and rugosity of leaves of virus-infected Hood soybeans. *Phytopathology* 60: 883-886.

Eastop, V. F. (1955). Selection of aphid species by different kinds of insect traps. *Nature (London)* 176: 936.

Fehr, W. R., Caviness, C. E., Burmood, D. T., and Pennington, J. S. (1971). Stage of development descriptions for soybeans, *Glycine max* (L.) Merr. *Crop Sci.* 11: 929-931.

Frosheiser, F. I. (1974). Alfalfa mosaic virus transmission to seed through alfalfa gametes and longevity in alfalfa seed. *Phytopathology* 64: 102-105.

Galvez, G. E. (1963). Host-range, purification, and electon microscopy of soybean mosaic virus. *Phytopathology* 53: 388-393.

Gardner, N. W., and Kendrick, J. B. (1921). Soybean mosaic. *J. Agr. Res.* 22: 111-114.

Ghanekar, A. M., and Schwenk, F. W. (1974). Seed transmission and distribution of tobacco streak virus in six cultivars of soybeans. *Phytopathology* 62: 112-114.

Gibbs, A. J., and Harrison, B. D. (1978). "Plant Virology: The Principles." John Wiley and Sons, New York.

Gibson, R. W., and Plumb, R. T. (1977). Breeding plants for resistance to aphid infestation. *In* "Aphids as Virus Vectors" (K. F. Harris and K. Maramorosch, eds.), pp. 473-500. Academic Press, New York. 559 pp.

Goodman, R. M., and Oard, J. A. (1980). Seed transmission and yield losses in tropical soybeans infected by soybean mosaic virus. (In preparation).

Goodman, R. M., Bowers, G. R., Jr., and Paschal, E. H., II. (1979). Identification of soybean germplasm lines and cultivars with low incidence of soybean mosaic virus transmission through seed. *Crop Sci.* 19: 264-267.

Gray, L. E. (1980). Personal communication.

Griffiths, D. C., Greenway, A. R., and Lloyd, S. L. (1978). The influence of repellent materials and aphid extracts on settling behavior and larviposition of *Myzus persicae* (Sulzer) (Hemiptera: Aphididae). *Bull. Entomol. Res.* 68: 613-619.

Gupta, B. M. (1977). Inhibition of plant virus infections by antiviral agents. *In* "Aphids as Virus Vectors" (K. F. Harris and K. Maramorosch, eds.), pp. 455-471. Academic Press, New York. 559 pp.

Halbert, S. E., and Irwin, M. E. (1980). Effect of soybean canopy closure on landing rates of aphids with implications for restricting spread of soybean mosaic virus. *Ann. Appl. Biol.* Submitted for publication.

Halbert, S. E., Irwin, M. E., and Goodman, R. M. (1980a). Alate aphid species and their relative importance as field vectors of soybean mosaic virus. *Ann. Appl. Biol.* (in press).

Halbert, S. E., Irwin, M. E., and Goodman, R. M. (1980b). Unpublished data.

Han, Y. H., and Murayama, D. (1970). Studies on soybean mosaic virus. I. Separation of virus strains by differential host. *J. Faculty Agr. Hokkaido Univ. Sapporo* 56: 303-310.

Heinze, K., and Köhler, E. (1941). Die mosaikkrankheit der Sojabohne und ihre Ubertragung durch Insekten. *Phytopathologische Zeitschrift* 13: 207-242.

Hepperly, P. R., and Sinclair, J. B. (1978). Quality losses in Phomopsis-infected soybean seeds. *Phytopathology* 68: 1684-1687.

Hepperly, P. R., Bowers, G. R., Jr., Sinclair, J. B., and Goodman, R. M. (1979). Predisposition to seed infection by *Phomopsis sojae* in soybean plants infected by soybean mosaic virus. *Phytopathology* 69: 846-848.

Hill, J. H., and Benner, H. I. (1979a). Properties of soybean mosaic virus ribonucleic acid. *Phytopathology* (in press).

Hill, J. H., and Benner, H. I. (1979b). Properties of soybean mosaic virus and its isolated protein. *Phytopathologische Zeitschrift* (in press).

Hymowitz, T., Newell, C. W., and Carmer, S. G. (1977). "Pedigrees of Soybean Cultivars Released in the United States and Canada." College of Agriculture, University of Illinois, Urbana. (INTSOY Series 13).

Iark, F., and Smith, J. C. (1976). Efeito dos espaamentos dos tomateiros ao ataque do *Macrosiphum euphorbiae* (Thomas, 1878). (Homoptera, Aphididae). *Anais da Sociedade Entomologica do Brasil* 5: 152-156.

Iizuka, N. (1973). Seed transmission of viruses in soybeans. *Bull. Tohoku Natl. Agr. Expt. Sta.* 46: 131-141.

Irwin, M. E. (1980a). Sampling aphids in soybean fields. *In* "Sampling Methods in Soybean Entomology" (M. Kogan and D. C. Herzog, eds.), pp. 239-259. Springer-Verlag, New York. 587 pp.

Irwin, M. E. (1980b). Unpublished observations.

Irwin, M. E., and Goodman, R. M. (1978). Factors affecting the field spread of soybean mosaic virus in north central United States. *Proc. N. Cent. Br. Entomol. Soc. Amer.* 33:

39-40.

Irwin, M. E., and Goodman, R. M. (1980a). Effect of soybean mosaic virus on leaf dry mass, yield and seed quality in soybeans inoculated at different growth stages. (In preparation).

Irwin, M. E., and Goodman, R. M. (1980b). Unpublished data.

Irwin, M. E., and Halbert, S. E. (1980). Unpublished data.

Irwin, M. E., Schultz, G. A., and Goodman, R. M. (1980). Unpublished data.

Kendrick, J. B., and Gardner, M. W. (1924). Soybean mosaic: Seed transmission and effect on yeild. *J. Agr. Res.* 27: 91-98.

Kennedy, B. W., and Cooper, R. L. (1967). Association of virus infection with mottling of soybean seed coats. *Phytopathology* 57: 35-37.

Khamala, C. P. M. (1978). Pests of grain legumes and their control in Kenya. *In* "Pests of Grain Legumes: Ecology and Control" (S. R. Singh, H. F. van Emden, and T. Ajibola Taylor, eds.), pp. 127-134. Academic Press, London. 454 pp.

Kiihl, R. A. S. (1976). Inheritance studies of two characteristics in soybeans (*Glycine max* (L.) Merrill): I. Resistance to soybean mosaic virus. II. Late flowering under short-day conditions. Ph.D. Diss. Mississippi State University, Mississippi State, Mississippi. 56 pp.

Kiihl, R. A. S., and Hartwig, E. E. (1979). Inheritance of reaction to soybean mosaic virus in soybeans. *Crop Sci.* 19: 372-375.

Koshimizu, V., and Iizuka, N. (1963). Studies on soybean virus diseases in Japan. *Bull. Tohoku Agric. Exp. Sta.* 27: 1-103.

Kring, J. B. (1954). New ways to repel aphids. *Frontiers of Science, Connecticut Agr. Exp. Sta.* 17: 607.

Kwon, S. H., and Oh, J. H. (1979). Inheritance of resistance to necrotic strain of SMV in soybean. *Soybean Genetics Newsl.* 6: 66-68.

Lee, Y., and Ross, J. P. (1972). Top necrosis and cellular changes in soybean doubly infected by soybean mosaic and bean pod mottle viruses. *Phytopathology* 62: 839-845.

Litsinger, J. A., Quirino, C. B., Lumahan, M. D., and Bandong, J. P. (1978). The grain legume pest complex of rice-based cropping systems at three locations in the Philippines. *In* "Pests of Grain Legumes: Ecology and Control" (S. R. Singh, H. F. van Emden, and T. Ajibola Taylor, eds.), pp. 309-320. Academic Press, London. 454 pp.

Lu, J. (1970). Essai in vivo de répression de la mosaïque du soja par l'huile. *Phytoprotection* 51: 149.

McDonald, J. G., and Hamilton, R. I. (1972). Distribution of southern bean mosaic virus in the seed of *Phaseolus vulgaris. Phytopathology* 62: 387-389.

McKinney, H. H., and Greely, L. W. (1965). "Biological Characteristics of Barley Stripe Mosaic Virus Strains and Their Evolution." *U.S. Dept. Agr. Tech. Bull.*, No. 1324.

McLean, D. M. (1957). Effect of insecticide treatments of beets on transmission of yellows virus by *Myzus persicae. Phytopathology* 47: 557-559.

Milbrath, G. M., and McLaughlin, M. R. (1980). Personal communication.

Muraveva, M. R. (1973). Soybean mosaic in the Khabarovsk region. *Trudy Dal'nerost N. II 5Kh* 13: 156-158.

Nariana, T. K., and Pingaley, K. V. (1960). A mosaic disease of soybeans (*Glycine max* (L.) Merr.). *Indian Phytopathology* 13: 130-136.

Nault, L. R., and Montgomery, M. E. (1977). Aphid pheromones. *In* "Aphids as Virus Vectors" (K. F. Harris and K. Maramorosch, eds.), pp. 527-545. Academic Press, New York. 559 pp.

Nawrocka, B. Z., Eckenrode, C. J., Uyemoto, J. K., and Young, D. H. (1975). Reflective mulches and foliar sprays for suppression of aphid-borne viruses in lettuce. *J. Econ. Entomol.* 68: 694-698.

Ovchinnikova, A. M. (1973). Diseases of soybean in Far East. *Trudy Dal'nerost N. II 15Kh* 13: 135-142.

Owen, F. V. (1927). Hereditary and environmental factors that produce mottling in soybeans. *J. Agr. Res.* 34: 559-587.

Paschal, E. H., II, and Goodman, R. M. (1978). A new source of resistance to soybean mosaic virus. *Soybean Genetics Newsl.* 5: 28-30.

Phelan, P. L., Montgomery, M. E., and Nault, L. R. (1976). Orientation and locomotion of apterous aphids dislodged from their hosts by alarm pheromones. *Ann. Entomol. Soc. Amer.* 69: 1153-1156.

Porto, M. D., and Hagedorn, D. J. (1975). Seed transmission of a Brazilian isolate of soybean mosaic virus. *Phytopathology* 65: 713-716.

Redborg, A., Milbrath, G. M., and Goodman, R. M. (1980). Unpublished data.

Roach, S. H., and Agee, H. R. (1972). Trap colors: Preference of alate aphids. *Environ. Entomol.* 1: 797-798.

Ross, J. P. (1967). Purification of soybean mosaic virus for antiserum production. *Phytopathology* 57: 465-467.

Ross, J. P. (1968). Effect of single and double infections of soybean mosaic virus and bean pod mottle virus on soybean yield and seed characters. *Plant Dis. Reptr.* 52: 344-348.

Ross, J. P. (1969). Pathogenic variation among isolates of soybean mosaic virus. *Phytopathology* 59: 829-832.

Ross, J. P. (1970). Effect of temperature on mottling of soybean seed caused by soybean mosaic virus. *Phytopathology* 60: 1798-1800.

Ross, J. P. (1975). A newly recognized strain of soybean mosaic virus *Plant Dis. Reptr.* 59: 806-808.

Ross, J. P. (1977). Effect of aphid-transmitted soybean mosaic virus on yields of closely related resistant and susceptible soybean lines. *Crop Sci.* 17: 869-872.

Schultz, G. A., Irwin, M. E., and Goodman, R. M. (1980). Unpublished data.

Shands, W. A., and Simpson, G. W. (1972). Effects of aluminum foil mulches upon abundance of aphids on, and yield of potatoes in northeastern Maine. *J. Econ. Entomol.* 65: 507-510.

Shepherd, R. J. (1972). Transmission of viruses through seed and pollen. *In* "Principles and Techniques in Plant Virology" (C. I. Kado and H. O. Agrawal, eds.), pp. 267-292. Van Nostrand Reinhold, New York.

Singh, B. B. (1980). Personal communication.

Smith, F. F., and Webb, R. E. (1969). Repelling aphids by reflective surfaces, a new approach to the control of insect-transmitted viruses. *In* "Viruses, Vectors, and Vegetation" (K. Maramorosch, ed.), pp. 631-639. Interscience Publ., New York. 666 pp.

Smith, F. F., Johnson, G. V., Kahn, R. P., and Bing, A. (1964). Repellency of reflective aluminum to transient aphid virus vectors. *Phytopathology* 54: 748.

Taylor, L. R. (1960). The distribution of insects at low levels in the air. *J. Anim. Ecol.* 29: 45-63.

Taylor, L. R., and Palmer, J. M. P. (1972). Aerial sampling. *In* "Aphid Technology" (H. F. van Emden, ed.), pp. 189-234. Academic Press, London. 454 pp.

Taylor, R. H., Grogan, R. G., and Kimble, K. A. (1961). Transmission of tobacco mosaic virus in tomato seed. *Phytopathology* 51: 837-842.

Toba, H. H., Kishaba, A. N., Bohn, G. W., and Hield, H. (1977). Protecting muskmelons against aphid-borne viruses. *Phytopathology* 67: 1418-1423.

Usman, K. M., Ranganattan, K., Kandaswamy, T. K., Sorjini Damodaran, A. P., and Ayyavoo, R. (1973). Studies on a mosaic disease of soybean. *Madras Agr. J.* 60: 472-474.

Vanderveken, J. J. (1977). Oils and other inhibitors of nonpersistent virus transmission. *In* "Aphids as Virus Vectors" (K. F. Harris and K. Maramorosch, eds.), pp. 435-454. Academic Press, New York. 559 pp.

Wallen, V. R., and Seaman, W. L. (1963). Seed infection of soybean by *Diaporthe phaseolorum* and its influence on host development. *Can. J. Bot.* 41: 13-21.

Wang, C. S. (1980). Personal communication.

Weiss, M. G. (1953). Registration of soybean varieties. IV. Ogden. *Agron. J.* 45: 570-571.

Whigham, D. K. (1980). Personal communication.

Wilcox, J. R., and Laviolette, F. A. (1968). Seedcoat mottling response of soybean geno-types to infection with soybean mosaic virus. *Phytopathology* 58: 1446-1447.

Yang, A. F., and Hamilton, R. I. (1974). The mechanism of seed transmission of tobacco ringspot virus in soybean. *Virology* 62: 26-37.

Yang, S. L., and Zettler, F. W. (1975). Effects of alarm pheromones on aphid probing be-havior and virus transmission efficiency. *Plant Dis. Reptr.* 59: 902-905.

Chapter 7

EARLY EVENTS IN PLANT VIRUS INFECTION

G. A. de Zoeten

Department of Plant Pathology
University of Wisconsin
Madison, Wisconsin

7.1 INTRODUCTION

Since virus infections that become localized tend to be self-eliminating, those that become systemic are potentially the most devastating and permanent of all plant virus diseases. Our knowledge of the infection process, virus-host specificity, and the establishment of systemic infection in plants is extremely limited and mostly confined to a knowledge of some of the biochemical processes in the replication cycle of the single-stranded (ss) RNA-containing-plant viruses. In this

Copyright © 1981 by Academic Press, Inc.
All rights of reproduction in any form reserved.
ISBN 0-12-470240-6

treatise, I will limit the discussion of virus infection to those caused by ss-RNA viruses. Because of space constraints, the plant Rhabdo viruses and the multiplication of double-stranded (ds) RNA-containing plant viruses will not be discussed. Furthermore, the biochemical aspects of plant virus RNA translation and transcription will be discussed in depth only where needed to develop the arguments presented. The readers are, therefore, referred to reviews that have appeared in the last few years and deal in depth with some of the aspects of plant virus replication that are not covered here (Hamilton, 1974; Kaesburg, 1977; Van Vloten-Doting et al., 1977; Zaitlin, 1977; Zaitlin et al., 1976).

A number of different steps can be distinguished in the replication of single-stranded (ss) RNA viruses, e.g., release of the viral genome, an as yet unknown but essential host response [inhibited by actinomycin D (AMD)], the recognition of the genome by its host, translation and transcription of the genome, assembly of the virus particle, and possibly maturation (La Fléche and Bové, 1971; La Fléche et al., 1972). Probably the most important events in virus infection are the release of the viral genome from its protein shell prior to replication, the AMD inhibitable step in infection and virus-host recognition phenomena. The initiation of infection (virus uncoating prior to multiplication and the AMD inhibitable step), the virus-host recognition and virus spread within the plant are the least understood processes involved in plant virus infection. The scarcity of information concerning the infection process is probably caused by the additional complexity of virus-plant host systems as compared to other virus-host systems. The plant cell wall and the chloroplasts complicate host-virus interaction and are features of plant cells not encountered in cells of organisms outside the plant kingdom.

Understanding the mechanism of invasion subsequent to the infection of the first cell, that is, the transport of the infectious entity and its nature during the progression of infection, is basic to the understanding of the development of a disease. In order to provide the necessary background to the concept that will be developed concerning the nature of the infectious entity during the different stages of infection and the virus-host recognition phenomena, the following review of this process is presented.

The phenomenon of systemic infection can be divided into three steps: (1) virus genome release and cell ingress (infection); (2) the host-virus recognition and the virus replication in the first cell entered; and (3) the transport of the infectious entity from cell to cell and over long distances within the plant.

7.2 INFECTION

The beginning of infection involves the release of the viral genome from its protein (uncoating) and its entrance into the cell to initiate multiplication. Plant viruses can be introduced into many different plant tissues by abrasion of the outer cell layers of plant organs (mechanical inoculation), by insects (with suck-ing as well as with biting mouthparts), by nematodes, by fungi, by grafting, by dodder (*Cuscuta* sp.), and by pollen (Das *et al.*, 1961). All of these except pos-sibly pollen transmission require wounding of the host plant to establish the intimate contact between host and virus necessary for infection. There is no un-equivocal proof that plant viruses enter through wounds made during natural or artificial mechanical inoculation or that these wounds provide specific virus at-tachment sites. Although virus has been found in abrasion wounds made during mechanical inoculation (Kontaxis and Schlegel, 1962), susceptibility involves other factors according to Herridge and Schlegel (1962).

7.2.1   Attachment and Genome Release

Specific attachment sites, as in the case of bacterial viruses (Luria and Darnell, 1967) have not been identified for plant viruses. Since the same forces involving plant virus assembly are involved in disassembly processes, infection sites (attach-ment sites?) should increase electrostatic repulsions between protein subunits and/or weaken their hydrophobic bonds. Caspar (1963) suggested that lipid-containing structures could provide the environment affecting the stabilizing interactions within a virus particle in a way similar to nonpolar solvents and de-tergents, resulting in uncoating and genome release.

The two lipid-containing barriers encountered in mechanical inoculations of plant viruses are the cuticle and the plasmalemma of the cell. All other means of inoculation have the potential of bringing the virus directly into contact with the plasmalemma.

We will first discuss evidence that the cell wall with its cuticle provides for virus attachment and uncoating. Shaw, in 1967 and 1969, has shown that an un-coating process with regard to tobacco mosaic virus (TMV) occurs on the leaf epidermis and can be looked upon as a two-stage process. The first step is a bio-physical, nonenzyme mediated step in which 25% of the virus becomes uncoated within 10 minutes after inoculation. The second step proceeds over a longer period and is temperature-sensitive. Kurtz-Fritsch and Hirth (1972), using high ionic strength buffers instead of the low ionic strength buffer used by Shaw (1967, 1969), showed that for turnip yellow mosaic virus (TYMV) and bromegrass mo-saic virus (BMV), two polyhedral viruses, uncoating proceeds to 60% of the irre-versibly leaf surface-bound virus and no second step is observed after 10 minutes.

Ultrastructural studies (Merkens *et al.*, 1972) revealed attachment of potato virus X (PVX) to the epidermal cuticle of inoculated leaves. Furthermore, it was shown (Gerola *et al.*, 1969), by means of replicate techniques as well as by scan-ning electron microscopy (Duafala and Nemania, 1974), that TMV attaches end-on to inoculated leaf surfaces. Gaard and de Zoeten (1979) showed that end-on wall attachment of rod-shaped viruses is not peculiar to the epidermal layers, but occurs also on the cell walls delimiting the intercellular spaces. They showed both by direct measurement and by autoradiography that virus [tobacco rattle virus (TRV)] dissociation of cell wall-attached particles occurred. These observations are in agreement with those of Kassanis and Kenten (1978) on TMV attachment and uncoating on leaf surfaces and in intercellular spaces and with those of Skreczkowski (1977) on the effect of adsorption on *in vitro* inactivation of PVX in extreme resistant and susceptible potato leaves.

Direct observation of the passage of viral genomic material through the cell wall or the plasmalemma is currently beyond our experimental capabilities. An unequivocal answer to the question whether or not the surface dissociated virus particles provide the infecting entity can, therefore, not be given at this time.

There is currently, however, no experimental evidence available to suggest that plant virus uncoating is initiated at an intracellular site as postulated by others (Zaitlin, 1977). On the contrary, evidence accumulated to date suggest strongly that the virus dissociative phase occurring at the plant cell surface represents uncoating and that it should be looked upon as an early step of the multifaceted entrance process of the viral genomic material into the plant cell. It is of interest here to note that cell surface mediated destabilization of picorna viruses is essentially similar to the cell surface-virus interactions in plants and is considered to be an essential part of the genome release process (uncoating) of these viruses during the infection of animal cells (de Sena and Mandel, 1977).

7.2.2 Ingress of Viral Genomic Material

The cell wall itself offers a rather formidable barrier to the passage of viral genomic material and should be considered here. The cell wall of expanded leaf cells is not a rigid structure of cellulose micelles, but a dynamic and pliable portion of a living cell in which ectodesmata appear and disappear without observable change in the cuticle overlaying these structures. Ectodesmata have been implicated as possible infection sites for plant viruses (Brants, 1964; Thomas and Fulton, 1968). They interrupt the continuity of the cell wall and are formed in outer epidermal walls and in the walls of cells bordering intercellular spaces merely by a parting and expanding of the cellulose fibrils, resulting in a local elevation of the surface. They do not contain membranes and, therefore, do not offer infecting inoculated plant viruses direct points of contact with the plasmalemma. If virus uncoating, upon mechanical inoculation, is accomplished on the cuticle, ectodesmata may only facilitate transport of viral RNA. Merkens *et al.* (1972) have not been able to locate and distinguish morphologically distinct PVX particles either in ectodesmata or in cell walls of inoculated plants, and concluded that PVX must uncoat on the cuticle of inoculated leaf surfaces.

The plasmalemma is the next obstacle to be overcome by the viral genomic material on its way into the cell.

During the infection by polio virus (a simple RNA-containing animal virus), an eclipse period is observed. On the basis of virus inactivation studies, the physical location of the infectious entity at the time of eclipse must be on or close to the cell surface (Chan and Black, 1970; Holland, 1962). Mandel (1967) in his studies of this virus came to the conclusion that uncoating must be initiated at the cell surface, but taken to completion intracellularly by means of pinocytotic movement. Mandel, however, left the possibility open that mechanisms of infection other than viropexis (Cocking, 1970; Dales, 1965) (pinocytotic movement of virus) were operable as well. Recently, de Sena and Mandel (1977) presented evidence that destabilization of the polio virus particle as a prelude to genome release occurred at the cell surface. Direct penetration of polio virus has been observed by others (Dunnebacke *et al.*, 1969), but has not been confirmed.

The situation in plant virus infection is even more unsettled. If uncoating takes place on the cuticle, the viral RNA has to get into the cell proper by direct (active) transport through the cell wall followed by transport through the plasmalemma, possibly by viropexis. Viropexis has been inferred to exist by Cocking and his co-workers (1970) for TMV moving into tomato fruit protoplasts, as well as for TMV, PVX, and cucumber mosaic virus (CMV) moving into tobacco protoplasts (Otsuki *et al.*, 1972). Motoyoshi *et al.* (1973) reported successful tobacco protoplast infection, presumably by viropexis of cowpea chlorotic mottle virus (CCMV) and its RNA. Whether the presumed pinocytotic vesicles are, indeed, a result of pinocytosis and whether the visualized virus is, indeed, being uncoated at this site is, in light of later work by Burgess *et al.* (1973), in question.

The subject of pinocytosis by plant cells as well as that by protoplasts seems to be controversial (Burgess *et al.*, 1973). Where Bradfute and co-workers in 1964 failed to find evidence for pinocytosis in plant cells, others have taken invaginations of the plasmalemma, which sometimes contained marker substances observed in fixed material, as proof of pinocytosis (Mahlberg 1972a, 1972b). Burgess *et al.* (1973) concluded that a requisite for good virus infection of protoplasts was stress applied to the cell membrane, whether mechanical, biochemical, or chemical. Damaged portions of the plasma membrane seemed to be favored binding sites for negatively charged particles. They stated furthermore that regions of vesiculate cytoplasm with macromolecular (virus) content may only represent a transient repair mechanism at work.

Although macromolecular transport across membranes is known (Kent 1966), the intricacies of virus genome transport across the plasmalemma remain to be elucidated.

7.3   HOST VIRUS RECOGNITION AND VIRUS REPLICATION

We have not considered the role which virus-host specificity plays in virus infection. Host specificity is extremely important in infection and if its molecular basis was understood, host specificity could be of great importance in the control of plant diseases. In the case of plant viruses, we can only guess at the reason for success or failure of infection until we can evaluate the contribution of the viral and host cell genomes to the process of virus replication.

Specificity may be centered in two broadly defined areas of interaction between plant hosts and viruses: (l) adsorption of the virus, uncoating of the RNA, and the penetration of the cell and (2) translation and transcription of viral RNA and the assembly and maturation of the virus. Atabekov (1975) recently described the present status of research in the area of host specificity of plant viruses. After reviewing past work and current ideas in regard to the mechanisms of virus-host specificity, he came to the following conclusion: "These observations are indicative of the existence of one (or many) intracellular barrier(s) operating at *later* stages of virus replication. Such barriers may constitute mechanisms controlling transportation or replication of the viral RNA, responsible for the specific defense of the plant cell against foreign genetic material."

Gaard and de Zoeten (1979) observed that TRV attached to cell walls of hosts as well as nonhosts, and that dissociation of virus proceeded in a similar fashion in tobacco (host) and corn (nonhost). Thus, it seems that the uncoating is not the source of specificity in plant virus interactions. Therefore, Atabekov's suggestion that specificity resides at a site involved in either transcription or translation of the viral RNA is appealing.

7.3.1 Transcription

Once the viral genome enters the cell, replication starts, mediated by a replicase enzyme possibly coded for, in part, by the viral genome and, in part, by the host genome (Kumarasamy and Symons, 1979).

This RNA-dependent RNA polymerase proceeds with the formation of (−) RNA complementary to the genomic (+) RNA forming a so-called replicative form (RF). The negative strand serves as template for the production of progeny (+) RNA by means of the replicase enzyme forming a partially double-stranded, partially single-stranded structure called the replicative intermediate (RI).

Controversies regarding the nature of the plant virus RNA polymerase are at this time unresolved. Since no pure replicase is available, no virus gene can be identified as the origin of a single polypeptide chain present in highly purified preparations (Zabel, 1978). Another difficulty is the presence of low levels of RNA polymerase in some healthy plant species used for experimentation (Leroy *et al.*, 1977; Stussi-Garaud *et al.*, 1977; White and Murakishi 1977; Zabel, 1978). Romaine and Zaitlin (1978) showed, for instance, that the RNA-dependent RNA polymerase from noninfected and TMV-infected tobacco plants was essentially identical. The appearance of soluble RNA polymerase activity following infection would then be due to increased activity of a pre-existing host enzyme rather than to *de novo* synthesis of a virus-specific enzyme; a point of view promoted by Fraenkel-Conrat (1976). However, Kumarasamy and Symons (1979) did not isolate any replicase activity when healthy cucumber plants were subjected to their replicase isolation procedures. This difference noted between plant species in regard to native replicase may mirror adaptations of virus replication strategies to host species or alternatively may only mean that the techniques employed were not sensitive enough to detect extremely low amounts of native replicase activity.

A third matter of concern regarding viral replicases isolated from plant virus systems (Bové, 1967; Bradley and Zaitlin, 1971; Leroy *et al.*, 1977; May *et al.*, 1970; Semal and Hamilton, 1968; Semal, 1970; Stussi-Garaud, 1977; Weening and Bol 1975; Zabel 1978) or from picorna virus systems (Clegg *et al.*, 1976; Lundquist *et al.*, 1974; Röder and Koschel, 1974, 1975) is the lack of template specificity of these enzymes. The pea enation mosaic virus (PEMV)-pea system as developed in our laboratory is unique in regard to its template specificity of the viral replicase.

Like the membrane-bound animal virus RNA replicases, the PEMV polymerase induced in healthy nuclei by PEMV-RNA is able to produce single-stranded RNAs (60% of total produced) as well as double-stranded viral specific RNA intermediates (40% of total produced) (Powell and de Zoeten, 1977). This implies the production of (−) strand RNA. In contrast, the membrane-bound plant virus polymerases produce mostly double-stranded RNA from existing (−) stranded RNA (Bové, 1967). Small amounts of single-stranded plant viral (+) RNA have been reported from crude polymerase preparations of brome mosaic virus (BMV) (Mouchés *et al.*, 1974, 1975; Semal and Kummert, 1971), broad bean mottle virus (BBMV) (Jacquemin 1972), cowpea mosaic virus (CPMV) (Zabel *et al.*, 1974), and cowpea chlorotic mottle virus (CCMV) (White and Dawson, 1978) infected tissues.

The template specificity for PEMV-RNA of the nuclear replicase triggered in healthy isolated pea nuclei is unique among the plant and picorna viruses. Although Zabel *et al.* (1979) recently showed that strong binding specificity existed for CPMV-RNA and CPMV-replicase, no template specificity could be found for this replicase. The enzyme utilized quite efficiently a variety of nonrelated viral templates.

It has been known for some time now that the replication of some plant viruses (Smith and Schlegel, 1965; Semal, 1967; Lockhart and Semancik, 1969; Dawson and Schlegel, 1976; Dawson, 1978), but not all plant viruses under all conditions (Otsuki and Takebe, 1973) can be inhibited by actinomycin D if administered early in infection. Recent studies by Rottier *et al.* (1979) yielded detailed biochemical evidence for the inhibition of CPMV replication in cowpea protoplasts if the antibiotic was administered at the time of inoculation. It was concluded that functional host DNA is a crucial factor in successful infection of a number of plant viruses. The elucidation of this host DNA-dependent step seems important in light of the possibility that for some virus-host combinations, specificity may be associated with this early step in replication.

Alternatively, the interaction of virus (+) RNA with the replicase enzyme may trigger recognition phenomena. It is of interest in this regard that a number of plant virus RNAs exhibit the t-RNA-like properties of selective amino acid acylation of their 3' termini (Bastin *et al.*, 1976; Busto *et al.*, 1976; Chen and Hall, 1973; Kohl and Hall, 1974). The fact that all members of the TYMO virus group contain RNA that will accept valine (Pinch *et al.*, 1972) argues for a biological function for aminoacylated viral RNAs. Furthermore, it has been shown (Kohl and Hall, 1977) that chemical alteration of both the 5' and 3' ends of BMV-RNA decreases RNA infectivity drastically. This underscores the importance of the integrity of the 3' and 5' ends of plant viral RNAs for infection.

Hall and Wepprich (1976) have postulated that the t-RNA behavior of viral RNA in aminoacylation may set up a recognition signal for the eukaryotic elongation factor 1 (EF 1) to commence transcription. The basis for this hypothesis is found in the fact that eukaryotic EF 1 will bind to aminoacylated, but not to uncharged RNA from TMV and TYMV (Litvak *et al.*, 1973). Furthermore, the prokaryotic counterparts of EF 1 are identical to two of the four subunits of QB replicase (Blumenthal *et al.*, 1972; Landers *et al.*, 1974). It has been postulated that virus-host specificity may be determined in the transcriptional phase of multiplication rather than at any of the preceding steps. If host elongation factors cannot interact with aminoacylated viral RNA, infection may not occur and conversely if aminoacylation does not occur *in vivo* in certain hosts, infection may not occur either.

This hypothesis for virus-host specificity may hold true for some viruses, but it cannot explain host specificity of other viruses with different replication strategies. Lack of aminoacylation of CPMV-RNA, and its binding specificity for its own replicase combined with the evidence for the need of the polyadenylated 3' end of CPMV-RNA for replicase binding, is an example of such a different strategy.

The failure of PEMV-RNA to become aminoacylated *in vitro* (German *et al.*, 1978), and its lack of polyadenylated sequences, combined with its template specificity is yet another example of a different replication strategy with its own possibilities for conferring host specificity. It should be clear, therefore, that virus-host specificity can easily reside in one of the steps in replication, but that generalizations as to its mechanism for single-stranded plant RNA viruses cannot be made.

7.3.2 Translation

For the purpose of this treatise, the translational events leading to the production of the replicase enzyme and a host membrane system as part of a replication complex seems of the greatest importance. However, not much is known about these early translational events.

The subject of virus RNA translation cannot be done justice in this presentation, but warrants separate review. The reader is, therefore, referred to earlier reviews on the subject (Hamilton, 1974; Kaesburg, 1977; Van Vloten-Doting *et al.*, 1977; Zaitlin, 1977; Zaitlin *et al.*, 1976) and cautioned that the area is a fast-moving one, with new data becoming available almost monthly. It is interesting to note that although the translation strategies among virus groups may seem similar, they are not identical (Zaitlin, 1977) in detail. Furthermore, there seem to be major differences in translational events between some virus groups. Post-translational processing of large polypeptides produced by means of near complete translation of whole CPMV-RNA has been reported (Pelham, 1979) for this viral RNA translated in a reticulocyte cell-free system. This would indicate a different translation stragegy than that of TMV (Zaitlin *et al.*, 1976) and some of the multicomponent viruses (Davies, 1976).

Suffice it to say that generally coat protein is produced in larger amounts than any of the presently known other virus proteins. The production of this protein is followed normally by a self-assembly process which can be mimicked *in vitro*. It is a process goverened by biophysical parameters and leads to the production of virus particles.

7.4   TRANSPORT OF THE INFECTIOUS ENTITY

7.4.1   Intracellular Transport

Virus and/or virus products, (+) and (−) ssRNA, dsRNA, and coat protein can move in the cell with the cytoplasmic streaming and enter organelles. There is a distinct possibility that virus can enter nuclei without the need for either virus multiplication or assembly in the nucleus to explain the presence of virus particles there. Not only did Esau (1967, 1968) show how TMV could be entrapped in nuclei during cell division, but de Zoeten and Gaard (1969) showed that no great obstacles have to be overcome for spherical plant viruses to move through nuclear pores into nuclei under the influence of diffusion pressure buildup in the cytoplasm.

Virus protein synthesis can be sustained by chloroplasts *in vitro* (Sela and Kaesberg, 1969) and other evidence has also associated TMV multiplication with chloroplasts (Ralph and Wojcik, 1969, Ralph *et al.*, 1971a; Ralph *et al.*, 1971b; Zaitlin and Boardman, 1958). However, cytological studies done with TMV failed to implicate the chloroplasts in *in vivo* synthesis of TMV protein. In the case of turnip yellow mosaic virus (TYMV), on the other hand, the chloroplasts have been implicated in the most fundamental step of virus replication, that of the production of dsRNA (RF) (La Fléche and Bové, 1971; La Fléche *et al.*, 1972). We have to conclude, therefore, that in that system, the viral message can be transferred into the chloroplasts.

Mitochondria have not been implicated as the sites of synthesis of virus, thus offering no evidence at this time that the viral message in one form or another can be transported across the membranes of this organelle. However, there have been reports of peculiar virus-mitochondria associations in the case of tobacco rattle virus (Harrison and Roberts, 1968) which are reminiscent of the associations of barley stripe mosaic virus with chloroplasts. Whether or not these associations are fortuitous remains to be established.

7.4.2 Cell to Cell Transport

An important question with regard to cell to cell transport of the infectious entity is whether it travels in the form of nucleic acid or as an assembled virus particle. Both intact virus and naked viral RNA can infect healthy cells so that both should be considered. Cytology of infected cells has shown conclusively that plasmodesmata do not form obstacles to virus movement from cell to cell (de Zoeten and Gaard, 1969; Esau, 1968). Kitajima and Lauritis (1969) showed that virus infection could even lead to modification of certain plasmodesmata in order to accommodate transport of virus particles. The frequency and localization of plasmodesmata on cell walls may be of importance in the rate as well as the direction of the transport, whether virus or viral-RNA is spread as the infectious entity.

In the light of our earlier statement on the necessity of uncoating of virus upon infection and the contention that this occurred at a lipid virus interface outside the cell proper, it seems doubtful that transported virus particles are important in the spread of virus infections from cell to cell. One may argue the possibility of enzymatic cytoplasm mediated uncoating or that the intracellular membranes could present the nonpolar environment needed to uncoat a virus particle. However, there is no evidence for an intracellular regulatory mechanism that would discriminate between infecting nucleoproteins that ought to be uncoated and the virus particles produced as a result of infection that should remain assembled. Rather than postulating a cellular mechanism by which cells selectively uncoat some virus particles (infecting) and not others (produced), it seems that there are precedents for movement of the infectious entity in the form of naked nucleic acid. An example of infectious RNA moving from cell to cell is the spread of the "unstable" form of tobacco rattle virus, the infectious principle of which was shown to be the RNA of the large particle (Lister, 1968). In 1962, Cadman showed that both cell to cell and long distance transport of the "NM" variant of tobacco rattle virus (now known to be the long particle RNA) occurred. The speed of systemic spread, however, was three to five times slower than that of the particulate variant of this virus. Although an attractive idea, there is no compelling reason (proof) to assume that this phenomenon (slow movement) is caused by an absence of assembled virus particles in "NM" variant-infected plants.

The infectious RNAs of the spindle tuber and the exocortis agents (Diener, 1972; Semancik *et al.*, 1973) which are probably more structured than viral ssRNA and less structured than the replicative forms of RNA viruses also spread through plants. These RNAs spread slowly, but are highly infectious. Moreover, we have established (Powell and de Zoeten, 1977) that pea nuclei isolated from healthy pea plants do not offer the uncoating environment necessary for infection with pea enation mosaic virus (PEMV). PEMV-RNA could initiate infection as measured by actinomycin D (AMD)-insensitive viral polymerase activity in isolated healthy pea nuclei, whereas PEMV particles could not. We concluded, therefore, that although the nucleus is the first organelle to become involved in replication after inoculation, it does not offer the necessary uncoating environment for the virus.

If cell to cell spread of the infectious entity occurs as ssRNA, the fact that cytology reveals so little of the actual process is due to the inability of the techniques of electron microscopy to visualize ssRNA in sections. If the infectious entity is either (+) RNA or (−) RNA, it must be assumed that systemic infections induce minimal decompartmentalization of RNases detrimental to these RNAs or that the viral genome or its transcription products are protected from degradation during cell to cell transport.

It has been assumed for some time that plant viral polymerases capable of transcribing plant viral RNAs were intrinsic proteins of some host membrane (Bradley and Zaitlin, 1971; Brishammer, 1970; Brishammer and Juntti, 1974; Fraenkel-Conrat, 1976; Zabel *et al*, 1974; Zabel *et al.*, 1976; Zaitlin *et al.*, 1973). The chaotropic substances used in the isolation procedures of these enzymes are an indication of their membrane-bound nature. Moreover, the allotypic behavior of most isolated viral polymerases when released from membranes is a well known phenomenon of intrinsic membrane enzymes and could account for the loss of some template specificity of viral polymerases when isolated (Bradley and Zaitlin, 1971; Brishammer, 1970; Brishammer and Juntti, 1974; Fraenkel-Conrat, 1976; Zabel *et al.*, 1974; Zabel *et al.*, 1976; Zaitlin *et al.*, 1973). Recently, however, Butterworth *et al.* (1976) showed that the fluidity of the phospholipid membrane did not affect polymerase activitiy suggesting that the polymerase is possibly a peripheral rather than an integral membrane protein.

The actinomycin D sensitivity of the infection process when AMD is administered early in infection [either before, at, or within 2 hours after inoculation with TMV (Dawson, 1978a; 1978b; Dawson *et al.*, 1975; Dawson and Schlegel, 1976a, 1976b) and cowpea mosaic virus (CPMV) (Lockhart and Semancik, 1969; Rottier *et al.*, 1979)] is indicative of a necessary host DNA-dependent step in the infection process. Whether this step is related to the host membrane manufacture or to the production of a presumed host elongation factor (Hall and Wepprich, 1976) combining with virus coded polymerase subunits on a membrane, or both, is at this time not known.

The occurrence of membrane-associated dsRNA has been established for TMV (Brishammer, 1970; Brishammer and Juntti, 1974; Zaitlin *et al.*, 1973),

CPMV (Assink *et al.*, 1973; de Zoeten *et al.*, 1974), and TYMV (La Fléche *et al.*, 1972). Moreover, it was shown that in the case of TYMV the membranes associated with the viral dsRNA were also associated with the viral polymerase. The RNase resistance of CPMV-dsRNA associated with vesicular material in the cytoplasm (autoradiographically determined) suggests the form in which either (–) RNA or RF and RI could be transported from cell to cell; namely, membrane-bound. Passage of membranes (ER) through plasmodesmata or, at least, membrane association with these structures has been observed (de Zoeten and Gaard unpublished). Actual observations, however, of the passage of these infection-associated vesicles (replication complexes) through plasmodesmata are lacking at this moment.

Nuclear involvement in the multiplication of PEMV-RNA was established earlier (de Zoeten *et al.*, 1972, 1976). Recently, we have been able to show that both the PEMV-specific dsRNA and the viral polymerase are associated with a fraction containing the nuclei (isolated from PEMV-infected tissue) and another fraction containing most of the vesicles generated by the nuclei and recognizable as part of the cytopathological structure described earlier for PEMV-infected cells (Powell *et al.*, 1977). The association of the virus-specific dsRNA and a viral polymerase with a host-generated membrane system as in the case of PEMV has also been reported for TYMV-infected plants. In the latter case, however, the chloroplasts (polyplasts), rather than the nuclei, serve as the source of the membraneous backbone (peripheral vesicles) for the RNA replication (La Fléche *et al.*, 1972). Since (+) RNA, (–) RNA, dsRNA, and virus particles are not themselves mobile, transport of the infectious entity (the replication complex) must be achieved in a passive manner following the modes of transport present in the cytoplasm.

The differential speed observed for the spread of TMV and its PM mutants which move extremely slow does not argue against the view that systemic virus spread is achieved through the transport of the replication complex as the infectious entity. There is, indeed, no proof that these mutants did not undergo other mutations than those involved in coat protein production or virus assembly, characters easy to observe. It, therefore, seems reasonable to assume that more mutations occurred, one of which affected the systemic spread of these mutants by the production of unstable replication complexes.

7.4.3 Long Distance Transport

More data than those presented in the reviews of Schneider (1965) and Esau (1967, 1968) in which long distance transport of virus was treated in depth have not become available. At that time, the most acceptable hypothesis for intercellular virus movement was that it followed the movement of the normal cell constituents. The velocity of long distance movement of virus is several orders of magnitude larger than cell to cell movement and much greater than the velocities of cytoplasmic streaming.

Although phloem transport of virus is most common, xylem transport has been reported for lettuce necrotic yellows virus (Chambers and Francki, 1966) and for Southern bean mosaic virus (Schneider and Worley, 1959). Long distance transport does not seem to be dependent on multiplication of the virus in the conducting tissues. There is no doubt, however, that some of the transported virus is produced in young, undifferentiated phloem elements. The continuity of the plasmalemma of fully differentiated phloem elements, and that of the plasmodesmata of companion and other phloem parenchyma cells, seems to indicate that plasmodesmata can serve as entry as well as exit ports for the infectious agent. The form in which the infectious entity travels in long distance transport has not been established. If the virus particles were the infectious entity, some uncoating mechanism should be operable at the interface between phloem and companion cells or the phloem parenchyma. The endoplasmic reticulum usually associated with plasmodesmata in phloem elements may provide the nonpolar uncoating environment needed for infection. On the other hand, the viral genome could travel in some other forms through conductive tissue which does not require uncoating. Since enzyme compartmentalization in the phloem must have broken down during the ontogeny of the phloem elements, protection of the infectious entity from RNases is imperative. This could be offered if the viral genome was packaged in a protein or protected by a membrane (dsRNA membrane association). Although it was shown that in the case of TMV infection the first progeny particle may be observed by electron microscopy within 15 hours after leaf inoculation, it has been known since the early 1930s (Kunkel, 1939; Samuel, 1934) that a minimum of 42 hours is needed for the TMV infection to move out of inoculated leaves. Thus, enough time has elapsed for the formation of replication complexes to allow for their involvement in systemic spread of the infection.

The conclusions of the review presented here are that (1) uncoating of the viral genome takes place on the outside of the cellular boundary offered by the plasmalemma; (2) host specificity of plant viruses is an important but largely unknown entity in the infection process that may be operational at the transcriptional level; (3) cell to cell and long distance transport of the infectious entity could occur as membrane associated virus-specific RNA [dsRNA, (−) RNA, or (+) RNA]. In this scheme of infection, constituting a working hypothesis, progeny virus particles do not play a role in systemic spread of the infection in the plant. The uncoating is a single event at the initiation of infection and is not repeated during the spread of virus infection, thus limiting the time the viral genetic material is most vulnerable to digestion by nucleases. Moreover, the infectious entity is protected during long distance transport by reason of its association with a membrane system.

As a consequence of the concept developed here, progeny virus can only express its infectivity after release from the infected plant and arrival at an extracellular uncoating site.

In closing this discussion, a word of caution seems appropriate. The concepts developed here are meant to be food for thought and experimentation. They represent the author's view on the subject of early events in plant virus infection.

7.5 ACKNOWLEDGMENT

The support of the College of Agricultural and Life Sciences, University of Wisconsin-Madison, during a study leave spent at the University of California-Davis, is gratefully acknowledged.

7.6 REFERENCES

Assink, A. M., Swaans, H., and van Kammen, A. (1973). The localization of virus-specific double-stranded RNA of cowpea mosaic virus in subcellular fractions of infected Vigna leaves. *Virology* 53: 384-391.

Atabekov, J. G. (1975). Host specificity of plant viruses. *Ann. Rev. Phytopathol.* 13: 127-145.

Blumenthal, T., Landers, T. A., and Weber, K. (1972). Bacteriophage Qβ replicase contains the protein biosynthesis elongation factors EF TU and EF T_s. *Proc. Nat. Acad. Sci.* 69: 1313-1317.

Bové, J. M. (1967). Thèse de Doctor et d'Etat. Fac. des Sci. de Paris # d'enregistrement au C. N. R. S.: A01289.

Bradfute, O. E., Chapman-Andresen, C., and Jensen, W. A. (1964). Concerning morphological evidence of pinocytosis in higher plants. *Exp. Cell Res.* 36: 207-210.

Bradley, D. W., and Zaitlin, M. (1971). Replication of tobacco mosaic virus. II. The *in vitro* synthesis of high molecular weight virus-specific RNAs. *Virology* 45: 192-199.

Brants, D. H. (1974). The susceptibility of tobacco and bean leaves to tobacco mosaic virus infection in relation to the condition of ectodesmata. *Virology* 23: 588-594.

Bastin, M., Dasgupta, R., Hall, T. C., and Kaesberg, P. (1976). Similarity in structure and function of the 3'-terminal region of the four brome mosaic virus RNAs. *J. Mol. Biol.* 103: 737-745.

Brishammer, S. (1970). Identification and characterization of an RNA replicase from TMV-infected tobacco leaves. *Biochem. Biophys. Res. Comm.* 41: 506-511.

Brishammer, S., and Juntti, N. (1974). Partial purification and characterization of soluble TMV replicase. *Virology* 59: 245-253.

Burgess, J., Motoyoshi, F., and Fleming, E. N. (1973). Effect of poly-L-ornithine on isolated tobacco mesophyll protoplasts: evidence against stimulated pinocytosis. *Planta 111*: 199-208.

Busto, P., Carriquiry, E., Tarragõ-Litvak, L., Castroviejo, M., and Litvak, S. (1976). Interactions of plant viral RNAs and tRNA nucleotidyl transferase. *Ann. Microbiol. (Inst. Pasteur)* 127A: 39-46.

Butterworth, B. E., Shimshick, E. J., and Yin, F. H. (1976). Association of polioviral RNA polymerase complex with phospholipid membranes. *J. Virol.* 19: 457-466.

Cadman, C. H. (1962). Evidence for association of tobacco rattle virus nucleic acid with a cell component. *Nature* 193: 49-52.

Caspar, D. L. D. (1963). Assembly and stability of the tobacco mosaic virus particle. *Advan. Protein Chem.* 18: 87-121.

Chambers, T. C., and Francki, R. I. B. (1966). Localization and recovery of lettuce necrotic yellows virus from xylem tissue of *Nicotiana glutinosa. Virology* 29: 673-676.

Chan, V. F., and Black, F. L. (1970). Uncoating of polio virus by isolated plasma membranes. *J. Virol.* 5: 309-312.

Chen, J. M., and Hall, T. C. (1973). Comparison of tyrosyl transfer ribonucleic acid and brome mosaic virus tyrosyl ribonucleic acid as amino acid donors in protein synthesis. *Biochemistry* 12: 4570-4574.

Clegg, J. C. S., Brzeski, H., and Kennedy, S. I. T. (1976). RNA polymerase components in Semliki forest virus-infected cells: synthesis from large precursors. *J. Gen. Virol.* 32: 413-430.

Cocking, E. C. (1970). Virus uptake, cell wall regeneration and virus multiplication in isolated plant protoplasts. *Intern. Rev. Cytol.* 28: 89-124.

Dales, S. (1965). Penetration of animal viruses into cells. *Progr. Med. Virol.* 7: 1-43.

Das, C. R., Milbrath, J. A., and Swenson, K. G. (1961). Seed and pollen transmission of prunus ringspot virus in buttercup squash. *Phytopathology* 51: 64.

Davies, J. W. (1976). The multipartite genome of brome mosaic virus: Aspects of *in vitro* translation and RNA structure. *Ann. Microbiol.* 127A: 131-142.

Dawson, W. O. (1978a). Time-course of cowpea chlorotic mottle virus RNA replication. *Intervirology* 9: 119-128.

Dawson, W. O. (1978b). Time-course of actinomycin D inhibition of tobacco mosaic virus multiplication relative to the rate of spread of the infection. *Intervirology* 9: 304-309.

Dawson, W. O., and Schlegel, D. E. (1976a). The sequence of inhibition of tobacco mosaic virus synthesis by actinomycin D, 2-thiouracil, and cycloheximide in a synchronous infection. *Phytopathology* 66: 177-181.

Dawson, W. O., and Schlegel, D. E. (1976b). Time-course of tobacco mosaic virus-induced RNA synthesis in synchronously infected tobacco leaves. *Phytopathology* 66: 437-442.

Dawson, W. O., and Schlegel, D. E. (1976c). Synchronization of cowpea chlorotic mottle virus replication in cowpea leaves. *Intervirology* 7: 284-291.

Dawson, W. O., Schlegel, D. E., and Lung, M. C. Y. (1975). Synthesis of tobacco mosaic virus in intact tobacco leaves systemically inoculated by differential temperature treatment. *Virology* 65: 565-573.

De Sena, J., and Mandel, B. (1977). Studies on the *in vitro* uncoating of polio virus. II. Characteristics of the membrane modified particle. *Virology* 78: 554-566.

de Zoeten, G. A., and Gaard, G. (1969). Possibilities for inter- and intracellular translocation of some icosahedral plant viruses. *J. Cell. Biol.* 40: 814-823.

de Zoeten, G. A., Gaard, G., and Diez, F. B. (1972). Nuclear vesiculation associated with pea enation mosaic virus-infected plant tissue. *Virology* 48: 638-647.

de Zoeten, G. A., Assink, A. M., and van Kammen, A. (1974). Association of cowpea mosaic virus induced double stranded RNA with a cytopathological structure in infected cells. *Virology* 59: 341-355.

de Zoeten, G. A., Powell, C. A., and Gaard, G. (1976). *In situ* localization of pea enation mosaic virus double stranded ribonucleic acid. *Virology* 70: 459-469.

Diener, T. O. (1972). Viroids. *Advan. Virus Res.* 17: 295-319.

Duafala, T. H., and Nemanic, M. K. (1974). Tobacco mosaic virus particles on inocualted leaves observed with SEM. *Proc. 7th Ann. Scann. Electr. Microsc. Symposium Partu.* pp. 622-628.

Dunnebacke, T. H., Levinthal, J. D., and Williams, R. C. (1969). Entry and release of polio virus as observed by electron microscopy of cultured cells. *J. Virol.* 4: 505-513.

Esau, K. (1967). Anatomy of plant virus infection. *Ann. Rev. Phytopathol.* 5: 45-76.

Esau, K. (1968). "Viruses in Plant Hosts: Form, Distribution and Pathological Effects." Univ. of Wisconsin Press, Madison.

Fraenkel-Conrat, H. (1976). RNA polymerase from tobacco necrosis virus infected and uninfected tobacco. Purification of the membrane-associated enzyme. *Virology* 72: 23-32.

Gaard, G., and de Zoeten, G. A. (1979). Plant virus uncoating as a result of virus-cell wall interactions. *Virology* 96: 21-31.

German, T. L., and de Zoeten, G. A. (1975). Purification and properties of the replicative forms and replicative intermediates of pea enation mosaic virus. *Virology* 66: 172-184.

German, T. L., de Zoeten, G. A., and Hall, T. C. (1977). Pea enation mosaic virus genome RNA contains no polyadenylate sequences and cannot be aminoacylated. *Intervirology*

9: 226-230.

Gerola, F. M., Bassi, M., Favali, M. A., and Betto, E. (1969). An electron microscopy study of the penetration of tobacco mosaic virus into leaves following experimental inoculation. *Virology* 68: 380-386.

Hamilton, R. I. (1974). Replication of plant viruses. *Ann. Rev. Phytopathol.* 12: 223-245.

Hall, T. C., and Wepprich, R. K. (1976). Functional possibilities for aminoacylation of viral RNA in transcription and translation. *Ann. Microbiol. (Inst. Pasteur)* 127A: 143-152.

Harrison, B. D., and Roberts, I. M. (1968). Association of tobacco rattle virus with mitochondria. *J. Gen. Virol.* 3: 121-124.

Herridge, E. A., and Schlegel, D. E. (1962). Autoradiographic studies of tobacco mosaic virus inoculations on host and nonhost species. *Virology* 18: 517-523.

Holland, J. J. (1962). Irreversible eclipse of polio virus by Hela cells. *Virology* 16: 163-176.

Jacquemin, J. M. (1972). *In vitro* product of an RNA polymerase induced in broad bean by infection with broad bean mosaic. *Virology* 49: 379-384.

Kaesberg, P. (1977). Translation of the RNAs of brome mosaic virus. *In* "Beltsville Symposia in Agricultural Research I. Virology in Agriculture" (J. A. Romberger, ed.), pp. 267-272. Allanheld, Osmun & Co., Montclair, New Jersey.

Kassanis, B., and Kenten, R. H. (1978). Inactivation and uncoating of TMV on the surface and intercellular spaces of leaves. *Phytopath. Z.* 91: 329-339.

Kent, S. P. (1966). Intracellular plasma protein: a manifestation of cell injury in myocardial ischaemia. *Nature* 210: 1279-1281.

Kitajima, E. W., and Lauritis, J. A. (1969). Plant virions in plasmodesmata. *Virology* 37: 681-685.

Kohl, R. J., and Hall, T. C. (1977). Loss of infectivity of brome mosaic RNA after chemical modification of the 3' and 5' terminus. *Proc. Nat. Acad. Sci.* 74: 2682-2686.

Kohl, R. J., and Hall, T. C. (1979). Amino acylation of RNA from several viruses: Amino acid specificity and differential activity of plant, yeast and bacterial synthetases. *J. Gen. Virol.* 25: 257-261.

Kontaxis, D. G., and Schlegel, D. E. (1962). Basal septa of broken trichomes in Nicotiana as possible infection sites for tobacco mosaic virus. *Virology* 16: 244-247.

Kumarasamy, R., and Symons, R. H. (1979). Extensive purification of the cucumber mosaic virus-induced RNA replicase. *Virology* 96: 622-632.

Kunkel, L. O. (1939). Movement of tobacco-mosaic virus in tomato plants. *Phytopathology* 29: 684-700.

Kurtz-Fritsch, C., and Hirth, L. (1972). Uncoating of two spherical plant viruses. *Virology* 47: 385-396.

La Flèche, D., and Bové, J. M. (1971). Virus de la mosaïque jaune du navet: site cellulaire de la replication du RNA viral. *Physiol. Veg.* 9: 487-503.

La Flèche, D., Bové, C., Dupont, C., Mouchès, C., Astier, T., Garnier, M., and Bové, J. M. (1972). Site of viral RNA replication in the cells of higher plants: TYMV-RNA synthesis on the chloroplast outer membrane system. *Proc. 8th Mtg. Fed. Euro. Biochem. Soc.* pp. 43-71.

Landers, T. A., Blumenthal, T., and Weber, K. (1974). Function and structure in ribonucleic acid phage $Q\beta$ ribonucleic acid replicase. *J. Biol. Chem.* 249: 5801-5808.

Le Roy, C., Stussi-Garaud, C., and Hirth, L. (1977). RNA-dependent RNA polymerases in uninfected and in alfalfa mosaic virus-infected tobacco plants. *Virology* 82: 48-62.

Lister, R. M. (1968). Functional relationships between virus specific products of tobacco rattle virus. *J. Gen. Virol.* 2: 43-58.

Lockhart, B. E. L., and Semancik, J. S. (1969). Differential effect of actinomycin D on plant virus multiplication. *Virology* 39: 362-365.

Litvak, S., Tarragô-Litvak, L., and Allende, J. E. (1973). Elongation factor-viral genome interaction dependent on the aminoacylation of TYMV and TMV RNAs. *Nature New Biol.* 241: 88-90.

Lundquist, R. E., Ehrenfeld, E., and Maizel, J. V. (1974). Isolation of a viral polypeptide associated with polio virus RNA polymerase. *Proc. Nat. Acad. Sci.* 71: 4773-4777.

Luria, S. E., and Darnell, J. E. (1967). "General Virology." John Wiley & Sons, New York. pp. 195-197.

Mahlberg, P. (1972a). Further observations on the phenomenon of secondary vacuolation in living cells. *Am. J. Botany* 59: 172-179.

Mahlberg, P. (1972b). Localization of neutral red lysosome structures in hair cells of *Tradescantia virginiana. Canad. J. Botany* 50: 857-859.

Mandel, B. (1967). The relationship between penetration and uncoating of polio virus in Hela cells. *Virology* 31: 701-712.

May, J. T., Gilliland, J. M., and Symons, R. H. (1970). Properties of a plant virus-induced RNA polymerase in particulate fractions of cucumbers infected with cucumber mosaic virus. *Virology* 41: 653-664.

Merkens, W. S. W., de Zoeten, G. A., and Gaard, G. (1972). Observations on ectodesmata and the virus infection process. *J. Ultrastruct. Res.* 41: 397-405.

Motoyoshi, F., Bancroft, J. B., Watts, J. W., and Burgess, J. (1973). The infection of tobacco protoplasts with cowpea chlorotic mottle virus and its RNA. *J. Gen. Virol.* 20: 177-193.

Mouchès, C., Bové, C., and Bové, J. M. (1974). Turnip yellow mosaic virus-RNA replicase: partial purification of the enzyme from solubilized enzyme-template complex. *Virology* 58: 409-423.

Otsuki, Y., Takebe, I., Honda, Y., and Matsui, C. (1972). Ultrastructure of infection of tobacco mesophyll protoplasts by tobacco mosaic virus. *Virology* 49: 188-194.

Otsuki, Y., Takebe, I., Honda, Y., Kajita, S., and Matsui, C. (1974). Infection of tobacco mesophyll protoplasts by potato virus X. *J. Gen. Virol.* 22: 375-385.

Pelham, H. R. B. (1979). Synthesis and proteolytic processing of cowpea mosaic virus proteins in reticulocyte lysates. *Virology* 96: 463-477.

Pinck, A. M., Chan, S. K., Genevaux, M., Hirth, L., and Duranton, H. (1972). Valine specific tRNA-like structure in RNAs of two viruses of the turnip yellow mosaic virus group. *Biochimie* 54: 1093-1094.

Powell, C. A., and de Zoeten, G. A. (1977). Replication of pea enation mosaic virus RNA in isolated pea nuclei. *Proc. Nat. Acad. Sci.* 74: 2919-2922.

Powell, C. A., de Zoeten, G. A., and Gaard, G. (1977). The localization of pea enation mosaic virus-induced RNA dependent RNA polymerase in infected peas. *Virology* 78: 135-143.

Ralph, R. K., and Wojcik, S. J. (1969). Double stranded tobacco mosaic virus RNA. *Virology* 37: 276-282.

Ralph, R. K., Bullivant, S., and Wojcik, S. J. (1971a). Cytoplasmic membranes; a possible site of tobacco mosaic virus RNA replications. *Virology* 43: 713-716.

Ralph, R. K., Bullivant, S., and Wojcik, S. J. (1971b). Evidence for the intracellular site of double stranded turnip yellow mosaic virus RNA. *Virology* 44: 473-479.

Röder, A., and Koschel, K. (1974). The reversible inhibition of polio virus RNA synthesis *in vivo* and *in vitro* by viral products. *J. Virol.* 3: 846-852.

Röder, A., and Koschel, K. (1975). Virus-specific proteins associated with the replication complex of poliovirus RNA. *J. Gen. Virol.* 28: 85-98.

Romaine, C. P., and Zaitlin, M. (1978). RNA-dependent RNA polymerases in uninfected and tobacco mosaic virus-infected tobacco leaves: Viral-induced stimulation of a host polymerase activity. *Virology* 86: 241-253.

Rottier, P. J. M., Rezelman, G., and van Kammen, A. (1979). The inhibition of cowpea mosaic virus replication by actinomycin D. *Virology* 92: 299-309.

Samuel, G. (1934). The movement of tobacco mosaic virus within the plant. *Ann. Appl. Biol.* 21: 90-111.

Schneider, I. R. (1965). Introduction, translocation, and distribution of viruses in plants. *Adv. Virus Res.* 11: 163-221.

Schneider, I. R., and Worley, J. F. (1959). Rapid entry of infectious particles of southern bean mosaic virus into living cells following transport of the particles in the water stream. *Virology* 8: 243-249.

Sela, I., and Kaesberg, P. (1969). Cell-free synthesis of tobacco mosaic virus coat protein and its combination with ribonucleic acid to yield tobacco mosaic virus. *J. Virol.* 3: 89-91.

Semal, J. (1967). Effects of actinomycin D in plant virology. *Phytopathol. Z.* 59: 55-71.

Semal, J. (1970). Properties of the products of UTP incorporation by cell-free extracts of leaves infected with bromegrass mosaic virus or with broad bean mottle virus. *Virology* 40: 244-250.

Semal, J., and Hamilton, R. (1968). RNA synthesis in cell-free extracts of barley leaves infected with bromegrass mosaic virus. *Virology* 36: 293-302.

Semal, J., and Kummert, J. (1971). *In vitro* synthesis of a segment of bromegrass mosaic virus ribonucleic acid. *J. Gen. Virol.* 11: 189-192.

Semancik, J. S., Morris, T. J., and Weathers, L. G. (1973). Structure and conformation of low molecular weight pathogenic RNA from exocortis disease. *Virology* 53: 448-456.

Shaw, J. C. (1967). *In vivo* removal of protein from tobacco mosaic virus after inoculation of tobacco leaves. *Virology* 31: 665-675.

Shaw, J. C. (1969). *In vivo* removal of protein from tobacco mosaic virus after inoculation of tobacco leaves. II. Some characteristics of the reaction. *Virology* 37: 109-116.

Skrzeczkowski, L. J. (1977). Biochemiciczna charakterystyka niektorych elementow krancowej odpornosci ziemniaka na wirus X. Doctoral thesis, University of Warsaw.

Smith, S., and Schlegel, D. E. (1965). The incorporation of ribonucleic acid precursors in healthy and virus-infected plant cells. *Virology* 26: 180-189.

Stussi-Garaud, C. Lemius, J., and Fraenkel-Conrat, H. (1977). RNA polymerase from tobacco necrosis virus-infected and uninfected tobacco. *Virology* 81: 224-236.

Thomas, P. E., and Fulton, R. W. (1968). Correlation of ectodesmata number with non-specific resistance to initial virus infection. *Virology* 34: 459-469.

van Vloten-Doting, L., Bol, J. F., and Jaspars, E. M. J. (1977). Plant viruses with multipartite genomes. *In* "Beltsville Symposia in Agricultural Research I. Virology in Agriculture" (J. A. Romberger, ed.), pp. 47-61. Allanheld, Osmun & Co., Montclair, New Jersey.

Weening, C. J., and Bol, J. F. (1975). Viral RNA replication in extracts of alfalfa mosaic virus-infected *Vicia faba*. *Virology* 63: 77-83.

White, J. L., and Murakishi, H. H. (1977). *In vitro* replication of tobacco mosaic virus RNA in tobacco callus cultures: solubilization of membrane-bound replicase and partial purification. *J. Virol.* 21: 484-492.

White, J. L., and Dawson, W. O. (1978). Characterization of RNA-dependent RNA polymerase in uninfected and cowpea chlorotic mottle virus-infected cowpea leaves: selective removal of host RNA polymerase from membranes containing CCMV RNA replicase. *Virology* 88: 33-43.

Zabel, P., Weenen-Swaans, H., and van Kammen, A. (1974). *In vitro* replication of cowpea mosaic virus RNA. I. Isolation and properties of the membrane-bound replicase. *J. Virol.* 14: 1049-1055.

Zabel, P., Jongen-Neven, I., and van Kammen, A. (1976). *In vitro* replication of cowpea mosaic virus RNA. II. Solubilization of membrane-bound replicase and the partial purification of the solubilized enzyme. *J. Virol.* 17: 679-685.

Zabel, P. (1978). Purification and properties of cowpea mosaic virus RNA replicase. Dissertation. Wageningen, The Netherlands.

Zabel, P., Jongen-Neven, I., and van Kammen, A. (1979). *In vitro* replication of cowpea mosaic RNA. III. Template recognition by cowpea mosaic virus RNA replicase. *J. Virol.* 29: 21-33.

Zaitlin, M. (1977). Replication of plant viruses: An overview. *In* "Beltsville Symposia in Agricultural Research I. Virology in Agriculture" (J. A. Romberger, ed.), pp. 33-46. Allanheld, Osmun & Co., Montclair, New Jersey.

Zaitlin, M., and Boardman, N. K. (1958). The association of tobacco mosaic virus with plastids. I. The isolation of virus from the chloroplast fraction of diseased-leaf homogenates. *Virology* 6: 743-757.

Zaitlin, M., Duda, C. T., and Petti, M. A. (1973). Replication of tobacco mosaic virus. V. Properties of the bound and solubilized replicase. *Virology* 53: 300-311.

Zaitlin, M., Beachy, R. N., Bruening, G., Romaine, C. P., and Scalla, R. (1962). Translation of tobacco mosaic virus RNA. *In* "Animal Virology" (D. Baltimore, A. S. Huang, and C. F. Fox, eds.), pp. 567-581. Academic Press, New York.

Chapter 8

VIRUS TRANSMISSION THROUGH SEED AND POLLEN

C. L. Mandahar

Department of Botany
Panjab University
Chandigarh, India

8.1 INTRODUCTION

Transmission of plant viruses by seeds was earlier considered to be a rare and an insignificant factor in the epidemiology of virus diseases of plants. The picture has now changed for the worst. The number of viruses reported to be trans-

TABLE I. Viruses Transmitted Through Seeds

Virus	Host	Percent[1,2] transmission	Reference
Abutilon mosaic	Abutilon spp.	< 1	Keur, 1933
	Abutilon thompsonii x	0.7	Keur, 1934
	A. mulleri		
Alfalfa mosaic	Capsicum annum	1-5	Sutic, 1959
	Medicago sativa	up to 6	Belli, 1962
	M. sativa	55	Zschan & Janke, 1962
	M. sativa	14	Frosheiser, 1964
	M. sativa	0.2-6	Frosheiser, 1970
	M. sativa	0.6-10.3	Hemmati & McLean, 1977
	M. sativa	4	Ekbote & Mali, 1978
	Nicandra physaloides	-	Galla & Ciampor, 1977
Andean potato latent[3]	Nicotiana clevelandii	-	Gibbs et al., 1966
	Petunia hybrida	-	Gibbs et al., 1966
	Solanum tuberosum	< 1[4]	Jones & Fribourg, 1977
Apple chlorotic leafspot	Rubus spp. (raspberry)	30-40	Cadman, 1965; Converse, 1967
Apple mosaic[5]	Vigna unguiculata	2	fide Neergaard, 1977
Apricot gummosis[6]	Prunus americana	-	Fridlund, 1966
	P. avium	15	Fridlund, 1966
	P. domestica	-	Traylor et al., 1963
	P. serrulata	-	Fridlund, 1966
Arabis mosaic	Beta vulgaris	13	Lister & Murant, 1967
	Capsella bursa-pastoris	33	Lister & Murant, 1967
	Chenopodium album	80	Lister & Murant, 1967
	Fragaria x ananassa	6.9	Lister & Murant, 1967
	Glycine max	6.3	Lister, 1960
	Humulus spp.	10	Kriz, 1959
	Lactuca sativa	60-100	Walkey, 1967
	Lamium amplexicaule	1.2-25	Lister & Murant, 1967

242

	Lycopersicon esculentum	1.8	Lister & Murant, 1967
	Myosotis arvensis	19-25	Lister & Murant, 1967
	Petunia hybrida	20	Lister, 1960
	P. violacea	up to 34	Phatak, 1974
	Plantago major	5.4-28	Lister & Murant, 1967
	Poa annua	4	Taylor & Thomas, 1968
	Polygonum persicaria	21-100	Lister & Murant, 1967
	Rheum rhaponticum	10-24	Tomlinson & Walkey, 1967
	Senecio vulgaris	2.2	Lister & Murant, 1967
	Stellaria media	57	Lister & Murant, 1967
Asparagus bean mosaic	*Vigna sesquipedalis*	35	*fide* Mandahar, 1978
Asparagus latent	*Asparagus officinalis*	65	Paludan, 1964
Avocado sunblotch	*Persea americana*	76	Wallace & Drake, 1953
Barley mottle mosaic	*Hordeum vulgare*	2-45	Dhanraj & Raychaudhuri, 1969
Barley stripe mosaic	*Aegilops* spp.	-	Nitzany & Gerechter, 1962
	Agropyron elongatum	22	Inouye, 1962
	Avena fatua	0-9.5	Chiko, 1975
	A. sativa	-	McKinney & Greeley, 1965
	Avena spp.	8	Nitzany & Gerechter, 1962
	Bromus inermis	-	Inouye, 1962
	Bromus spp.	4	Nitzany & Gerechter, 1962
	Commelina communis	3	Inouye, 1962
	Hordeum depressum	2	Inouye, 1962
	H. glaucum	58	Inouye, 1962
	H. vulgare	up to 90	McKinney, 1951a
	H. vulgare	50-100	McKinney, 1953
	H. vulgare	4-64	Gold *et al.*, 1954
	H. vulgare	38-86	Eslick & Afanasiev, 1955
	H. vulgare	3-53	Inouye, 1962
	H. vulgare	38	McKinney & Greeley, 1965
	H. vulgare	5-60	Catherall, 1972
	H. vulgare	61-70	Phatak, 1974
	H. vulgare		Carroll & Mayhew, 1976b

TABLE I. Continued

Virus	Host	Percent[1,2] transmission	Reference
	Poa exilis	-	Nitzany & Gerechter, 1962
	Triticum aestivum	71	Hagborg, 1954
Barley yellow dwarf	*T. aestivum*	6.7-80	McNeal & Afanasiev, 1955
Bean common mosaic	*Hordeum vulgare*	-	*fide* Mandahar, 1978
	Macroptilum lathyroides	-	Kaiser & Mossahebi, 1974;
			Provvidenti & Braverman, 1976
	Phaseolus acutifolius var. *latifolius*	7-34	Lockhart & Fischer, 1974
	P. acutifolius var. *latifolius*	7-22	Provvidenti & Cobb, 1975
	P. mungo	-	Nene, 1972; Agarwal *et al.*, 1977
	P. vulgaris	50	Reddick & Stewart, 1919
	P. vulgaris	50	Burkholder & Muller, 1926
	P. vulgaris	20-60	Harrison, 1935
	P. vulgaris	2-66	Smith & Hewitt, 1938
	P. vulgaris	25-86	Medina & Grogan, 1961
	P. vulgaris	7-20	Phatak, 1974
	Vigna sesquipedalis	37	Snyder, 1942
	V. sinensis	25-40	Sachchidananda *et al.*, 1973
Bean southern mosaic	*Phaseolus vulgaris*	21	Crowley, 1959
	P. vulgaris	5	Smith, 1972
	Vigna sinensis	3-4	Shepherd & Fulton, 1962
	V. unguiculata	1-3	Shepherd & Fulton, 1962;
			Lamptey & Hamilton, 1974
Bean western mosaic	*Phaseolus vulgaris*	2-3	Skotland & Burke, 1961
Bean yellow mosaic	*P. vulgaris*	-	Uyemoto & Grogan, 1977
	Lupinus albus	-	Blaszczak, 1963, 1965
	L. luteus	6.2	Corbett, 1958
	L. luteus	7-21	Porembskaya, 1964
	Melilotus alba	3-5	Phatak, 1974

244

Beet cryptic	*Phaseolus aureus*	-	Bengno & Favali-Hedayat, 1977
	P. vulgaris	-	Hino, 1962
Beet 41 yellows[7]	*Pisum sativum*	10-30	Inouye, 1967
Black gram leaf-crinkle	*Trifolium pratense*	12-15	Hampton, 1967
Black raspberry latent[8]	*Vicia faba*	-	Quantz, 1954; Bos, 1970
(see raspberry latent)	*V. faba*	0.1-2.4	Kaiser, 1972b, 1973; Evans, 1973
	Vigna sinensis	1-37	Snyder, 1942
	Beta vulgaris	-	Kassanis *et al.*, 1978
	Beta vulgaris	47	Clinch & Loughnane, 1948
	Black gram	20 & 41.9	Naryanaswamy & Jaganathan, 1975
Brinjal mosaic	*Solanum melongena*	-	Mayee & Khatri, 1975
Broad bean mild mosaic	*Vicia faba*	-	Devergne & Cousin, 1966
Broad bean mosaic	*Vicia faba*	-	Brunt, 1969; *fide* Neergaard, 1977
Broad bean mottle	*Phaseolus mungo*	6-7	Phatak, 1974
Broad bean stain	*Vicia faba*	1	Lloyd *et al.*, 1965
	V. faba	1	Varma & Gibbs, 1967
	V. faba	up to 10	Gibbs & Smith, 1970
	V. faba	7.3	Cockbain *et al.*, 1976
	V. faba	1.3, 16	Vorra-urai & Cockbain, 1977
Broad bean true mosaic (see EAMV)			
Carrot red leaf	*Daucus carrota*	25	Watson & Sergent, 1962
Cauliflower mosaic[9]	*Capsella bursa-pastoris*		Tomlinson & Walker, 1973
	Raphanus raphanistrum		Tomlinson & Walker, 1973
Cherry leaf roll	*Chenopodium amaranticolor*	100	Lister & Murant, 1967
	Glycine max	100	Lister & Murant, 1967
	Nicotiana tabaccum	1	Schmelzer, 1965, 1966
	Phaseolus vulgaris	12-40	Lister & Murant, 1967
	Rheum rhaponticum	72	Tomlinson & Walkey, 1967
	Sambucus racemosa	13-44	Schimanski & Schmelzer, 1972
			fide Neergaard, 1977
	Viola tricolor	1.2-6.1	Lister & Murant, 1967

TABLE I. Continued

Virus	Host	Percent[1,2] transmission	Reference
Cherry necrotic ringspot (see Prunus necrotic ringspot)			
Cherry necrotic rusty mottle	Prunus avium	-	Nyland, 1962
	P. cerasus	-	Nyland, 1962
Cherry rasp leaf	Chenopodium amaranticolor	-	Hanson et al.,1974
	Cherry	-	Hanson et al.,1974
	Taraxacum officinale	-	Hanson et al.,1974
Cherry ring mottle [10]	Cherry	33.3	Ramaswamy & Posnette, 1971
Cherry yellows	Prunus cerasus	9.0	Cation, 1952
	P. mahaleb	8.7	Cation, 1949
Citrus psorosis	Citrus sinensis x Ponicrus trifoliata	19	Childs & Johnson, 1966
	Citrus spp.	Trace	Wallace, 1957
Citrus xyloporosis	Citrus aurantifolia	66	Childs, 1956
Clover (white) mosaic	Trifolium pratense	6.0	Hampton, 1963
	T. pratense	24	Hampton & Hanson, 1968
Clover yellow mosaic	T. pratense	7.6	Hampton, 1963
Cocoa necrosis[11]	Phaseolus vulgaris	1-24	Kenten, 1972
Coffee ringspot	Coffea excelsa	10-11	Reyes, 1961
Cowpea aphid-borne mosaic	Phaseolus angularis	-	Tsuchizaki et al., 1970
	Vigna sinensis	about 1	Lovisolo & Conti, 1966
	V. sinensis	35	Phatak fide Neergaard, 1977
	V. unguiculata	27	Snyder, 1942
	V. unguiculata	0-30	Khatri & Chohan, 1972; Phatak, 1974
	V. unguiculata	0-3	Bock, 1973a
	V. unguiculata	up to 20.9	Ladipo, 1977
Cowpea banding mosaic	V. sinensis	15-31	Sharma & Verma, 1975
Cowpea chlorotic spot virus	V. sinensis	3-16	Sharma & Verma, 1975

Cowpea isometric mosaic	V. sinensis	5-16	Chenulu et al., 1968
Cowpea mild mottle	V. unguiculata	90	Brunt & Kenten, 1973
	Glycine max	90	Brunt & Kenten, 1973
	Phaseolus vulgaris	6	Brunt & Kenten, 1973
Cowpea mosaic	Vigna catjang	17	Capoor & Verma, 1956
	V. sinensis	17.5	Diwakar & Mali, 1977
	V. sinensis	23	Capoor & Verma, 1956
	V. sinensis	0-55	Anderson, 1957
	V. sinensis	10	Shepherd, 1964
	V. sesquipedalis	8	Dale, 1949
Cowpea mottle	V. unguiculata	3-10.3	Shoyinka et al., 1978
Cowpea ringspot	V. unguiculata	15-20	Phatak et al., 1976
	V. unguiculata	10-30	Phatak, 1974
Cucumber green mottle mosaic	Cucumis sativum	44	Yakovlva, 1965
	Lagenaria siceraria	-	Komuro et al., 1971; Sharma & Chohan, 1974
Cucumber mosaic	Benincasa hispida	1	Sharma & Chohan, 1973
	Cerastium holosteoides	2	Tomlinson & Carter, 1970b
	Cucumis melo	2.1	Kendrick, 1934
	C. pepo	-	Reddy & Nariani, 1963; Sharma & Chohan, 1974
	C. sativus	1.4	Doolittle, 1920
	Cucurbita moschata	0.7	Sharma & Chohan, 1974
	Echinopsis lobata	9.1	Doolittle & Gilbert, 1919
	E. lobata	55	Doolittle & Walker, 1925
	E. lobata	15	Lindberg et al., 1956
	Lupinus luteus	-	Troll, 1957
	L. luteus	14	Porembskaya, 1964
	Lycopersicon esculentum	0.2	Van Koot, 1949
	Phaseolus aureus	5	Phatak, 1974
	P. vulgaris	up to 30	Marchoux et al., 1977
	Stellaria media	1-30	Hani, 1971; Hani et al., 1970

TABLE I. Continued

Virus	Host	Percent[1,2] transmission	Reference
	S. media	10	Hani et al., 1970
	S. media	5-8	Tomlinson & Carter, 1970a
	S. media	3-40	Tomlinson & Carter, 1970b
	S. media	14	Tomlinson & Walker, 1973
	Lamium purpureum	4	Tomlinson & Carter, 1970b
	Spergula arvensis	2	Tomlinson & Carter, 1970b
	Vigna cylindrica		Brantley et al., 1965
	V. radiata	0.61	Purivirojkul et al., 1978
	V. radiata	11	Iwaki, 1978
	V. sesquipedalis	–	Anderson, 1957
	V. sinensis	4-28	Anderson, 1957
	V. sinensis	10	Iwaki, 1978
	V. unguiculata	4-28	Anderson, 1957
Datura quercina[12]	Datura stramonium	79	Blakeslee, 1921
Desmodium trifolium mottle	Desmodium trifolium	–	Suteri & Joshi, 1978
Dodder latent mosaic	Cascuta californica	2.4	Bennett, 1944a
	C. campestris	4.9	Bennett, 1944a
Dulcamara mottle	Solanum dulcamara	2-3	Gibbs et al., 1966
Echte Ackerbohnenmosaik virus (EAMV)[13]	Vicia faba	3.2	Cockbain & Bowen (quoted in Vorra-urai & Cockbain, 1977)
	V. faba	1-3	Quantz, 1953
	V. faba	up to 15	Gibbs & Smith, 1970
	V. faba	0.8	Cockbain et al., 1976
	V. faba	5	Vorra-urai & Cockbain, 1977
	V. faba	15	Gibbs & Paul, 1970
Eggplant mosaic[14]	Nicotiana clevelandii	–	Gibbs et al., 1966
	Petunia hybrida	–	Gibbs et al., 1966
Elm mosaic	Ulmus americana	1-3.5	Bretz, 1950
	U. americana	48	Callahn, 1957

248

Virus	Host	Value	Reference
Elm mottle	*U. glabra*	-	Jones & Mayo, 1973
Euonymus mosaic[15]	*Euonymus* sp.	-	Bojnansky & Kosljarova, 1968
Grape decline[16]	Grapevine		Dias & Cation, 1976
Grapevine Bulgarian latent virus	Grapevine	4.54	Uyemoto *et al.*, 1977
	Chenopodium quinoa	12.50	Uyemoto *et al.*, 1977
Grapevine fanleaf	*C. amaranticolor*	0.7	Dias, 1963
	C. amaranticolor	1.3	Dias, 1963
	C. quinoa	3	Cory & Hewitt, 1968
	Vitis vinifera[17]	-	Cory & Hewitt, 1968
Grape yellow mosaic	*Chenopodium amaranticolor*	0.7	Dias, 1963
Grape yellow vein[18]	*C. amaranticolor*	57	Cory & Hewitt, 1968
Guar symptomless virus	*Cyamompsis tetragonoloba*	12-28	Hansen & Leseman, 1978
Hippeastrum mosaic	*Hippeastrum hybridum*	-	Brants & Van Den Heuvel, 1965
Hop chlorosis	*Humulus lupulus*	27	Salmon & Ware, 1935
Lettuce mosaic	*Chenopodium quinoa*	0-1	Phatak, 1974
	Lactuca sativa	3	Newhall, 1923
	L. sativa	10	Ogilvie *et al.*, 1935
	L. sativa	6-15	Kramer *et al.*, 1945
	L. sativa	1-8	Grogan & Bardin, 1950
	L. sativa	up to 11	Herold, 1956
	L. sativa	up to 13	Rohloff, 1962
	L. sativa	5	Ryder, 1964
	L. serriola	0.2-6.2	van Hoff, 1959
Lettuce yellow mosaic	*L. sativa*	30	*fide* Mandahar, 1978
Lima bean mosaic	*Phaseolus limensis*	25	*fide* Mandahar, 1978
Loganberry degeneration	*Rubus loganobaccus*	-	Ormerod, 1970
Lychnis ringspot[19]	*Beta vulgaris*	9.5	Bennett, 1959
	Capsella bursa-pastoris	9.4	Bennett, 1959
	Cerastium viscosum	27	Bennett, 1959
	Lychnis divaricata	58	Bennett, 1959
	Silene gallica	28	Bennett, 1959
	S. noctiflora	41	Bennett, 1959

TABLE I. Continued

Virus	Host	Percent[1,2] transmission	Reference
Maize dwarf mosaic (see sugarcane mosaic)			
Melon necrotic spot	*Cucumis melo*	-	Gonzalez *et al.*, 1979
Mungbean mosaic[20]	*Phaseolus aureus*	8-32	Kaiser *et al.*, 1968; Kaiser & Mossahebi, 1974
	P. aureus	3	Phatak, 1974
Muskmelon mosaic	*Cucumis melo*	12-93	Rader *et al.*, 1947
	Cucurbita flexuous	-	Rader *et al.*, 1947
	C. moschata	-	Rader *et al.*, 1947
	C. pepo	-	Rader *et al.*, 1947
Nicotiana glutinosa	*Nicotiana glutinosa*	72	Randles *et al.*, 1976
Nicotiana veluntina mosaic	*Nicotiana* spp.	-	Randles *et al.*, 1976
Onion mosaic	*Allium cepa*	-	Cheremushkina, 1977
Onion yellow dwarf	*A. cepa*	-	Härdtl, 1964, 1972
Pea early browning	*Pisum sativum*	37	Bos & van der Want, 1962
	P. sativum	1-2	Harrison, 1973
Pea enation mosaic	*P. sativum*	40	Kovachevski, 1978
Pea false leaf roll	*P. sativum*	2-15	Thottappilly & Schmutterer, 1968
Pea leaf rolling	*P. sativum*	55	Musil, 1970
			Musil, 1970
Pea mosaic	*P. sativum*	-	Quantz, 1958
	Lathyrus odoratus	0.7	*fide* Mandahar, 1978
	Trifolium hybridum	47	*fide* Mandahar, 1978
	T. pratense		*fide* Mandahar, 1978
Pea seed-borne mosaic	Field pea	0.5, 5.8	Chiko & Zimmer, 1978
	Lens culinaris	32-44	Hampton & Muetibauer, 1977
	Pisum sativum	8-30	Inouye, 1967
	P. sativum	0-88	Stevenson & Hagedorn, 1969, 1973

250

Peach latent	*P. sativum*	20-80	Hampton, 1969
	P. sativum	65-90	Mink *et al.*, 1969
	P. sativum	4-32	Hampton, 1972
Peach latent	Cherry	2.4	*fide* Mandahar, 1978
Peach necrotic leafspot	*Prunus persica*	3-9	Wagnon *et al.*, 1960
Peach ringspot	*Prunus americana*	-	*fide* Mandahar, 1978
	P. persica	3-9	*fide* Mandahar, 1978
Peach rosette mosaic	*Chenopodium quinoa*	90	Dias & Cation, 1976
	Taraxacum officinale	3.6	Ramsdell & Meyers, 1978
	Vitis labrusca	9.5	Ramsdell & Meyers, 1978
Peanut bunchy top	*Arachis hypopea*	-	*fide* Mandahar, 1978
Peanut chlorosis	*A. hypogea*	-	*fide* Mandahar, 1978
Peanut marginal chlorosis	*A. hypogea*	30-100	van Velsen, 1961
Peanut mottle	*A. hypogea*	2	Kuhn, 1965
	A. hypogea	20	Bock, 1973b
	A. hypogea	3.7	Paguio & Kuhn, 1974
	A. hypogea	0-8.5	Adams & Kuhn, 1977
Peanut stunt	*A. hypogea*	0.2	Troutman *et al.*, 1967
	Lycopersicon esculentum	2-11	Singh, 1970
Potato spindle tuber viroid	*Physalis peruviana*	29	McClean, 1948 *fide* Phatak, 1974
	Solanum incanum	53	McClean, 1948 *fide* Phatak, 1974
	S. tuberosum	87-100	Hunter *et al.*, 1969
	S. tuberosum	6-12	Singh, *fide* Phatak, 1974
Potato virus T	*Datura stramonium*	5	Salazar & Harrison, 1978
	Solanum demissum	39	Salazar & Harrison, 1978
	Nicandra physaloides	28	Salazar & Harrison, 1978
Potato virus X	*Solanum tuberosum*	0.6-2.3	Darozhkin & Grabenshchykava, 1974
Potato virus Y	*S. tuberosum*	14, 16	*fide* Mandahar, 1978
Prune dwarf	*Prunus cerasus*	3-30	Gilmer & Way, 1960
	P. cerasus	15	Gilmer & Way, 1960
	P. mahaleb	9	Cation, 1949, 1952
	P. persica	-	Cochran, 1950
	Cherry	17, 77.7	*fide* Mandahar, 1978

TABLE I. Continued

Virus	Host	Percent[1,2] transmission	Reference
Prunus necrotic ringspot	Cucurbita maxima	2.7	Das et al., 1961
	Prunus americana	-	Hobart, 1956
	P. avium	6.0	Cochran, 1946
	P. avium	37	Megahed & Moore, 1967
	P. cerasus	20-56	Cation, 1949, 1952
	P. cerasus	up to 91	Megahed & Moore, 1967
	P. cerasus	-	Davidson, 1976
	P. mahaleb	10	Cation, 1949
	P. mahaleb	70	Megahed & Moore, 1967
	P. pennsylvanica	37	Megahed & Moore, 1967
	P. persica	3-9	Cochran, 1950
	P. persica	16	Millikan, 1959
	P. persica	1-12	Wagnon et al., 1960
Raspberry bushy dwarf	Rubus idaeus	40-50	Cadman, 1965
Raspberry latent[8] (Black raspberry latent)	Rubus spp.	10	Converse & Lister, 1969
Raspberry ringspot	Stellaria media	8-26	Murant et al., 1968
	Beta vulgaris	50-55	Lister & Murant, 1967
	Capsella bursa-pastoris	2-10	Lister & Murant, 1967
	Fragaria x ananassa	35	Lister, 1960
	Fragaria spp.	35	Lister & Murant, 1967
	Glycine max	7.2	Lister, 1960
	G. soja	20	fide Mandahar, 1978
	Petunia violacea	8-20	Phatak, 1974
	Rubus idaeus	18	Lister & Murant, 1967
	Stellaria media	29	Lister & Murant, 1967
Red clover mosaic	Trifolium pratense	12-18	Hampton, 1967
	Vicia faba	-	Sander, 1959
Red clover vein mosaic	Trifolium pratense	up to 100	Matsulevich, 1957

Disease	Host	%	Reference
Runner bean mosaic	*Vicia faba*	up to 100	Sander, 1959
	Phaseolus coccineus	42	Vashisth & Nagaich, 1965
Sincomas mosaic	*Pachyrrhizus erosus*	40-80	Fajardo & Maranon, 1932
Sour cherry yellows (see prune dwarf)			
Sowbane mosaic	*Atriplex pacifica*	21	Bennett & Costa, 1961
	Chenopodium album	30	Bennett & Costa, 1961
	C. amaranticolor	14-62	Kado, 1967; Dias & Waterworth, 1967
	C. murale	20-70	Bennett & Costa, 1961
	C. quinoa	2-46	Bancroft & Tolin, 1967; Dias & Waterworth, 1967
Soybean mild mosaic[21]	*Glycine max*	22-76	Takahashi *et al.*, 1974
	G. max	34	Iizuka, 1973
	G. max	0-68	Kendrick & Gardner, 1924
	G. max	1-18	Ross, 1963
	G. max	1-24	Kennedy & Cooper, 1967
	G. max	4-30	Phatak, 1974
	G. soja	10-25	*fide* Mandahar, 1978
Soybean stunt	*G. max*	50	Koshimizu & Iizuka, 1963
	G. max	95	Iizuka, 1973
	Vigna unguiculata	5	Iizuka, 1973
Squash mosaic	*Citrullus vulgaris*	1.5	Nelson & Knuhtsen, 1969
	Cucumis melo	1-27	Mahoney, 1935
	C. melo	12-93	Rader *et al.*, 1947
	C. melo	9	Kemp *et al.*, 1972
	C. melo	4	Phatak, 1974
	C. melo	0-34.6	Alvarez & Campbell, 1978
	Cucubita flexuous	—	Rader *et al.*, 1947
	C. maxima	0.2-1.5	Grogan *et al.*, 1959
	C. melo	6.6-21	Grogan *et al.*, 1959
	C. melo	3	Nelson & Knuhtsen, 1973a
	C. mixta	0.3	Grogan *et al.*, 1959;
	C. moschata	-	Rader *et al.*, 1947; Campbell, 1971

TABLE I. Continued

Virus	Host	Percent[1,2] transmission	Reference
Stone fruit ringspot	C. pepo	2.2	Middleton, 1944
	C. pepo	up to 5	Grogan et al., 1959; Nelson & Knuhtsen, 1973a
	Squash		fide Mandahar, 1978
Strawberry latent ringspot	Apium graveolens	98-100	Walkey & Whittingham-Jones, 1970
	Capsella bursa-pastoris	4	Schmelzer, 1969; Allen et al., 1970
	Chenopodium quinoa	65-100	Allen et al., 1970; Schmelzer, 1969
	C. quinoa	74	Hanson & Campbell, 1979
	Lamium amplexicaule	–	Murant & Goold, 1969
	Mentha arvensis	6	Taylor & Thomas, 1968
	Petroselinum crispum var. neopolitianum	6.6	Hanson & Campbell, 1979
	Stellaria media	97	Murant & Goold, 1969
	Senecio vulgaris	20	Murant & Goold, 1969
Sugarcane mosaic[22]	Zea mays	up to 1	Shepherd & Holdeman, 1965; Williams et al., 1968; Baudin, 1969
Tobacco mosaic	Apple cv. McIntosh	up to 21	Allen, 1969
	Capsicum frutescens[23]	22	McKinney, 1952
	Carthamus tinctorius[24]	–	Lockhart & Goethals, 1977
	Lycopersicon esculentum[23]	6	Doolittle & Beecher, 1937
	Malus platycarpa	38	Gilmer & Wilks, 1967
	M. sylvestris	3-37	Gilmer & Wilks, 1967
	Plantago major	–	Prochazkova, 1977
	Pyrus communis	35	Gilmer & Wilks, 1967
	Vitis vinifera	20	Gilmer & Kelts, 1968
	Vigna unguiculata	1.4	Phatak, 1974
Tobacco rattle	Capsella bursa-pastoris	1.9	Lister & Murant, 1967
	Lamium amplexicaule	2.2	Lister & Murant, 1967
	Myosotis arvensis	6.0	Lister & Murant, 1967

Disease	Species		Reference
Tobacco ringspot	Papaver rhoeas	1.1	Lister & Murant, 1967
	Viola arvensis	3	Lister & Murant, 1967
	Cucumis melo	3-7	McLean, 1962
	Glycine max	54-78	Desjardins et al., 1954
	G. max	78-82	Kahn, 1956
	G. max	100	Athow & Bancroft, 1959
	G. max	40	Kahn et al., 1962
	G. max	up to 100	Owusu et al., 1968
	G. max	100	Iizuka, 1973
	G. max	94-97	Yang & Hamilton, 1974
	Gomphrena globosa	25-50	Kahn et al., 1962
	G. globosa	46	Iizuka, 1973
	Lactuca sativa	3.0	Grogan & Schnathorst, 1955
	L. sativa	21	Iizuka, 1973
	Nicotiana glutinosa	—	Stace-Smith, fide Phatak, 1974
	N. tabacum	17	fide Mandahar, 1978
	N. tabacum	4.9	Valleau, 1932a
	Pelargonium hortorum	4-6	Scarborough & Smith, 1975
	Petunia hybrida	20	Henderson, 1931
	Senecio vulgaris	52	Tomlinson & Carter, 1971
	Taraxicum officinale	9-36	Tuite, 1960
	Vigna sinensis	82	Kahn, 1956
	Zinnia elegans	5	Iizuka, 1973
Tobacco streak[25]	Chenopodium quinoa	—	fide Neergaard, 1977
	Datura stramonium	79	Blakeslee, 1921
	D. stramonium	up to 94	Edwardson & Purcifull, 1974
	Glycine max	3.31	Ghanekar & Schwenk, 1974
	Phaseolus vulgaris	1-26	Thomas & Graham, 1951
Tomato aspermy	Stellaria media	—	Noordam et al., 1965
Tomato black ring	Beta vulgaris	3-27	Gibbs & Harrison, 1964
	B. vulgaris	56	Lister & Murant, 1967
	Capsella bursa-pastoris	90	Lister, 1960
	Cerastium vulgatum	33-100	Lister & Murant, 1967

TABLE I. Continued

Virus	Host	Percent[1,2] transmission	Reference
	Chenopodium album	84	Lister & Murant, 1967
	Fragaria x ananassa	40	Lister, 1960
	Fumaria officinalis	100	Lister & Murant, 1967
	Glycine max	83	Lister, 1960
	Lamium amplexicaule	10-48	Lister & Murant, 1967
	Lingustrum vulgare	5.7-8.3	Lister & Murant, 1967
	Lycopersicon esculentum	19	Lister & Murant, 1967
	Myosotis arvensis	100	Lister & Murant, 1967
	Nicotiana rustica	4.4-8.8	Lister & Murant, 1967
	Petunia violacea	9-30	Phatak, 1974
	Poa annua	2.7	Lister & Murant, 1967
	Polygonum persicaria	21-100	Lister & Murant, 1967
	Rubus idaeus	1-6	Lister & Murant, 1967
	Rubus spp.	5-19	Lister & Murant, 1967
	Senecio vulgaris	14	Lister, 1960
	S. vulgaris	3-40	Lister & Murant, 1967
	Spergula arvensis	63	Lister & Murant, 1967
	Stellaria media	65	Lister & Murant, 1967
	Vigna sinensis	23	Lister, 1960
Tomato bunchy top[26] (see potato spindle tuber viroid)			
Tomato bushy stunt	Malus pumila	up to 17	Allen, 1969
Tomato ringspot	Fragaria vesca	68	Mellore & Stace-Smith, 1963; Kahn, 1956
	Glycine max	76	Kahn, 1956
	Gomphrena globosa	76	Keplinger & Braun, 1973
	Lycopersicon esculentum	3	Hollings et al., 1972
	Nicotiana tabacum	11	Hollings et al., 1972
	Pelargonium hortorum	11	Hollings et al., 1972

	Rubus idaeus	—	Stace-Smith, *fide* Phatak, 1974; Braun & Keplinger, 1973
	Sambucus spp.	11	Uyemoto *et al.*, 1971
	Trifolium pratense	3-7	Hampton, 1967
Tomato spotted wilt	*Senecio* spp.	96	*fide* Mandahar, 1978
	S. cruentus	up to 70	Jones, 1944
Tomato streak	*Lycopersicon esculentum*	66	*fide* Mandahar, 1978
Turnip mosaic[9]	*Raphanus raphanistrum*	4	Tomlinson & Walker, 1973
Urd bean leaf crinkle	*Phaseolus mungo*	18.30	Kolte & Nene, 1972
Watermelon mosaic	*Echinocystis lobata*	2	Lindberg *et al.*, 1956
White clover mosaic			
(see Clover white mosaic)			

[1] Variable estimates or a range of per cent seed transmission, where given, are due in the majority of cases either to the use of different cultivars of the same host species or of different strains of the same virus. [2] Per cent seed transmission has not been given in some cases becuase either it has not been reported or due to the nonavailability of original paper. [3] Strain of eggplant mosaic virus (Gibbs and Harrison, 1973). [4] Seeds collected from infected plants were sown after a storage of 12 months. [5] Synonym of rose mosaic virus (*CMI/AAB Descriptions of Plant Viruses*, No. 83). [6] Probably a strain of *Prunus necrotic* ringspot virus (Fridlund, 1966). [7] Not clearly demonstrated to be a virus. [8] Strain of tobacco streak virus (Jones and Mayo, 1975). [9] Present in testa only of seeds of both the hosts, but was not transmitted to the seedlings. [10] Strain of prune dwarf virus (Ramaswamy and Posnette, 1971). [11] Serotype of tomato black ring virus (Kenten, 1972). [12] Also known as tobacco streak virus (Neergaard, 1977). [13] Isonym of broad bean true mosaic virus. [14] Strain of potato (Andean) latent virus (Neergaard, 1977). [15] Strain of tobacco necrosis virus (Mali, 1976). [16] This disease is caused by peach rosette mosaic virus (Dias and Cation, 1976). [17] Hevin *et al.* (1973) failed to obtain seed transmission in this host. [18] Strain of tomato ringspot virus (Martyn, 1968). [19] Serologically distantly related to barley stripe mosaic virus (Gibbs *et al.*, 1963). [20] Earlier considered as bean common mosaic virus but now given an independent ranking with present name by Phatak (1974). [21] Synonym of soybean chlorotic spot virus. [22] Its maize-infecting strain is maize dwarf mosaic virus. [23] Embryo transmission absent but surface-borne virus infects seedlings, particularly during transplanting. [24] Combined samples of its pericarp and integuments contained the virus which, however, was not transmitted to seedlings (Lockhart and Goethals, 1977). [25] Also known as *Datura quercina* virus (Neergaard, 1977; *CMI/AAB Descriptions of Plant Viruses*, No. 44). [26] May be the same as potato spindle tuber viroid (O'Brien and Raymer, 1964; Bensen *et al.*, 1965).

mitted through seed has been on the increase over the years: from 8 (Smith, 1951) to about 20 (Crowley, 1957) to 36 (Fulton, 1964) to 53 (Bennett, 1969) to 64 (Carter, 1973) to 85 (Phatak, 1974) and finally to 119 (excluding strains) (Table I). Similarly, the number of plant species reported to seed-transmit plant viruses has increased from about 40 (Crowley, 1957) to 63 (Fulton, 1964) to about 82 (Bennett, 1969) and finally to 162 (Table I). The picture slowly seems to be getting worse; but, keeping all the information in view (Section 8.4.1), it is not expected to become so alarming in the near future—at least not for the majority of seed-borne viruses. In some cases, however, unequivocal evidence shows that transmission of virus through seed is a positive threat to crop husbandry (Section 8.4).

Earlier reviews on seed transmission of viruses are by Fulton (1964), Baker and Smith (1966), Noble and Richardson (1968), Bennett (1969), Shepherd (1972), Carter (1973), Phatak (1974), and Neergaard (1977).

8.1.1 Terminology

The following terms used in the text are defined according to Neergaard (1977). *Seed transmitted* viruses are those which are transferred from one place to the other through the agency of the seed and cause infection of the plant produced by germination of such seed. *Seed-borne* viruses are carried in, on, or otherwise within the seed. *Infected seeds* are those in which the virus has penetrated into the tissues of the seed and is often in a resting stage. *Contaminated (infested) seeds* are those in which the virus is carried adhering to the external surface of the seeds.

8.2 SEED TRANSMISSION

Only those viruses which are present *in situ* at the time of differentiation of male and female sporogenous cells and before the callose walls are laid down will be able to infect microspores, megaspores, and embryos. For example, in barley infected by barley stripe mosaic virus, virus was present in floral meristems, subsequently in female and male archesporial cells and, still later, in megaspore mother cells and microspore mother cells before their separation from parental tissues, nucellus and tapetum, respectively, by the callose wall (Carroll and Mayhew, 1976a, 1976b). Similarly, bean common mosaic virus was detected in ovule primordia formed on placentae of ovaries of *Phaseolus vulgaris* cv. Beka (Schippers, 1963). There was hardly any question of the presence of callose wall at that time. This confirms the postulate of Bennett (1969) that seed transmission of a virus may depend upon its ability to invade the floral meristem early in its development and differentiation. Its failure to do so will permit the pollen-

and megaspore mother cells to escape infection, because of the callose wall, which will result in the formation of virus-free pollen and embryo sac. Non-seed and non-pollen transmissible isolate of barley stripe mosaic virus, for example, rarely invades anthers, and then only in a very late stage of their development (Carroll and Mayhew, 1976a).

Table I lists the viruses reported by the middle of 1979 to be seed transmitted. Extent and frequency of seed transmission of viruses varies over the whole range from 0 to 100%. Several plant viruses show 100% seed transmission; but percentage transmission of most seed-borne viruses rarely exceeds 50% (Crowley, 1957; Table I) and, in many cases, it is much less. This in no way decreases the epidemiological significance of seed transmission as shown by lettuce mosaic in lettuce where disease control becomes a losing proposition if the per cent seed transmission exceeds 0.1% (Broadbent et al., 1951; Grogan et al., 1952; Zink et al., 1956).

A few seed-borne viruses are present in immature seeds or seeds nearing maturity but are eliminated from mature, ripened seeds by viral inactivation during seed maturation. Such viruses are potato virus X in *Datura* and potato (Stelzner, 1942), southern bean mosaic virus in bean seeds (Cheo, 1955; Crowley, 1959; McDonald and Hamilton, 1972; Uyemoto and Grogan, 1977), pea streak virus in pea seeds (Ford, 1966), probably brinjal mosaic virus in brinjal seed (Mayee and Khatri, 1975; Mayee, 1977), broad bean stain virus and Echtes Ackerbohnen-mosaik-Virus (EAMV) in broad bean seeds (Cockbain et al., 1976; Vorra-urai and Cockbain, 1977), and peanut mottle virus strain M2 in peanut seed (Adams and Kuhn, 1977). Such viruses are transmitted to a high degree by immature seeds but are either not transmitted or transmitted to a very low level by mature seeds.

This loss of virus during ripening has been traditionally ascribed to the presence of some seed inhibitors or inactivating substances or systems during maturation (Cheo, 1955) but proof for the alleged *in vivo* activity of these inhibitors is lacking. On the other hand, many leaves can support virus replication even when they contain large quantities of polyphenols. However, Mandahar et al. (1979) found that the amount of phenols in virus-infected *Euphorbia pulcherrima* was influenced by virus infection which, in turn, lead to the break in dormancy of buds on infected stem cuttings.

8.2.1   Factors Affecting Seed Transmission

Bennett (1969) and Shepherd (1972) have exhaustively dealt with this topic; hence these factors are treated here in brief. Environmental factors (temperature), host plants, virus strains, and time of infection are all factors affecting seed transmission of viruses. Seed transmissibility of a particular virus varies greatly in different species or even varieties and cultivars of a particular host plant (Bennett, 1969; Shepherd, 1972; Khatri and Chohan, 1972; Stevenson and Hagedorn,

1969, 1973; Takahashi *et al.*, 1974; Ghanekar and Schwenk, 1974; Mayee and Khatri, 1975; and Ladipo, 1977). This has been correlated with susceptibility or resistance, and therefore, with the severity of disease of the mother plant: greater numbers of seeds of severely infected, therefore susceptible, plants will transmit virus than will seeds of less severely infected, therefore tolerant or resistant, plants. This close relationship can be explained on the basis of the mechanism of disease resistance. Resistance of a host to virus infection (*fide* Mandahar, 1978) may be due either to suppression of virus multiplication and, therefore, lower virus concentration in resistant and partially resistant varieties (Pring, 1974; Levy *et al.*, 1974; Pilowsky and Cohen, 1974; Wyatt and Kuhn, 1979), or to a reduced rate of systemic movement of virus in a plant or its failure to move out from inoculated leaves of resistant varieties (Tu and Ford, 1970; Jones and Tolin, 1972; Jensen, 1973; Wyatt and Kuhn, 1979). An isolate of peanut mottle virus had the highest concentration in leaves but was not seed transmissible, whereas another isolate, which had the lowest concentration in leaves, was seed transmissible at significant rates. The former situation may have been due to a decreased capacity of the virus to move out of infected leaves (Adams and Kuhn, 1977). It appears, therefore, that the ability of virus to move out of infected leaves is a more important determinant of seed transmissibility than is virus concentration in leaves.

Seed transmissibility of different strains or isolates of a virus can differ even in the same cultivar (Bennett, 1969; Shepherd, 1972; Nelson and Knuhtsen, 1973a, 1973b; Adams and Kuhn, 1977). Much evidence indicates that infected tissue harbors, in addition to the dominant variant producing typical disease symptoms, many other strains or isolates which seemingly remain restricted to the particular infected plant (*fide* Mandahar, 1978). It is conceivable, therefore, that in nature any virus may exist in an infected plant as a conglomeration of strains or isolates differing slightly from each other in various characteristics. Thus, it should be within the biological capability of a virus to preferentially "utilize" strains depending upon directional selection pressures and its particular environmental circumstance (Zitter and Murakishi, 1969; Wilkins and Catheral, 1974). This could also be true of the different strains of a virus with respect to its being or not being seed transmissible. Indeed, seed transmissibility may have been an evolutionary necessity for the epidemiology of some viruses in certain ecological situations. This appears to be true of viruses with narrow host ranges, of viruses which are carried from one season to the next exclusively or primarily in infected seed, and of nematode-transmitted, seed-borne viruses. The existence of such viral conglomerates is confirmed by the conclusion of Hanada and Harrison (1977) that seed transmission of different strains of raspberry ringspot, tomato black ring, and strawberry latent ringspot viruses at different temperatures in different geographical areas (eastern Scotland, England and Continental Europe) was adapted to the climatic conditions of the region. Viruses, like cellular organisms, apparently are governed by the processes of natural selection and survival of the fittest.

8.2.2 Genetics of Seed Transmission

As already mentioned, both host and virus genotypes control the capacity and extent of seed transmissibility of a virus in a host. Recently, some new and more pertinent evidence based on crossing experiments has been reported which clearly shows that this is so.

In crosses between resistant and susceptible barley cultivars, Carroll *et al.* (1979) found that resistance to seed transmission of barley stripe mosaic virus isolate MI-3 in barley was controlled by a single recessive gene in resistant barley and that this gene is probably specific against seed transmission of the MI-3 isolate and isolates identical to it. Seed transmission of bean common mosaic virus is dependent upon the dominance or recessiveness of resistance to transmission in the host plant (Medina and Grogan, 1961).

Harrison and coworkers (Harrison *et al.*, 1974; Harrison and Hanada, 1976; Hanada and Harrison, 1977; Harrison, 1977) have built the first genetic map, although still incomplete, of the nepovirus raspberry ringspot virus by using pseudo-recombinant isolates of this multicomponent virus. They found that the property of seed transmissibility of this virus is controlled by a gene located in RNA-1. Certain other characters of this virus which directly or indirectly aid in its seed transmission are also determined largely by RNA-1.

8.3 POLLEN TRANSMISSION

8.3.1 Methods Employed to Detect Pollen-Borne Viruses

8.3.1.1 Pollen Trituration. Pollen triturate is prepared in buffer and then inoculated to some indicator plant or subjected to conventional serology to test for the presence of virus. The following points are noteworthy in this connection. (i) Pollen should be washed before trituration to remove any external contaminating virus particles. Rader *et al.* (1947) had reported the presence of squash mosaic virus in triturates of presumably unwashed pollen, whereas Alvarez and Campbell (1978) later failed to detect the same virus in triturates of washed pollen. All such other reports on triturates of unwashed pollen are suspect, although, here, as in the original papers, no such distinction is made. (ii) The presence of viruses in triturates of washed or unwashed pollen has been interpreted as implying pollen transmission of these viruses without, in many cases, actually performing experiments to see if such pollen does lead to the production of infected seed. Thus both broad bean stain virus (BBSV) and *Echtes Ackerbohnenmo saik* virus (EAMV) were detected in pollen from infected plants but glasshouse experiments showed that EAMV was not transmitted through pollen to seeds, whereas

BBSV was (Vorra-urai and Cockbain, 1977). Similarly, potato virus T was detected in the pollen of *Datura stramonium, Nicandra physaloides* and *Solanum demissum* but was transmitted from pollen to seed only in *S. demissum* (Salazar and Harrison, 1978). Nevertheless, the detection of viruses in triturates of washed or unwashed pollen is here correlated, as in the majority of the original papers, with pollen transmission of these viruses.

8.3.1.2 Pollination and Fertilization. Pollen from infected plants is used for pollination and fertilization of an emasculated flower borne on healthy plants. Seed produced in this way, if infected, indicates the presence of virus in pollen. Pollen is generally mechanically carried to the stigma. Desjardins *et al.* (1979) employed bees for bringing about pollination, since mechanical pollination failed.

8.3.1.3 Conventional Electron Microscopy. It has often been employed to detect viruses in pollen and in various cells produced by germination of pollen.

8.3.2 Pollen Transmitted Viruses

Reddick and Stewart (1918) were probably the first to suggest the possible transmission of a virus, bean common mosaic virus, through pollen. Since then 37 viruses have been reported to be pollen transmitted (Table II). Per cent transmission of viruses through pollen may vary greatly depending upon the variety of the plant used and the strains of the virus employed (Medina and Grogan, 1961; Ramaswamy and Posnette, 1971).

A few of the non-seed transmitted viruses are carried by pollen as contaminant on the external surface of the pollen produced in anthers of infected plants. Such viruses are sowbane mosaic virus in *Atriplex coulteri* (Bennett and Costa, 1961); apple chlorotic leaf spot virus in *Chenopodium quinoa* and *C. amaranticolor* (Cadman, 1965); potato virus X in *Petunia hybrida* (Phatak, *fide* Neergaard, 1977), and bromegrass mosaic virus, southern bean mosaic virus and TMV on the exine of mature pollen of bean and cowpea (Hamilton *et al.*, 1977). About 40% of anthers of TMV-infected bean and cowpea contained infested pollen (Hamilton, *et al.*, 1977). None of these viruses have been proved to be seed transmitted because pollen was never infected (Hamilton *et al.*, 1977). The last three viruses— bromegrass mosaic, southern bean mosaic and TMV—are highly infectious and easily mechanically transmissible and, therefore, may easily spread in the natural state to plants pollinated by infested pollen. However, the exact epidemiological role of the infested pollen is not yet known.

TABLE II. Viruses Transmitted Through Pollen and/or Ovule

Virus	Host	Per cent transmission through			Reference
		Pollen	Ovule	Pollen & ovule	
Alfalfa mosaic	*Medicago sativa*	0.5-26.5	0-9.5	-[1]	Frosheiser, 1974
	M. sativa	1.0-14.0	0.5-6.0	-	Hemmati & McLean, 1977
Apple mosaic[2]	*Betula pendula*	x[3]	x	-	Gotlieb & Berbee, 1973; *fide* Neergard, 1977
A virus from decline infected pear	*Pyrus communis*	x	x	-	Williams & Smith, 1967
A virus of olive	*Olea europaea*	x	-	-	Pacini & Cresti, 1977
Avocado sunblotch	*Persea americana*	1.8-3.12	-	-	Desjardins *et al.*, 1979
Barley stripe mosaic	*Hordeum vulgare*	10.0	-	-	Gold *et al.*, 1954
	H. vulgare	35.0	-	-	Inouye, 1962
	H. vulgare	68.4	-	-	Carroll, 1974
	H. vulgare	45-59	61-70	-	Carroll & Mayhew, 1976 a and b
Bean common mosaic	*Phaseolus vulgaris*	25.0[4]	25.0[4]	50.0[4]	Nelson & Down, 1933
	P. vulgaris	18-76	8-78	21-86	Medina & Grogan, 1961
Bean southern mosaic	*P. vulgaris*	21.3	-	-	Crowley, 1959
Bean yellow mosaic	*Melilotus alba*	x	-	-	Frandsen, 1952
Beet cryptic	*Beta vulgaris*	x	x	-	Kassanis *et al.*, 1978
Black raspberry latent[5]	Black raspberry	x	-	-	Converse & Lister, 1969
Broad bean stain	*Vicia faba*	0.52	x	-	Vorra-urai & Cockbain, 1977
Cherry chlorotic leafspot	*Chenopodium quinoa*	x	-	-	*fide* Mandahar, 1978
Cherry necrotic ringspot (see under *Prunus* necrotic ringspot)					
Cherry rasp leaf	Cherry	x	-	-	Hansen *et al.*, 1974
Cherry ring mottle	Cherry	25, 33	-	-	Ramaswamy & Posnette, 1971
Cowpea aphid-borne mosaic	Cowpea	x	x	-	Tsuchizaki *et al.*, 1970
Cucumber mosaic	*Stellaria media*	x	-	-	Tomlinson & Carter, 1970b
Datura quercina	*Datura stramonium*	x	-	-	Blakeslee, 1921

TABLE II. Continued

Virus	Host	Per cent through transmission			Reference
		Pollen	Ovule	Pollen & ovule	
EAMV[6]	*Vicia faba*	x	x	-	Vorra-urai & Cockbain, 1977
Elm mosaic	*Ulmus americana*	30.5	75.0	48.0	Callahn, 1957
	Sambucus racemosa	x	-	-	Schmelzer, 1966
Elm mottle	*Syringa vulgaris*	x	-	-	Schmelzer, 1969
Lettuce mosaic	*Lactuca sativa*	0.48	5.8	5.2	Ryder, 1964
Loganberry degeneration	*Rubus loganbaccus*	x	-	-	Ormerod, 1970
Lychnis ringspot	*Lychnis divaricata*	18.6	30.7	36.6	Bennett, 1959
	Silene noctiflora	18.6	28.7	37.6	Bennett, 1959
Onion mosaic	*Allium cepa*	x	-	-	Cheremushkina, 1979
Onion yellow dwarf	*A. cepa*	x	-	-	Louie and Lorbeer, 1966
Peach rosette mosaic[7]	*Vitis labrusca*	-	x	-	Ramsdell & Meyers, 1978
Peanut mottle strain M3	*Arachis hypogea*	-	-	8.5	Adams & Kuhn, 1977
Potato spindle tuber	*Lycopersicon esculentum*	x	-	-	Singh, 1970
	Solanum toberosum	x	x	-	Singh, 1970
Potato virus T[8]	*Datura stramonium*	x	-	-	Salazar & Harrison, 1978
	Nicandra physaloides	x	-	-	Salazar & Harrison, 1978
	Solanum demissum	x	-	-	Salazar & Harrison, 1978
Prune dwarf	*Prunus avium*	x	-	-	Schimanski & Schade, 1974
	P. cerasus	19.3	-	-	Gilmer & Way, 1960
	P. mahaleb	x	-	-	Schimanski & Schade, 1974
Prunus necrotic ringspot	*Cucurbita maxima*	0.0	0.8	0.5	Das et al., 1961
	Prune	x	-	-	Cameron et al., 1978
	Prunus avium	x	-	-	Schimanski & Schade, 1974
	P. cerasus	27.8	-	-	Way & Gilmer, 1958
	P. cerasus	21	-	-	Cameron et al., 1973
	P. cerasus	x	-	-	Davidson, 1976
	P. mahaleb	x	-	-	Schimanski & Schade, 1974
Raspberry bushy dwarf	*Rubus idaeus*	x	-	-	Cadman, 1965; Barnett & Murant, 1970

Raspberry ringspot	*Frageria virginiana*	5.5	10.5	44.0	Lister & Murant, 1967
	Rubus sp.	5.2	13.5	-	Lister & Murant, 1967
Squash mosaic virus[9]	*Cucumis melo*	-	x	-	Alvares & Campbell, 1978
Stone fruit ringspot	Squash	x	-	-	*fide* Mandahar, 1978
Tobacco ringspot	*Glycine max*	x	x	-	Yang & Hamilton, 1974
Tomato black ring	*Beta vulgaris*	x	-	-	Gibbs & Harrison, 1964
	Rubus sp.	12.9	3.2	10.3	Lister & Murant, 1967
Tomato ringspot	*Pelargonium hortorum*	1	-	-	Scarborough & Smith, 1975
	Red raspberry	-	1.25	-	Braun and Keplinger, 1973

[1] Sign (-) means that transmission of the relevant virus through that organ was reported not to occur. [2] Strain of rose mosaic virus (*fide* Neergaard, 1977). [3] Sign (x) means that transmission of the relevant virus through that organ was reported but the per cent transmission was not found out by the investigator. [4] Approximate. [5] Strain of tobacco streak virus (Jones and Mayo, 1975). [6] Detected in pollen from infected plants but could not be transmitted through pollen to seed in glasshouse tests (Vorra-urai and Cockbain, 1977). [7] Appears to be seed-borne only via maternal tissue (Ramsdell and Myers, 1978). [8] This virus was detected in the pollen from each of the three plants (*D. stramonium*, *N. physaloides* and *S. demissum*) but was transmitted from pollen to seed only in *S. demissum* (Salazar and Harrison, 1978). [9] Rader *et al.* (1947) recovered it earlier from pollen but Alvarez and Campbell (1978) failed to transmit it through pollen to seed or to detect it in washed triturates of pollen. It is transmitted only through ovule (Alvarez and Campbell, 1978).

8.3.3   Relative Efficiency of Ovule and Pollen Transmitted Viruses to Become Seed Transmitted

Transmission of viruses through pollen as compared to ovule transmission is relatively an inefficient method for the viruses to become seed transmitted. Numbers of barley stripe mosaic virus-infected seeds produced on infected barley plants often are greater by 50% or more than those produced on healthy plants pollinated by infected pollen (McKinney, 1951a, 1951b). About 30.5% of elm seeds were infected with elm mosaic virus when pollen from infected plants was used to pollinate flowers on healthy plants, whereas 75% of the seeds produced by infected female plants harbored the virus (Callahn, 1957). Seed transmission of *Lychnis* ringspot virus was 18.6% in the seeds of *Lychnis divaricata* when only the male plant was infected, 30.7% when only the female plant was infected, and 33.6% when both parents were infected (Bennett, 1959). Efficiency of pollen transmission of lettuce mosaic virus was about one-tenth that of ovule transmission (Ryder, 1964). Infected pollen played no significant role in transmission of tobacco ringspot virus in soybean, so that seed transmission of this virus depended on ovule infection (Yang and Hamilton, 1974). And fewer broad bean seeds were infected with broad bean stain virus through pollen than on BBSV-infected mother plants in glasshouse experiments (Vorra-urai and Cockbain, 1977). However, pollen transmission of bean common mosaic virus is more efficient (18-21%) than ovule transmission (8%) in one bean variety, but almost equally efficient (67% pollen transmission compared to 60-76% ovule transmission) in another variety (Medina and Grogan, 1961); whereas Nelson and Down (1933) report that both pollen and ovule transmission of this virus in bean seeds was 25%. Low efficiency of seed transmission of pollen-borne viruses can be accounted for in one or more of the following ways.

(i) Pollen of plants infected by some viruses aborts. This is true in *Datura stramonium* infected by tobacco streak virus (*Datura quercina* virus) (Blakeslee, 1921), in tobacco infected by certain strains of tobacco ringspot virus (Valleau, 1932b, 1939), in barley infected by barley stripe mosaic virus (Yamamoto, 1951; Inouye, 1962; Carroll and Mayhew, 1976a), in lettuce infected by lettuce mosaic virus (Ryder, 1964); in *Pelargonium hortorum* infected by tomato ringspot virus (Murdock *et al.*, 1976) and in onion infected by onion mosaic virus (Cheremushkina, 1979). This will lead to pollen- and male-sterility.

(ii) The number of pollen grains produced per flower or anther is considerably reduced in virus-infected plants as compared to healthy plants. The number of pollen grains per flower produced by tobacco ringspot virus-infected soybean plants ranged from 0-2000 compared to 4000-6500 produced by flowers on healthy plants (Yang and Hamilton, 1974). Earlier, Valleau (1939) had reported reduced production of pollen in tobacco infected by this very virus. Carroll and Mayhew (1976a) found that the average number of pollen grains in a microscope field was 61.3 (range 50-75) in anthers of barley stripe mosaic virus, MI-1 strain, infected barley plants and 84.7 (range 57-111) in anthers of healthy barley plants.

The average pollen sterility was 10% in infected and 6% in healthy plants. Thus anthers on primary tillers of MI-1 infected barley contained less total pollen and more sterile pollen than the anthers of healthy plants. Yamamoto (1951) and Inouye (1962) had also reported that barley stripe mosaic virus infection caused a reduction in normal pollen.

(iii) Tomato black ring virus and raspberry ringspot virus infected pollen of raspberry is less competitive in fertilizing ovules than is healthy pollen (Lister and Murant, 1967). This could be due to either of the following causes reported by Yang and Hamilton (1974) for tobacco ringspot virus infected soybean plants: the germination rate of pollen from infected plants was 47%, compared to 77% for pollen from healthy plants; in addition, pollen from infected plants produced shorter germ tubes which also elongated more slowly than the normal germ tubes of healthy pollen. Pollen from tobacco ringspot virus infected soybean is, therefore, less viable and obviously less competitive than pollen from healthy plants.

(iv) The number of diseased plants in a field is generally much less than the number of healthy plants. Hence, the total amount of infected pollen produced is bound to be much less than the healthy pollen produced. This, together with the three aforementioned reasons and the fact that pollination and fertilization are largely random processes, will lead to an overall preponderance of healthy over infected seed.

(v) Much of the data listed in Table II about per cent seed transmission through pollen is based on experiments carried out under controlled glasshouse conditions in which pollen from infected plants is mechanically and deliberately carried to stigmas by the investigator. Even then pollen transmission competes poorly with ovule transmission of viruses. No such guiding hand exists in nature, where pollen transmission of viruses is bound to be still less effective.

(vi) Pollen is generally short-lived and, hence, can not be carried far from the source of its origin.

The foregoing discussion indicates that the epidemiological role of pollen transmission of viruses in the production of infected seed is likely to be slight. This role is bound to be still less important in exclusively or predominantly self-pollinating crops like broad bean, barley, and other cereals. Thus seed transmission of broad bean stain virus in broad bean and barley stripe mosaic virus in barley may not be greatly dependent upon pollen (Vorra-urai and Cockbain, 1977; Inouye, 1962).

8.3.4 Infection of the Ovule-Bearing Plant

Pollen transmitted viruses remain restricted to the developing embryo because of the surrounding callose layer. They normally cannot escape this isolation and cause systemic infection of the parent plant. However, about a dozen cases are known where virus brought by pollen to the flower-bearing plants becomes systemically distributed in the latter and incites disease: prunus (necrotic) ringspot virus in cherry (Way and Gilmer, 1958; George and Davidson, 1963; Cameron *et al.*, 1973; Davidson, 1976) and in squash (*Cucurbita maxima*) (Das and Milbrath, 1961); cherry yellows virus in cherry (Gilmer and Way, 1963; George and Davidson, 1963); prune dwarf virus in cherry (Gilmer and Way, 1963); apple chlorotic leaf spot virus (raspberry bushy dwarf virus) in raspberry (Cadman, 1965), and black raspberry latent virus in raspberry (Converse and Lister, 1969).

Cameron *et al.* (1973) and Davidson (1976) performed the following convincing experiment to establish firmly that infected pollen was the major means of spread of prunus (necrotic) ringspot virus in cherry orchards. In the case of Cameron *et al.* (1973), half of 68 cherry trees were debloomed (treated) each year over a 7-year period, while the remaining half of the trees were allowed to bloom normally (untreated). Davidson (1976) conducted an identical experiment, except that it involved a total of 64 trees and a 9-year test period. None of the debloomed cherry trees became infected during the test periods, whereas 87.5% (Davidson, 1976) and 73.5% (Cameron *et al.*, 1973) of the normally flowering trees had become infected with the virus. George and Davidson (1963) had earlier performed an identical experiment, lasting 3 years, on the spread of this virus in 93 cherry trees and arrived at the same results. Additionally, extensive attempts by Swenson and Milbrath (1964) and others to find any insect vector for the spread of this virus proved negative. Cadman (1965) also failed to find any vector for apple chlorotic leaf spot virus in raspberry, although the disease spreads rapidly in raspberry orchards after its initial appearance. How the trees are infected via pollination is not known, especially since the callose layer presumedly has already been formed much before the differentiation of egg and its fertilization.

8.3.5 Histological and Cytological Studies on Development of Infected Anther and Infected Pollen

8.3.5.1 Causes of Pollen Abortion. Abortion of pollen in anthers of infected plants has generally been found to be due to abnormal and underdeveloped anthers and abnormal anther dehiscence (Yamamoto, 1951; Inouye, 1962; Carroll and Mayhew, 1976a; Murdock *et al.*, 1976), degeneration of cytoplasm and collapse of walls of pollen grains (Murdock *et al.*, 1976), and abnormal mitosis and meiosis during anther and pollen development (Caldwell, 1952).

Anthers in flowers of tobacco ringspot virus infected soybean plants were necrotic and puffy, and they usually failed to dehisce (Yang and Hamilton, 1974). Anthers in flowers of *Pelargonium hortorum* infected by tomato ringspot virus were shrunken; anthers and filaments showed generalized tissue disintegration; and tapetum, apparently completely disintegrated, was not to be seen (Murdock *et al.*, 1976). Anthers of barley stripe mosaic virus infected barley were, on the whole, shorter than anthers of healthy plants; moreover, about 22% of the anthers of the former were abnormally dehiscent and had underdeveloped longitudinal slits at maturity, as compared to 10.4% abnormally dehiscent anthers in flowers of healthy barley (Carroll and Mayhew, 1976a).

Cytoplasm within the pollen grains of *Pelargonium hortorum* infected with tomato ringspot virus was either present in very small amounts or lacking. And many infected grains were distorted or crescent-shaped because of partial collapse of their walls (Murdock *et al.*, 1976).

Viruses are known to bring about cytological aberrations in meiotic and mitotic cell division during the formation of sex cells. Meiotic division of the mother cells leading to the formation of microspores and megaspores in chrysanthemums and tomato is stopped by aspermy virus (Caldwell, 1952).

8.3.5.2 Location of Viruses In Situ. Seed transmission of pollen-borne viruses implies that the transfer of virions takes place through pollen tube to the egg along with male gametes, that virions are present *in situ* at the time of fertilization, and that the virions are then also present within the developing embryo and endosperm. There was, however, no direct proof of this obviously true speculation till about 6-8 years ago. That direct proof is now available as a result of the work on barley stripe mosaic virus in barley by Carroll (1972, 1974), Mayhew and Carroll (1974a, 1974b) and Carroll and Mayhew (1976a), and on tobacco ringspot virus in soybean by Yang and Hamilton (1974).

The seed-borne strain of barley stripe mosaic virus (MI-1 strain) infected barley pollen (Carroll, 1972, 1974), and virions were seen in the cytoplasm of vegetative cells (Carroll, 1972; Gardner, 1967) and in sperms (Carroll, 1974). Carroll and Mayhew (1976a) found that MI-1 invaded the floral meristem in its earliest stages of initiation and development and that MI-1 virions were subsequently present in the cytoplasm of cells of all the developmental stages of anthers and pollen studied. Virions were present in interphase cells as well as in mitotically and meiotically dividing cells, i.e., in cells undergoing development and differentiation. They were present primarily in the cytoplasm as individual virions, virion aggregates, or virion-cytoplasmic tubule complexes in resting cells, among the chromosomes and in association with spindle microtubules in mitotically and meiotically dividing cells. MI-1 virions were seen in the cytoplasm of many of the cells of the walls of anther locules, archesporium, pollen mother cells, dyads, tetrads, developing and maturing microspores in various stages of vacuolation, mature microspores, vegetative cells, generative cells, and sperms. Some virions in sperms were also present among the contents of nuclei around which there were no envelopes. Carroll (1974) had earlier reported that MI-1 particles occurred within 68.4% of the pollen grains; within the cytoplasm of 41.2% of sperm cells, of which 14.4% also contained virus particles in their nuclei, and within the cytoplasm of 68.4% of vegetative cells, out of which 12.2% also contained virions in their nuclei. Virions were, however, never detected by Carroll (1974) and Carroll and Mayhew (1976a) in primary or callose walls of pollen mother cells and microspores, in intines and exines of pollen, and in cell walls of generative cells.

Yang and Hamilton (1974) detected tobacco ringspot virus particles within intines of pollen and within generative cells and their walls, but not in vegetative cells of pollen from infected soybean plants. No surface contaminating barley stripe mosaic virus and tobacco ringspot virus particles were carried externally on exines of pollen from their respective infected plants.

Pacini and Cresti (1977) found that virus particles were present in the cytoplasm, adjacent to the callose wall, of microspore mother cells and all subsequent stages of microspore development up to maturation. Later, they enter the germination pores since the plasmalemma harboring these particles evaginates into the pores. A layer of pectocellulosic material is laid down, soon after first haploid mitosis, between the plasmalemma and the germination pore so that the evagination gets separated from the cytoplasm and virus particles cannot return to it. Thus, mature pollen grains are free of the virus; virions present in evaginations also disappear, possibly because of the presence of some lytic enzyme.

8.3.5.3 Cell-to-Cell Transfer of Virus Particles.   The studies of Mayhew and Carroll (1974a) and Carroll and Mayhew (1976a) are also interesting in another way. It is widely agreed now that cell-to-cell movement of viruses occurs through plasmodesmata (Mandahar, 1978). Mayhew and Carroll (1974a) and Carroll and Mayhew (1976a), on the other hand, provide evidence that indicates that movement of MI-1 virions from parent to progeny cells in dividing and differentiating cells of anthers and pollen takes place via spindle microtubule transfer from parent to progeny cells. This conclusion is based on the following: virions were never detected in plasmodesmata of these cells; both long and short virus particles were mostly present on and intimately associated with or attached to spindle microtubules during meiosis as well as mitosis, including all the meiotic stages of pollen development. This was also true of ovule and megagametophyte development (Carroll and Mayhew, 1976b). This, however, does not appear to be true for cell-to-cell transfer of tobacco ringspot virus during ovule and megagametophyte development in infected soybean plants, since particles of this virus were present in plasmodesmata between nucellar cells and between egg and synergid cells (Yang and Hamilton, 1974).

8.4   ECOLOGY AND EPIDEMIOLOGY

Seed transmission of viruses has great potential ecological and epidemiological significance, as has been demonstrated in at least some cases. Some of seed transmitted viruses cause havoc to crops as is clear from some selected reports given here. There was a 17-24% reduction in barley yield because of seed transmission of barley stripe mosaic virus (Timian and Sisler, 1955); although yield reduction was much greater, 62%, in experiments (Catherall, 1972). A 0.1% level of seed transmission of lettuce mosaic virus in lettuce may result in a total crop loss (Broadbent et al., 1951; Grogan et al., 1952; Zink et al., 1956). Seed-borne bean common mosaic virus is one of the principal causes of yield loss of bean in Morocco, where growers report an estimated 50% reduction in yield when infection is highest (Lockhart and Fischer, 1974). The same seed-borne virus reduced bean yields in Iran by 80% or more (Kaiser fide Neergaard, 1977). Seed-transmitted bean yellow mosaic virus in *Vicia faba* resulted in 4-34% of plants becoming diseased after 15 weeks and 93-100% after 22 weeks (Kaiser, 1973).

Many viruses have restricted host ranges, others are confined to annual herbaceous plants and have no perennial hosts, whereas still others have insect vectors like nematodes which have very limited mobility. All such viruses would risk oblivion or become confined to very narrow ecological niches if they had not evolved the capacity to survive by seed transmission. This ability ensures the survival of viruses and their perpetuity from one season to the next or over several years—in fact, for as long as the virus carrying seeds remain viable.

Unequivocal evidence clearly indicates that seed transmission is the primary source of infection, and the only or the most important factor in the epidemiology, survival, annual recurrence and initiation of the following virus diseases: bean common mosaic virus in bean (Fajardo, 1930; Meiners et al., 1978), lettuce mosaic in lettuce (Broadbent et al., 1951; Grogan et al., 1952), tobacco mosaic in tomato (Taylor et al., 1961; Broadbent, 1965, 1976), barley stripe mosaic in barley and wheat (Slykhuis, 1976; Chiko, 1973), broadbean strain and EAMV in spring sown broad bean (Cockbain et al., 1975b, 1976), squash mosaic in cantaloupe (Alvarez and Campbell, 1978), brinjal mosaic in brinjal (Mayee and Khatri, 1975; Mayee, 1977), peanut mottle in peanut and soybean (Paguio and Kuhn, 1974; Demski, 1975; Adams and Kuhn, 1977) and nepoviruses in various hosts (Harrison, 1977; Hanada and Harrison, 1977; Lister and Murant, 1967; Murant and Lister, 1967).

Seedlings produced by germination of infected seeds serve as the initial sources of virus inoculum and establish foci of infection from which a virus is later transmitted by natural vectors to other plants in the field. These infected seedlings are randomly scattered in the crop which ensures optimum distribution of infection foci in the field. A crop may ultimately be extensively damaged. Lettuce mosaic virus is rarely transmitted through more than 5% of seeds of diseased plants (Broadbent et al., 1959); yet seed transmission of this virus is a major factor in its spread in England and California. In fact, as already mentioned, if its level of seed transmission exceeds 0.1%, severe outbreaks of the disease occur and control of the disease is likely to be unsatisfactory.

Long-distance dispersal of several plant viruses can be easily accounted for by their being seed transmitted. Establishment of such viruses in new, distant, geographically isolated regions, or over wide areas extending over even several countries is also a natural corollary of seed transmission. The primary factor responsible for the worldwide distribution of soybean mosaic virus in soybean is its being seed-borne (Kendrick and Gardner, 1924; Bowers, 1977, quoted in Cho and Goodman, 1979). Fajardo (1930) blamed infected commercial seed lots for the widespread geographical distribution of bean common mosaic virus. Spread of lettuce mosaic virus in England and California is mainly through infected seed (Broadbent *et al.*, 1951; Grogan *et al.*, 1952). The rapid worldwide distribution of certain strains of sugarcane mosaic virus in recent years has in all probability occurred through their being carried in the seeds of corn and perennial Johnson-grass (*Sorghum halepense*) (Shepherd and Holdeman, 1965).

Squash mosaic virus was introduced into the United States by seed from Iran (Leppik, 1964) and to New Zealand from the U.S. in seed of *Cucumis melo* (Thomas, 1973). Lettuce seed in Denmark meant for export was frequently found to be infected by lettuce mosaic (Anderson, 1974, *fide* Neergaard, 1977). Seed imported from Sweden gave rise to pea seed-borne virus-infected plants in quarantine in Tasmania, Australia (Munro, 1978). Broad bean stain virus and EAMV were detected in Australia in seedlings of *Vicia faba* raised from seeds imported from the United Kingdom (Randles and Duke, 1977). Guar symptomless virus appears to have been introduced into the U.S.A. from India in infected seeds of guar, *Cyamopsis tetragonoloba* (Hansen and Leseman, 1978). Strawberry latent ringspot virus has been introduced to the U.S. from Western Europe in imported seedlots of Plain parsley, *Petroselinum crispum* var. *neopolitanum* (Hansen and Campbell, 1979).

Plant breeding is providing disease resistant and higher yielding varieties. To this end, germplasm banks and plant introduction stations have been established in many countries of the world. These seed lots and lines, if already infected by viruses, can represent wasted time, money, and energy. For example, about 16% of the cherry seeds imported from France for use as root stocks were infected with cherry necrotic ringspot virus (Gilmer, 1955). Barley stripe mosaic virus occurs in PI lines and there is a danger of its spreading to breeders' plots through the pollen used in crossing experiments (Slykhuis, 1976). Cucumber mosaic virus was present in the PI lines of bean in Puerto Rico (Meiners *et al.*, 1977). Four of 172 pea breeding lines and 10 of 74 commercial seed lots contained low levels of pea seedborne mosaic virus (Mink and Parsons, 1978). Later it was found that 23% of the PI lines of *Pisum sativum* at the Northeastern Regional Plant Introduction Station, Geneva, New York, were infected by this virus (Hampton and Braverman, 1979). Seven strains of soybean mosaic virus were present in the soybean seeds of the United States Department of Agriculture's germplasm collections (Cho and Goodman, 1979).

Well documented research on the ecology and epidemiology of seed-borne plant viruses has been conducted using nematode-transmitted viruses (Lister and Murant, 1967; Murant and Lister, 1967; Harrison, 1977). Seeds of three common weeds, *Capsella bursa-pastoris*, *Senecio vulgaris*, and *Stellaria media*, in the U.K. transmit arabis mosaic, raspberry ringspot and tomato black ring viruses at fairly high rates (Table I) in nature. Subsequently, nematode vectors acquire these viruses from infected seedlings and transmit them to commercially important crops.

Persistence and survival of raspberry ringspot and tomato ringspot viruses in soil is not dependent upon the nematode vector since they can survive in it in fallow soil for only 9 weeks. Instead they survive in infected weed seeds in nematode infested soils and provide on germination a continuing source of virus for vectors for further transmission. These two viruses, therefore, survive the fallow period in weed seeds. Arabis mosaic and grapevine fanleaf viruses, on the other hand, are retained by *Xiphinema* spp. for 8 months or even more. Moreover, these nematode vectors have greater association with perennial hosts which act as virus reservoirs. These viruses should, therefore, be less dependent upon weed seeds for survival and transmission. It was actually found to be so in nature since these viruses were detected only occasionally in weed seeds in nematode-infested soil. Thus, seeds play a greater role in the ecology and epidemiology of some nematode transmitted viruses than others.

Many of the viruses that infect weeds are also transmitted through their seeds. Infected weed seedlings are the sources of primary virus inoculum for natural vectors. The importance of seed transmission in the epidemiology of nematode-borne viruses has already been emphasized. Johnsongrass is the primary overwintering host of maize dwarf mosaic virus in the U.S.A. *Viola arvensis* transmitted tobacco rattle virus through winter (Cooper and Harrison, 1973). Cucumber mosaic virus-infected seeds of *Stellaria media* germinated to produce infected seedlings even after being buried in soil for 21 months. Such seeds can serve as reservoirs of the virus and be a source of infection every year (Tomlinson and Walker, 1973). Kazda and Hevert (1977) found that cucumber mosaic virus passes autumn and winter in seeds of *S. media* which germinate in the following spring to produce infected seedlings from which virus is transmitted by aphids to cucumber.

8.4.1 Epidemiological Significance

The epidemiological significance of seed transmission of viruses has been convincingly established so far in only a few cases, such as the seed transmitted viruses already mentioned above. The epidemiological significance of seed transmission of most other viruses is seemingly not as great. That that is so is borne out by the following facts.

The total number of known plant viruses is much larger (Martyn, 1968, 1971; Fenner, 1976) than the viruses known to be seed transmitted (Table I), and the total number of plants reported to seed-transmit viruses (Table I) is only a fraction of the huge number of angiosperm species and cultivars. Seed-borne viruses may not now appear to be exceptional but they are certainly far less common than their non-seed-borne counterparts.

Only four or five pollen transmitted viruses are known to escape isolation from the embryo and spread systemically in blossom-bearing plants.

Some of the most widely distributed and most common viruses are either not seed transmitted or are seed transmitted in very few plants. The host range of some seed-borne viruses comprises several hundred plant species: more than 350 for tobacco rattle virus (Schmelzer, 1957), more than 300 for alfalfa mosaic virus (Neergaard, 1977; Smith, 1972), 165 species in 34 families for tomato spotted wilt virus (Best, 1968). Similarly, cucumber mosaic virus and TMV have very wide host ranges (Smith, 1972). However, all these viruses are transmitted through the seeds of comparatively very few plants (Table I). Thus, a compatible virus-host combination, with respect to seed transmission, is rather an occasional, if not rare, event for most seed transmitted viruses.

Even with seed transmitted viruses in their transmitting hosts, the number of infected seeds produced is generally less than 50% (Table I). Thus healthy seed normally predominates.

The transmission data given in Table I are based on three types of experiments: (i) Seeds collected from naturally infected plants are sown in a glasshouse to determine the number of infected seedlings produced. (ii) Seeds collected from manually inoculated plants in the field or glasshouse are sown in the glasshouse to determine the number of infected seedlings produced. (iii) Infected seeds are sown in the field or glasshouse, and the seeds produced from these infected plants are sown to determine the number of infected seedlings produced. These are quite different methodologies and, given the varied reactions of different cultivars in any one particular virus-host complex or of the same cultivar to different virus strains or isolates, they are bound to give widely different results.

Data collected in the glasshouse under controlled conditions and by manual inoculation cannot generally be considered as a fair indication of the efficiency of seed transmission or its relative epidemiological and ecological importance in nature. Virus-infected pollen is a poor competitor compared to healthy pollen (Lister and Murant, 1967). The rate of transmission of cucumber mosaic virus in seeds of *Stellaria media* varied from 21-40% in plants grown from infected seed, from 4-29% in naturally infected plants, and from 3-21% in inoculated plants (Tomlinson and Carter, 1970b). Arabis mosaic, raspberry ringspot, strawberry latent ringspot, and tomato black ring viruses were frequently seed transmitted in many weed and crop species in glasshouse experiments; however, seed transmission of two of these viruses, arabis mosaic and strawberry ringspot, was uncommon in the field (Lister and Murant, 1967; Murant and Goold, 1967; Hanada and Harrison, 1977).

Several of the reports of seed transmitted viruses are based solely on experiments carried out in the greenhouse to determine the host range of a virus or to specifically test its seed transmissibility in various hosts. No information is available about the seed transmission of such viruses in nature.

Thus, it would appear that seed transmission is only an occasional episode in the general drama of non-seed-transmission of viruses.

8.5 CONTROL

Seed treatment of TMV-infested tomato seeds has proved very effective. The virus is eliminated by fermenting the seed, by soaking it in HCl or by exposing it to trisodium-phosphate (Chamberlain and Fry, 1950; Alexander, 1960; Taylor et al., 1961; Gooding, 1975). Treating tomato seeds with 10% or higher concentrations of trisodium phosphate for 10 minutes or 20% HCl for 30 minutes has already become a routine practice (Neergaard, 1977; Brezhnev and Vlasov, 1977; Benott, 1977).

Treatment of seeds carrying virus internally in the embryo by heat and/or chemicals has not been totally successful. The viruses, in general, cannot be inactivated by seed treatment without also impairing seed viability (Reddick and Stewart, 1919; Owusu et al., 1968; Cockbain et al., 1976; Valerie and Walkey, 1978). However, some recent reports note the curative value of high temperature on seed infection (Howles, 1978; Cooper and Walkey, 1978). Also heat treatment of polyethylene glycol (PEG)-imbibed seeds has given very encouraging results (Cooper, 1976, quoted in Cooper and Walkey, 1978; Walkey and Dance, 1979).

Seed transmission of viruses mostly continues to take place at about the initial rate even after long storage. However, some reports demonstrate substantially decreased or no transmission of viruses by seeds after long storage periods of 3-6 years (Valleau, 1939; Middleton and Bohn, 1947; Rader et al., 1947; Fulton, 1964; Megahed and Moore, 1969). It is doubtful that reduction in per cent seed transmission or in the level of seed-borne viruses after such long-term storage can be of any practical help to the growers. But it is conceivable that those viruses which are inactivated within a few months during storage could be kept at bay by merely storing the seeds for that short duration. This is true of all viruses which are found in immature seeds but are subsequently inactivated during seed maturation (section 8.2). These viruses are not seed transmissible.

Tobacco mosaic virus occurs as a surface contaminant on roots and cotyledons of tomato seedlings produced from infected seed. These seedlings become infected with virus through injuries and abrasions received mostly during transplanting (Taylor et al., 1961; Broadbent, 1965). Thus the incidence of TMV infection increased 10-15 times in transplanted versus nontransplanted tomatoes; the yield of nontransplanted plants increased by 19% (Nikitina, 1966). Nezhnev (1977) also found that tomato plants grown without transplanting had lower incidences of TMV infection. Hence, raising the plants in fields directly from tomato seeds rather than by transplantation helps control TMV (Brezhnev and Vlasov, 1977).

Use of disease-free seed for control of plant virus diseases is now considered the most reliable method for controlling seed-borne virus diseases and producing healthy, reasonably disease-free crops. Government and private agencies in many countries now regularly index and certify stocks and seeds of various plants to be virus-free. This is particularly true of barley, bean, cherry, citrus, lettuce, pea, *Prunus*, strawberry, and sugarbeet (Quantz, 1958; Marrou and Messiaen, 1967; Hamilton, 1965, 1977; Childs, 1968; Ryder and Johnson, 1970; Phatak, 1974; Slack and Shepherd, 1975; Neergaard, 1977; Agrawal *et al.*, 1977; Hunnius, 1977). Use of certified and virus-free seeds and stocks has kept diseases of these important plants in check in India, the U.K., the U.S.A., Germany, Scotland and other countries. In a massive effort to control the spread of pea seed-borne mosaic virus in the U.S.A., 450,000 lbs of contaminated seed was not planted in 1977 (Mink and Parsons, 1978).

Two very useful new techniques for the detection of seed-borne viruses are ELISA and SSEM. Enzyme-linked immunosorbent assay (ELISA) has been used to detect viruses in seeds of cherry infected by prune dwarf virus (Caspar, 1977), in seeds of soybean infected by tobacco ringspot and soybean mosaic viruses (Lister, 1978), in soybean seeds infected by soybean mosaic virus (Bossennec and Maury, 1978), and in pea seeds infected by pea seed-borne mosaic virus (Hamilton and Nicholas, 1978). ELISA is a very sensitive test, its sensitivity being greater than that of other simple and convenient serological assays, and infectivity assays on indicator hosts. It is an economical, convenient and rapid method for detecting viruses in seed certification programs and for screening seed lots intended for germplasm pools.

Serologically specific electron microscopy (SSEM) has been employed for detecting and identifying seed-borne viruses in crude extracts of infected seeds. Various viruses detected in crude extracts of infected seed include tobacco ringspot, barley stripe mosaic, lettuce mosaic, pea seed-borne mosaic, and soybean mosaic (Bralansky and Derrick, 1976, 1979; Hamilton and Nicholas, 1978). It too is a very sensitive method, in some cases, even more so than ELISA. Therefore, SSEM can also be a very useful tool in quarantine measures and in other ways already mentioned in relation to ELISA.

The role of commercial seed lots in the worldwide distribution of some viruses and introduction of viruses into new countries through import of seed for germplasm pools has already been emphasized. Hence, rigorous application of quarantine measures for detection of seed-borne viruses is a necessity.

8.6 REFERENCES

Adams, D. B., and Kuhn, C. W. (1977). Seed transmission of peanut mottle virus. *Phytopathology* 67: 1126-1129.

Agarwal, V. K., Nene, Y. L., and Beniwal, S. P. S. (1977). Detection of bean common mosaic virus in urd bean (*Phaseolus mungo*) seeds. *Seed Sci. Technol.* 5: 619-625.

Alexander, L. J. (1960). Inactivation of tobacco mosaic virus from tomato seed. *Phytopathology* 50: 627.

Allen, W. R. (1969). Occurrence and seed transmission of tomato bushy stunt virus in apple. *Can. J. Pl. Sci.* 49: 797-799.

Allen, W. R., Davidson, T. R., and Briscoe, M. R. (1970). Properties of a strain of strawberry latent ringspot virus isolated from sweet cherry growing in Ontario. *Phytopathology* 60: 1262-1265.

Alvarez, M., and Campbell, R. N. (1978). Transmission and distribution of squash mosaic virus in seeds of cantaloupe. *Phytopathology* 68: 257-263.

Andersen, H. (1974). *fide* Neergaard, 1977.

Anderson, C. W. (1957). Seed transmission of three viruses in cowpea. *Phytopathology* 47: 515.

Anonymous. (1977). Annual Report 1977. Centro Internacional de Agricultura Tropica (CIAT), Cali, Colombia.

Athow, K. L., and Bancroft, J. B. (1959). Development and transmission of tobacco ring-spot virus in soybean. *Phytopathology* 49: 697-701.

Baker, K. F., and Smith, S. H. (1966). Dynamics of seed transmission of plant pathogens. *Annu. Rev. Phytopathol.* 4: 311-334.

Bancroft, J. B., and Tolin, S. A. (1967). Apple latent virus 2 is sowbane mosaic virus. *Phytopathology* 57: 639-640.

Barnett, O. W., and Murant, A. F. (1970). Host range, properties and purification of raspberry bushy dwarf virus. *Ann. Appl. Biol.* 65: 435-449.

Baudin, P. (1969). Transmission par graines de mais du virus de la mosaique de la canne a' sucre. *Revue Agric. Sucr. Ile Maurice* 48: 277-278.

Belli, G. (1962). Notes and experiments on the transmission of lucerne mosaic virus through the seed and demonstration of its exclusion from clones of virus infected vines. Reprinted from *Ann. Fac. Agr. Milano* 10(1961): 1-15 (*fide* Neergaard, 1977).

Bengno, D. A., and Favali-Hedayat, M. A. (1977). Investigations on previously unreported or noteworthy plant viruses and virus diseases in Philippines. *FAO Pl. Protection Bull.* 25: 78-84.

Bennett, C. W. (1944a). Latent virus of dodder and its effect on sugarbeet and other plants. *Phytopathology* 34: 77-91.

Bennett, C. W. (1959). *Lychnis* ringspot. *Phytopathology* 49: 706-713.

Bennett, C. W. (1969). Seed transmission of plant viruses. *Adv. Virus Res.* 14: 221-261.

Bennett, C. W., and Costa, A. S. (1961). Sowbane mosaic caused by a seed-transmitted virus. *Phytopathology* 51: 546-550.

Bennett, C. W., and Esau, K. (1936). Further studies on the relation of the curly top virus to plant tissues. *J. Agr. Res.* 53: 595-620.

Benott, M. A. (1977). Does tobacco seed transmit tobacco mosaic virus? *Planter* 53: 197-201.

Benson, A. P., Raymer, W. B., Smith, W., Jones, E., and Munro, J. (1965). Potato diseases and their control. *Potato Handbook* 1965: 32-38.

Best, R. J. (1968). Tomato spotted wilt virus. *Ad. Virus Res.* 13: 65-146.

Blakeslee, A. F. (1921). A graft infectious disease of datura resembling a vegetative mutation. *J. Genet.* 11: 17-36.

Blaszczak, W. (1963). Seed transmission of narrowleavedness of yellow lupin (NYL). *Genet. Polon.* 4: 65-77.

Blaszczak, W. (1965). Severe strain of yellow bean mosaic virus found on *Trifolium pratense* L. *Bull. Acad. Pol. Sci.* Cl. V. *Ser. Sci. Biol.* 13: 381-384.

Bock, K. R. (1973a). East African strains of cowpea aphid-borne mosaic virus. *Ann. Appl. Biol.* 74: 75-83.

Bock, K. R. (1973b). Peanut mottle virus in East Africa. *Ann. Appl. Biol.* 74: 171-179.

Bojannsky, V., and Koslarova, V. (1968). Euonymus mosaic. *Biol. Planta.* (Praha) 10: 322-324.

Bos, L. (1970). Bean yellow mosaic virus. *CMI/AAB Descriptions of Plant Viruses*, No. 20.

Bos, L., and van der Want, J. P. H. (1962). Early browning of pea, a disease caused by a soil and seed-borne virus. *Tijdschr. PlZiekt.* 68: 368-390.

Bossennec, J. M., and Maury, Y. (1978). Use of the ELISA technique for the detection of soybean mosaic virus in soybean seeds. *Ann. de Phytopathologie* 10: 263-268.

Brlansky, R. H., and Derrick, K. S. (1976). Detection of seed-borne plant viruses using serologically specific electron microscopy. *Proc. Amer. Phytopathol. Soc.* 3: 334.

Brlansky, R. H., and Derrick, K. S. (1979). Detection of seed-borne plant viruses using serologically specific electron microscopy. *Phytopathology* 69: 96-100.

Brantley, B. B., Kuhn, C. W., and Sowell, G., Jr. (1965). Effect of cucumber mosaic virus on southern pea (*Vigna sinensis*). *Proc. Am. Soc. Hort. Sci.* 87: 355-358.

Brants, D., and van der Heuvel, J. (1965). Investigation of *Hippeastrum* mosaic virus in *Hippeastrum hybridum*. *Neth. J. Plant Pathol.* 71: 145-151.

Braun, A. J., and Keplinger, J. A. (1973). Seed transmission of tomato ringspot virus in raspberry. *Pl. Dis. Reptr.* 57: 431-432.

Bretz, T. W. (1950). Seed transmission of the elm mosaic virus. *Phytopathology* 40: 3.

Brezhnev, D. D., and Vlasov, Yu. I. (1977). (The problem of seed production of tomato on a virus-free basis.) *Trudy po Prikladno i Botanike. Genetike i Selektsii* 61: 76-81.

Broadbent, L. (1965). The piedemiology of tomato mosaic. XI. Seed-transmission of TMV. *Ann. Appl. Biol.* 56: 177-205.

Broadbent, L. (1976). Epidemiology and control of tomato mosaic virus. *Annu. Rev. Phytopathol.* 14: 75-96.

Broadbent, L., Tinsley, T. W., Buddin, W., and Roberts, E. T. (1951). The spread of lettuce mosaic in the field. *Ann. Appl. Biol.* 38: 689-706.

Brunt, A. A., and Kenten, R. H. (1973). Cowpea mild mottle, a newly recognized virus infecting cowpeas (*Vigna unguiculata*) in Ghana. *Ann. Appl. Biol.* 74: 67-74.

Burkholder, W. H., and Muller, A. S. (1926). Hereditary abnormalities resembling certain infectious diseases in beans. *Phytopathology* 16: 731-737.

Cadman, C. H. (1965). Filamentous viruses infecting fruit trees and raspberry and their possible mode of spread. *Pl. Dis. Reptr.* 49: 230-232.

Caldwell, J. (1952). Some effects of plant viruses on nuclear division. *Ann. App. Biol.* 39: 98-102.

Callahn, K. L. (1957). Pollen transmission of elm mosaic virus. *Phytopathology* 47: 5.

Cameron, H. R., Milbrath, J. A., and Tate, L. A. (1973). Pollen transmission of prunus ringspot virus in prune and sour cherry orchards. *Pl. Dis. Reptr.* 57: 241-243.

Campbell, R. N. (1971). Squash mosaic virus. *CMI/AAB Descriptions of Plant Viruses*, No. 43.

Capoor, S. P., and Varma, P. M. (1956). Studies on a mosaic disease of *Vigna cylindrica* Skeels. *Ind. J. Agric. Sci.* 26: 95-103.

Carroll, T. W. (1972). Seed transmissibility of two strains of barley stripe mosaic virus. *Virology* 48: 323-336.

Carroll, T. W. (1974). Barley stripe mosaic virus in sperm and vegetative cells of barley pollen. *Virology* 60: 21-28.

Carroll, T. W., and Mayhew, D. E. (1976a). Anther and pollen infection in relation to the pollen and seed transmissibility of two strains of barley stripe mosaic virus in barley. *Can. J. Bot.* 54: 1604-1621.

Carroll, T. W., and Mayhew, D. E. (1976b). Occurrence of virions in developing ovules and embryo sacs of barley in relation to the seed transmissibility of barley stripe mosaic virus. *Can. J. Bot.* 54: 2497-2512.

Carroll, T. W., Gossel, P. L., and Hockett, E. A. (1979). Inheritance of resistance to seed transmission of barley stripe mosaic virus in barley. *Phytopathology* 69: 431-433.

Carter, W. (1973). Modes of plant virus transmission. Chapter 11 in "Insects in Relation to Plant Disease" (2nd edition). Wiley-Interscience, New York, 759 pp.

Caspar, R. (1977). Testung von *Prunus avium-samen* aus prune dwarf virus mit dem ELISA Verfahren. (Assay of *Prunus avium* seed for prune dwarf virus by ELISA method.) *Phytopathologische Zeitschrift* 90: 91-94.

Catherall, P. L. (1972). Barley stripe mosaic virus. *Rept. Welsh. Plant Breeding Station*, 1971: 62.

Cation, D. (1949). Transmission of cherry yellows virus complex through seeds. *Phytopathology* 39: 37-40.

Cation, D. (1952). Further studies on transmission of ringspot and cherry yellows viruses through seeds. *Phytopathology* 42:4.

Chamberlain, E. E., and Fry, P. R. (1950). Influence of method of tomato seed extraction on seed transmission of tobacco-mosaic and tomato-streak. *New Zealand J. Sci. Tech.* (A) 32: 19-23.

Chenulu, V. V., Sachchidananda, J., and Mehta, S. C. (1968). Studies on a mosaic disease of cowpea from India. *Phytopath. Z.* 63: 381-387.

Cheo, P. C. (1955). Effect of seed maturation on inhibition of southern bean mosaic virus in bean. *Phytopathology* 45: 17-21.

Cheremushkina, N. P. (1977). O rasprostranenii virusa mozaiki luka samenami. (The spread of onion mosaic virus by seeds.) *Tr. VNII Selektsii i Semenovod. Ovoshch. Kultur.* No. 6: 52-54.

Chiko, A. W. (1973). Failure to transmit barley stripe mosaic virus by aphids, leaf-hoppers, and grasshoppers. *Pl. Dis. Reptr.* 57: 639-641.

Chiko, A. W. (1975). Natural occurrence of barley stripe mosaic virus in wild oats (*Avena fatua*). *Can. J. Bot.* 53: 417-420.

Chiko, A. W., and Zimmer, R. C. (1978). Effect of pea seed-borne mosaic virus on two cultivars of field pea grown in Manitoba. *Can. J. Pl. Sci.* 58: 1073-1080.

Childs, J. F. L. (1956). Transmission experiments and xyloporosiscachexia relations in Florida. *Pl. Dis. Reptr.* 40: 143-145.

Childs, J. F. L. (1968). Indexing procedures for 15 virus diseases of citrus trees. *U.S.D.A. Agr. Handbook* No. 333. 96 pp.

Childs, J. F. L., and Johnson, R. E. (1966). Preliminary report of seed transmission of psorosis virus. *Pl. Dis. Reptr.* 50: 81-83.

Cho, E.-K., and Goodman, R. M. (1979). Strains of soybean mosaic virus: classification based on virulence in resistant soybean cultivars. *Phytopathology* 69: 467-470.

Clinch, P. E. M., and Loughnane, J. B. (1948). Seed transmission of virus yellows of sugar-beet (*Beta vulgaris* L.) and the existence of strains of this virus in Eire. *Proc. R. Dublin Soc.* 24: 307-318.

Cochran, L. C. (1946). Passage of the ringspot virus through Mazzard cherry seeds. *Science* 104: 269-270.

Cochran, L. C. (1950). Passage of the ringspot virus through peach seeds. *Phytopathology* 40: 964.

Cockbain, A. J., Cook, S. M., and Vorra-urai, S. (1973). Disease of field beans (*Vicia faba* L.). *Rothamsted Exp. Sta. Rep.* 1972: 142-144.

Cockbain, A. J., Bowen, R., and Etheridge, P. (1975a). Attempts to control the spread of BBSV/EAMV. *Rep. Rothamsted Exp. Sta.* 1974: 235-236.

Cockbain, A. J., Cook, S. M., and Bowen, R. (1975b). Transmission of broad bean stain virus and Echtes Ackerbohnenmosaik-Virus to field bean (*Vicia faba*) by weevils. *Ann. Appl. Biol.* 81: 331-339.

Cockbain, A. J., Bowen, R., and Vorra-urai, S. (1976). Seed transmission of broad bean stain virus and Echtes Ackerbohnenmosaik-Virus in field bean (*Vicia faba*). *Ann. Appl. Biol.* 84: 321-332.

Converse, R. H. (1967). Pollen- and seed-borne raspberry viruses. *Phytopathology* 57: 97-98.

Converse, R. H., and Lister, R. M. (1969). The occurrence and some properties of black raspberry latent virus. *Phytopathology* 59: 325-333.

Cooper, J. I., and Harrison, B. D. (1973). The role of weed hosts and the distribution and activity of vector nematodes in the ecology of tobacco rattle virus. *Ann. Appl. Biol.* 73: 53-66.

Cooper, V. C., and Walkey, D. G. A. (1978). Thermal inactivation of cherry leaf roll virus in tissue cultures of *Nicotiana rustica* raised from seeds and meristem-tips. *Ann. Appl. Biol.* 88: 273-278.

Corbett, M. K. (1958). A virus disease of lupines caused by bean yellow mosaic virus. *Phytopathology* 48: 86-91.

Cory, L., and Hewitt, W. B. (1968). Some grapevine viruses in pollen and seeds. *Phytopathology* 58: 1316-1320.

Crowley, N. C. (1957). Studies on the seed transmission of plant virus diseases. *Aust. J. Biol. Sci.* 10: 449-464.

Crowley, N. C. (1959). Studies on the time of embryo infection by seed-transmitted viruses. *Virology* 8: 116-123.

Dale, W. T. (1949). Observations on a virus disease of cowpea in Trinidad. *Ann. Appl. Biol.* 36: 327-333.

Darozhkin, M. A., and Grabenshchykava, S. I. (1974). Da pytannya peredachy X-virusa nasennem bul 'by. (Transmission of virus X by potato seeds.) *Vestsi Akademii Navuk BSSR. Biyalagichnykh Navuk* 5: 80-85.

Das, C. R., and Milbrath, J. A. (1961). Plant-to-plant transfer of stone fruit ringspot virus in squash by pollination. *Phytopathology* 51: 489-490.

Davidson, T. R. (1976). Field spread of prunus necrotic ringspot in sour cherries in Ontario. *Pl. Dis. Reptr.* 60: 1080-1082.

Demski, J. W. (1975). Source and spread of peanut mottle virus in soybean and peanut. *Phytopathology* 65: 917-920.

Desjardins, P. R., Latterell, R. L., and Mitchell, J. E. (1954). Seed transmission of tobacco ringspot virus in Lincoln variety of soybean. *Phytopathology* 44: 86.

Desjardins, P. R., Drake, R. J., Atkins, E. L., and Bergh, B. O. (1979). Pollen transmission of avocado sunblotch virus experimentally demonstrated. *Calif. Agr.* 33: 14-15.

Devergne, J. C., and Cousin, R. (1966). Le virus de la mosaique de la feve (MF) et les symptomes d'ornementation sur graines. *Annls. Epiphyt.* 17 (Hors-Ser.): 147-161.

Dhanraj, K. S., and Raychaudhuri, S. P. (1969). A note on barley mosaic in India. *Pl. Dis. Reptr.* 53: 766-767.

Dias, H. F. (1963). Host range and properties of grapevine fanleaf and grapevine yellow mosaic viruses. *Ann. Appl. Biol.* 51: 85-95.

Dias, H. F., and Cation, D. (1976). The characterization of a virus responsible for peach rosette mosaic and grape decline in Michigan. *Can. J. Bot.* 54: 1228-1239.

Dias, H. F., and Waterworth, H. E. (1967). The identity of a seed-borne mosaic virus of *Chenopodium amaranticolor* and *C. quinoa. Can. J. Bot.* 45: 1285-1295.

Diwakar, M. P., and Mali, V. R. (1977). Cowpea mosaic virus disease—a new record for Marathwada. *J. Maharas. Agr. Univ.* 1: 274-277.

Doolittle, S. P. (1920). The mosaic disease of cucurbits. *U.S. Dept. Agr. Bull.* 879: 1-69.

Doolittle, S. P., and Gilbert, W. W. (1919). Seed transmission of cucurbit mosaic by the wild cucumber. *Phytopathology* 9: 326-327.

Doolittle, S. P., and Walker, M. N. (1925). Further studies on the overwintering and dissemination of cucurbit mosaic. *J. Agr. Res.* 31: 1-58.

Edwardson, J. R., and Purcifull, D. E. (1974). Relationship of *Datura quercina* and tobacco streak viruses. *Phytopathology* 64: 1322-1324.

Ekbote, A. U., and Mali, V. R. (1978). Occurrence of alfalfa mosaic virus on alfalfa in India. *Ind. Phytopathol.* 31: 171-175.

Eslick, R. F., and Afanasiev, M. M. (1955). Influence of time of infection with barley stripe mosaic on symptoms, plant yield, and seed infection of barley. *Pl. Dis. Reptr.* 39: 722-724.

Evans, I. R. (1973). Seed-borne bean yellow mosaic virus of fababean in Canada. *Can. Pl. Dis. Surv.* 53: 123-126.

Fajardo, T. G. (1930). Studies on the mosaic disease of the bean (*Phaseolus vulgaris* L.). *Phytopathology* 20: 469-494.

Fajardo, T. G., and Maranon, J. (1932). The mosaic disease of sincomas, *Pachyrrhizus erosus* (Linnaeus) Urban. *Philipp. J. Sci.* 48: 129-142.

Fenner, F. (1976). Classification and nomenclature of viruses. *Intervirology* 7: 1-116.

Fischer, H. L., and Lockhart, B. E. L. (1978). Host range and properties of peanut stunt virus from Morocco. *Phytopathology* 68: 289-293.

Ford, R. E. (1966). Recovery of pea streak from pea seed parts and its transmission by immature seed. *Phytopathology* 56: 858-859.

Frandsen, N. O. (1952). Untersuchungen zur Virusresistenzzüchtung bei *Phaseolus vulgaris* L. I. Phytopathologische Untersuchungen *Z. Pfl. Züchtung* 31: 381-420.

Fridlund, P. R. (1966). Transmission and lack of transmission of seven viruses through Prunus seed. *Pl. Dis. Reptr.* 50: 902-904.

Frosheiser, F. I. (1964). Alfalfa mosaic virus transmitted through alfalfa seed. *Phytopathology* 54: 893.

Frosheiser, F. I. (1970). Virus-infected seeds in alfalfa seed lots. *Pl. Dis. Reptr.* 54: 591-594.

Frosheiser, F. I. (1974). Alfalfa mosaic virus transmission to seed through alfalfa gametes and longevity in alfalfa seed. *Phytopathology* 64: 102-105.

Fulton, R. W. (1964). Transmission of plant viruses by grafting, dodder, seed and mechanical inoculation. *In* "Plant Virology" (M. K. Corbett and H. D. Sisler, eds.). pp. 39-67. Univ. of Florida Press, Gainesville. 527 pp.

Gallo, J., and Ciampor, F. (1977). Transmission of alfalfa mosaic virus through *Nicandra physaloides* seeds and its location in embryo cotyledons. *Acta. Virologica* 21: 344-346.

Gardner, W. S. (1967). Electron microscopy of barley stripe mosaic virus: comparative cytology of tissues infected during different stages of maturity. *Phytopathology* 57: 1315-1326.

George, J. A., and Davidson, T. R. (1963). Pollen transmission of necrotic ringspot and sour cherry yellows viruses from tree to tree. *Can J. Pl. Sci.* 43: 276-288.

Ghanekar, A. M., and Schwenk, F. W. (1974). Seed transmission and distribution of tobacco streak virus in six cultivars of soybeans. *Phytopathology* 64: 112-114.

Gibbs, A. J., and Harrison, B. D. (1964). Nematode transmitted viruses in sugarbeet in East Anglia. *Pl. Pathol.* 13: 144-150.

Gibbs, A. J., and Harrison, B. D. (1973). Eggplant mosaic virus. *CMI/AAB Descriptions of Plant Viruses*, No. 124.

Gibbs, A. J., and Paul, H. L. (1970). Echtes Ackerbohnenmosaik-virus. *CMI/AAB Descriptions of Plant Viruses*, No. 20.

Gibbs, A. J., and Smith, H. G. (1970). Broad bean stain virus. *CMI/AAB Descriptions of Plant Viruses*, No. 29.

Gibbs, A. J., Kassanis, B., Nixon, H. L., and Woods, R. D. (1963). The relationship between barley stripe mosaic and Lynchnis ringspot viruses. *Virology* 20: 194-198.

Gibbs, A. J., Hecht-Poinar, E., Woods, R. D., and McKee, R. K. (1966). Some properties of three related viruses: Andean potato latent, dulcamara mottle, and ononis yellow mosaic. *J. Gen. Microbiol.* 44: 177-193.

Gilmer, R. M. (1955). Imported Mahaleb seeds as carriers of necrotic ringspot virus. *Pl. Dis. Reptr.* 39: 727-728.

Gilmer, R. M., and Kelts, L. J. (1968). Transmission of tobacco mosaic virus in grape seed. *Phytopathology* 58: 277-278.

Gilmer, R. M., and Way, R. D. (1960). Pollen transmission of necrotic ringspot, and prune dwarf viruses in sour cherry. *Phytopathology* 50: 624-625.

Gilmer, R. M., and Way, R. D. (1963). Evidence for tree-to-tree transmission of sour cherry yellow virus by pollen. *Pl. Dis. Reptr.* 47: 1051-1053.

Gilmer, R. M., and Wilks, J. M. (1967). Seed transmission of tobacco mosaic virus in apple and pear. *Phytopathology* 57: 214-217.

Gold, A. H., Suneson, C. A., Houston, B. R., and Oswald, J. W. (1954). Electron microscopy and seed and pollen transmission of rod-shaped particles associated with the false stripe disease of barley. *Phytopathology* 44: 115-117.

Gonzalez-Garza, R., Gumpf, D. J., Kishaba, A. N., and Bohn, G. W. (1979). Identification, seed transmission and host range pathogenecity of California isolate of melon necrotic spot virus. *Phytopathology* 69: 340-345.

Gooding, G. V. (1975). Inactivation of tobacco mosaic virus on tomato seed with trisodium orthophosphate and sodium hypochlorate. *Pl. Dis. Reptr.* 59: 770-772.

Grogan, R. G., and Bardin, R. (1950). Some aspects concerning seed transmission of lettuce mosaic virus. *Phytopathology* 40: 965.

Grogan, R. G., and Schnathorst, W. C. (1955). Tobacco ringspot virus–the cause of lettuce calico. *Pl. Dis. Reptr.* 39: 803-806.

Grogan, R. G., Welch, J. E., and Bardin, R. (1952). Common lettuce mosaic and its control by the use of mosaic free seed. *Phytopathology* 42: 573-578.

Grogan, R. G., Hall, D. H., and Kimble, K. A. (1959). Cucurbit mosaic viruses in California. *Phytopathology* 49: 366-376.

Hagborg, W. A. F. (1954). Dwarfing of wheat and barley by the barley stripe mosaic (false-stripe) virus. *Can. J. Bot.* 32: 24-37.

Hamilton, R. I. (1965). An embryo test for detecting seed-borne barley stripe mosaic virus in barley. *Phytopathology* 55: 798-799.

Hamilton, R. I. (1977). Detecting pea seed-borne mosaic. *Can. Agr.* 22: 15-17.

Hamilton, R. I., and Nichols, C. (1978).Serological methods for detection of pea seed-borne mosaic virus in leaves and seeds of *Pisum sativum*. *Phytopathology* 68: 539-543.

Hamilton, R. I., Leung, E., and Nichols, C. (1977). Surface contamination of pollen by plant viruses. *Phytopathology* 67: 395-399.

Hampton, R. O. (1963). Seed transmission of white clover mosaic and clover yellow mosaic viruses in red clover. *Phytopathology* 53: 1139.

Hampton, R. O. (1967). Seed transmission of viruses in red clover. *Phytopathology* 57: 98.

Hampton, R. O. (1969). Characteristics of virus particles associated with the seed-borne pea fizzletop disease. *Phytopathology* 59: 1029.

Hampton, R. O. (1972). Dynamics of symptom development of the seed-borne pea fizzletop virus. *Phytopathology* 62: 268-272.

Hampton, R. O., and Braverman, S. W. (1979). Occurrence of pea seed-borne mosaic virus and new virus-immune germplasm in the plant introduction collection of *Pisum sativum*. *Pl. Dis. Reptr.* 63: 95-99.

Hampton, R. O., and Hanson, E. W. (1968). Seed transmission of viruses in red clover; evidence and methodology of detection. *Phytopathology* 58: 914-920.

Hampton, R. O., and Muetilbauer, F. J. (1977). Seed transmission of the pea seed-borne mosaic virus in lentils. *Pl. Dis. Reptr.* 61: 235-238.

Hanada, K., and Harrison, B. D. (1977). Effects of virus genotype and temperature on seed transmission of nepoviruses. *Ann. Appl. Biol.* 85: 79-92.

Häni, A. (1971). Zur Epidemiologie des Gurkenmosaikvirus im Tessin. *Phytopath. Z.* 72: 115-144.

Häni, A., Pelet, F., and Kern, H. (1970). Zur Bedeutung von *Stellaria media* (L.) Vill. in der Epidemiologie des Gurkenmosaikvirus. *Phytopath. Z.* 68: 81-83.

Hansen, A.-J., and Leseman, D. E. (1978). Occurrence and characteristics of a seed-transmitted potyvirus from Indian, African, North American Guar. *Phytopathology* 68: 841-846.

Hansen, A. G., Nyland, G., McElroy, F. D., and Stace-Smith, R. (1974). Origin, cause, host range and spread of cherry raspleaf disease in North America. *Phytopathology* 64: 721-727.

Hanson, C. M., and Campbell, R. N. (1979). Strawberry latent ringspot virus from 'Plain' parsley in California. *Pl. Dis. Reptr.* 63: 142-146.

Harder, D. E., and Bakker, W. (1973). African cereal streak, a new disease of cereals in East Africa. *Phytopathology* 63: 1407-1411.

Härdtl, H. (1964). Zum Nachweis von Gesund-und Abbaulagen bei Zweibelkultue. *Gesunde Pfl.* 16: 218-221.

Härdtl, H. (1972). Die Übertragung der Zwiebelgelbstreifigkeit durch den Samen. *Z. PflKrank. PflSchutz.* 79: 694-701.

Harrison, A. L. (1935). Mosaic of the Refugee bean. *New York St. Agr. Exp. Sta. Bull.* 656: 19 pp.

Harrison, B. D. (1973). Pea early-browning virus. *CMI/AAB Descriptions of Plant Viruses*, No. 120.

Harrison, B. D. (1977). Ecology and control of viruses with soil-inhabiting vectors. *Annu. Rev. Phytopathol.* 15: 331-360.

Harrison, B. D., and Hanada, K. (1976). Competitiveness between genotypes of raspberry ringspot virus is mainly determined by RNA 1. *J. Gen. Virol.* 31: 455-457.

Harrison, B. D., Murant, A. F., Mayo, M. A., and Roberts, I. M. (1974). Distribution of determinants for symptom production, host range, and nematode transmissibility between the two RNA components of raspberry ringspot virus. *J. Gen. Virol.* 22: 233-247.

Hemmati, K., and McLean, D. L. (1977). Gamete-seed transmission of alfalfa mosaic virus and its effect on seed germination and yield in alfalfa plants. *Phytopathology* 67: 576-579.

Henderson, R. G. (1931). Transmission of tobacco ringspot by seed of petunia. *Phytopathology* 21: 225-229.

Herold, F. (1956). Ist eine wirksame Bekämpfung des Salatmosaikvirus möglich? *Saatgutwirtschaft* 1956: 307-309.

Hevin, M., Ottenwaelter, M. M., Doazan, J. P., and Rives, M. (1973). Investigating the transmission of marbrure and fan-leaf through the seed in the grapevine. *Rivista di Patologia Vegetale* (Ser. 4) 9: 253-258.

Hino, T. (1962). Studies on the Adzuki-bean mosaic virus. *Ann. Phytopathol. Soc. Japan* 27: 138-142.

Hobart, O. F. (1956). Introduction and spread of necrotic ringspot virus in sour cherry nursery trees. *Iowa State Coll. J. Sci.* 30: 381-382.

Hollings, M., Stone, O. M., and Dale, W. T. (1972). Tomato ringspot virus in Pelargonium in England. *Pl. Pathol.* 21: 46-47.

Howles, R. (1978). Inactivation of lettuce mosaic virus in lettuce seed by dry heat treatment. *J. Aus. Inst. Agr. Sci.* 44: 131-132.

Hunnius, W. (1977). Virustestung bei Pflanzkartoffeln. (Virus testing of seed potatoes.) *Zeitschrift für Pflanzenkrankheiten und Pflanzenschutz.* 84: 748-763.

Hunter, D. E., Darling, H. M., and Beale, W. L. (1969). Seed transmission of potato spindle tuber virus. *Am. Potato J.* 46: 247-250.

Iizuka, N. (1973). (Seed transmission of viruses in soybean.) *Bull. Tohoku. Nat. Agr. Exp. Sta.* 46: 131-141.

Inouye, T. (1962). Studies on barley stripe mosaic in Japan. *Ber. Ohara Inst. Landw. Biol.* 11: 413-496.

Inouye, T. (1967). A seed-borne mosaic virus of pea. *Ann. Phytopathol. Soc. Japan* 33: 38-42.

Iwaki, M. (1978). Seed transmission of cucumber mosaic virus in mungbean (*Vigna radiata*). *Ann. Phytopathol. Soc. Japan.* 44: 337-339.

Jensen, S. G. (1973). Systemic movement of barley yellow dwarf virus in small grains. *Phytopathology* 63: 854-856.

Jones, A. T. (1978). Incidence, field spread, seed transmission and effects of broad bean stain virus and Echtes Ackerbohnenmosaik-virus in *Vicia faba* in eastern Scotland. *Ann. Appl. Biol.* 88: 137-144.

Jones, A. T., and Mayo, M. A. (1973). Purification and properties of elm mottle virus. *Ann. Appl. Biol.* 75: 347-357.

Jones, A. T., and Mayo, M. A. (1975). Further properties of black raspberry latent virus and evidence of its relationship to tobacco streak virus. *Ann. Appl. Biol.* 79: 297-306.

Jones, L. K. (1944). Streak and mosaic of *Cineraria. Phytopathology* 34: 941-953.

Jones, R. K., and Tolin, S. A. (1972). Concentration of maize dwarf mosaic virus in susceptible and resistant corn hybrids. *Phytopathology* 62: 640-644.

Jones, R. A. C., and Fribourg, C. E. (1977). Beetle, contact and potato true seed transmission of Andean potato latent virus. *Ann. Appl. Biol.* 86: 123-128.

Kado, C. I. (1967). Biological and biochemical characterization of sowbane mosaic virus. *Virology* 31: 217-229.

Kahn, R. P. (1956). Seed transmission of tomato ringspot virus in the Lincoln variety of soybeans. *Phytopathology* 46: 295.

Kahn, R. P., Scott, H. A., and Monroe, R. L. (1962). Eucharis mottle strain of tobacco ringspot virus. *Phytopathology* 52: 1211-1216.

Kaiser, W. J. (1972). Diseases of food legumes caused by pea leaf roll virus in Iran. *FAO Pl. Prot. Bull.* 20: 127-132; 135.

Kaiser, W. J. (1973). Biology of bean yellow mosaic and pea leaf roll viruses affecting *Vicia faba* in Iran. *Phytopath. Z.* 78: 253-263.

Kaiser, W. J., and Mossahebi, G. H. (1974). Natural infection of mungbean by bean common mosaic virus. *Phytopathology* 64: 1209-1214.

Kaiser, W. J., Danesh, Dariush, Okhovat, Mahmoud, and Mossahebi, Hossein (1968). Diseases of pulse crops (edible legumes) in Iran. *Pl. Dis. Reptr.* 52: 687-691.

Karimov, T. M. (1978). (Infection of tomato seeds by TMV.) *Zashchita Rastenii* No. 6, 43.

Kassanis, B., Russell, G. E., and White, R. F. (1978). Seed and pollen transmission of beet cryptic virus in sugar beet plants. *Phytopathologische Zeitschrift* 91: 76-79.

Kazda, V., and Hevert, V. (1977). Epidemiologie viru mozaiky okurky u sklenikovych okurek. (Epidemiology of cucumber mosaic virus of glasshouse cucumber.) *Ochrana Rostlin* 13: 169-176.

Kemp, W. G., Weibe, J., and Patrick, Z. A. (1972). Squash mosaic virus in muskmelon seed distributed commercially in Ontario. *Can. Pl. Dis. Surv.* 52: 58-59.

Kendrick, J. B. (1934). Cucurbit mosaic transmitted by muskmelon seed. *Phytopathology* 24: 820-823.

Kendrick, J. B., and Gardner, M. W. (1924). Soybean mosaic: seed transmission and effect on yield. *J. Agr. Res.* 27: 91-98.

Kennedy, B. W., and Cooper, R. L. (1967). Association of virus infection with mottling of soybean seed coats. *Phytopathology* 57: 35-37.

Kenten, R. H. (1972). The purification and some properties of cocoa necrosis virus, a serotype of tomato black ring virus. *Ann. Appl. Biol.* 71: 119-126.

Keplinger, J. A., and Braun, A. J. (1973). Seed transmission of tomato ringspot virus in *Gomphrena globosa. Pl. Dis. Reptr.* 57: 433.

Keur, J. Y. (1933). Seed transmissions of the virus causing variegation of *Abutilon. Phytopathology* 23: 20.

Keur, J. Y. (1934). Studies on the occurrence and transmission of virus diseases in the genus *Abutilon. Bull. Torrey Bot. Club.* 61: 53-70.

Khatri, H. L., and Chohan, J. S. (1972). Studies on some factors influencing seed transmission of cowpea mosaic virus in cowpea. *Ind. J. Myc. Pl. Pathol.* 2: 40-44.

Kolte, S. J., and Nene, Y. L. (1972). Studies on symptoms and mode of transmission of the leaf crinkle virus of urd bean (*Phaseolus mungo*). *Ind. Phytopathol.* 25: 401-404.

Komuro, Y., Tochihara, H., Fukatsu, R., Nagai, Y., and Yoneyama, S. (1971). Cucumber green mottle mosaic virus (watermelon strain) in watermelon and its bearing on deterioration of watermelon fruit known as 'Konnyaku' disease. *Ann. Phytopath. Soc. Japan* 37: 34-42.

Koshimizu, Y., and Iizuka, N. (1963). Studies on soybean virus diseases in Japan. *Bull. Tohoku Nat. Agr. Expt. Sta.* 27, 38: 1-103.

Kramer, M., Orlando, A., and Silberschmidt, K. M. (1945). Estudos sobre uma grave doenca de virus, responsavel pelo deperecimento de nossas culturas de alface. (Studies on a serious virus disease responsible for the dying-off of our lettuce crops.) *Biologico* 11: 121-134.

Kriz, J. (1959). Transmission of curl disease and infectious sterility of hops to the progeny through the seeds. *Ann. Acad. Tchecosl. Agr.* 32: 951-970.

Kuhn, C. W. (1965). Symptomatology, host range, and effect on yield of seed-transmitted peanut virus. *Phytopathology* 55: 880-884.

Ladipo. J. L. (1977). Seed transmission of cowpea aphid-borne virus in some cowpea cultivars. *Nigerian J. Pl. Protect.* 3: 3-10.

Lamptey, P. N. L., and Hamilton, R. I. (1974). A new cowpea strain of southern bean mosaic virus from Ghana. *Phytopathology* 64: 1100-1104.

Leppik, E. E. (1964). Some epiphytotic aspects of squash mosaic. *Pl. Dis. Reptr.* 48: 41-42.

Levy, A., Loebenstein, G., Smookler, M., and Drori, T. (1974). Partial suppression by UV irradiation of the mechanism of resistance to cucumber mosaic virus in a resistant cucumber cultivar. *Virology* 60: 37-44.

Lindberg, G. D., Hall, D. H., and Walker, J. C. (1956). A study of melon and squash mosaic viruses. *Phytopathology* 46: 489-495.

Lister, R. M. (1960). Transmission of soil-borne viruses through seed. *Virology* 10: 547-549.

Lister, R. M. (1978). Application of enzyme-linked immunosorbent assay for detecting viruses in soybean seed and plants. *Phytopathology* 68: 1393-1400.

Lister, R. M., and Murant, A. F. (1967). Seed-transmission of nematode-borne viruses. *Ann. Appl. Biol.* 59: 49-62.

Lloyd, A. T. E., Smith, H. G., and Jones, L. H. (1965). Evesham stain—a virus disease of broad beans (*Vicia faba* L.) *Hort. Res.* 5: 13-18.

Lockhart, B. E. L., and Fischer, H. U. (1974). Chronic infection by seed-borne bean common mosaic virus in Morocco. *Pl. Dis. Reptr.* 58: 307-308.

Lockhart, B. E. L., and Goethals, M. (1977). Natural infection of safflower by a tobamovirus. *Pl. Dis. Reptr.* 61: 1010-1012.

Louie, R., and Lorbeer, J. W. (1966). Mechanical transmission of onion yellow dwarf virus. *Phytopathology* 56: 1020-1023.

Lovisolo, O., and Conti, M. (1966). Identification of an aphid-transmitted cowpea mosaic virus. *Neth. J. Pl. Pathol.* 72: 265-269.

Mahoney, C. H. (1935). Seed transmission of mosaic in inbred lines of muskmelon (*Cucumis melo* L.) *Proc. Ann. Soc. Hort. Sci.* 32: 477-480.

Mali, V. R. (1976). Studies on euonymus mosaic virus disease and its transmission by *Olpidium brassicae* in Czechoslovakia. *Ind. Phytopathol.* 29: 262-268.

Mandahar, C. L. (1978). Introduction to Plant Viruses. S. Chand & Co., New Delhi. 333 pp.

Mandahar, C. L., Nath, S., and Gulati, A. (1979). Role of growth inhibitors in bud dormancy in virus infected stem cuttings of *Euphorbia pulcherrima*. *Experientia* 35: 315-316.

Marchoux, G., Quiot, J. B., and Devergne, J. C. (1977). Caracterisation d'un isolat du virus de la mosaique du concombre transmis par les graines du karicot (*Phaseolus vulgaris* L.). *Ann. de. Phytopathologie.* 9: 421-434.

Marrou, J., and Messiaen, C. M. (1967). The *Chenopodium quinoa* test: a critical method for detecting seed transmission of lettuce mosaic. *Proc. Int. Seed Test Assoc.* 32: 49-57.

Martyn, E. B. (1968). Plant Virus Names. An Annotated List of Names and Synonyms of Plant Viruses and Diseases. Commonwealth Mycological Institute, Kew, Surrey, England. 204 pp.

Martyn, E. B. (1971). Plant Virus Names. Supplement No. 1. Additions and Corrections to Phytopathological Paper No. 9 (Martyn, 1968) and Newly Recorded Plant Viruses. Commonwealth Mycological Inst., Kew, Surrey, England. 41 pp.

Mayee, C. D. (1977). Storage of seed for pragmatic control of a virus causing mosaic disease of brinjal (eggplant). *Seed Sci. & Technol.* 5: 555-558.

Mayee, C. D., and Khatri, H. L. (1975). Seed transmission of brinjal mosaic virus in some varieties of brinjal. *Ind. Phytopathol.* 28: 238-240.

Mayhew, D. E., and Carroll, T. W. (1974a). Barley stripe mosaic virions associated with spindle microtubules. *Science* 185: 597-598.

Mayhew, D. E., and Carroll, T. W. (1974b). Barley stripe mosaic virus in the egg cell and egg sac of infected barley. *Virology* 58: 561-567.

McDonald, J. G., and Hamilton, R. I. (1972). Distribution of southern bean mosaic virus in the seed of *Phaseolus vulgaris. Phytopathology* 62: 387-389.

McKinney, H. H. (1951a). A seed-borne virus causing false stripe in barley. *Phytopathology* 41: 563-564.

McKinney, H. H. (1951b). A seed-borne virus causing false-stripe symptoms in barley. *Pl. Dis. Reptr.* 35: 48.

McKinney, H. H. (1952). Two strains of tobacco mosaic virus, one of which is seed-borne in an etch-immune pungent pepper. *Pl. Dis. Reptr.* 36: 184-187.

McKinney, H. H. (1953). New evidence on virus diseases in barley. *Pl. Dis. Reptr.* 37: 292-295.

McKinney, H. H., and Greely, L. W. (1965). Biological characteristics of barley stripe mosaic virus strains and their evolution. *U.S. Dep. Agr. Tech. Bull.* 1324: 84 pp.

McLean, D. M. (1962). Seed transmission of tobacco ringspot virus in cantaloupe. *Phytopathology* 52: 21.

McNeal, F. H., and Afanasiev, M. M. (1955). Transmission of barley stripe mosaic through the seed in 11 varieties of spring wheat. *Pl. Dis. Reptr.* 39: 460-462.

Medina, A. C., and Grogan, R. G. (1961). Seed transmission of bean common mosaic viruses. *Phytopathology* 51: 452-456.

Megahed, E. S., and Moore, J. D. (1967). Differential mechanical transmission of Prunus viruses from seed of various *Prunus* spp. and from different parts of the same seed. *Phytopathology* 57: 821-822.

Megahed, E. S., and Moore, J. D. (1969). Inactivation of necrotic ringspot and prune dwarf viruses in seeds of some *Prunus* spp. *Phytopathology* 59: 1758-1760.

Meiners, J. P., Waterworth, H. E., Smith, F. F., Alconero, R., and Lamson, R. H. (1977). A seed transmitted strain of cucumber mosaic virus isolated from bean. *J. Agr., Univ. Puerto Rico.* 61: 137-147.

Meiners, J. P., Gillaspie, A. G. Jr., Lowson, R. H., and Smith, F. F. (1978). Identification and partial characterization of a strain of bean common mosaic virus from *Rhynochosia minima. Phytopathology* 68: 283-287.

Mellor, F. C., and Stace-Smith, R. (1963). Reaction of strawberry to a ringspot virus from raspberry. *Can. J. Bot.* 41: 865-870.

Middleton, J. T. (1944). Seed transmission of squash-mosaic virus. *Phytopathology* 34: 405-410.

Middleton, J. T., and Bohn, G. W. (1953). Cucumbers, melons, squash. *U. S. Dep. Agr. Yearbook of Agr.*, 1953: 483-493.

Millikan, D. F. (1959). The incidence of the ringspot virus in peach nursery and orchard trees. *Pl. Dis. Reptr.* 43: 82-84.

Mink, G. I., and Parsons, J. L. (1978). Detection of pea seedborne mosaic virus in pea seed by direct seed-assay. *Pl. Dis. Reptr.* 62: 249-253.

Mink, G. I., Kraft, J., Knesek, J., and Jafri, A. (1969). A seed-borne virus of peas. *Phytopathology* 59: 1342-1343.

Munro, D. (1978). Pea seed-borne mosaic virus in quarantine. APPS *Newsletter* 7:10.

Murant, A. F., and Goold, R. A. (1969). Strawberry latent ringspot virus. *Scottish Hort. Res. Inst. Annu. Rep. 1968.* 48.

Murant, A. F., and Lister, R. M. (1967). Seed-transmission in the ecology of nematode-borne viruses. *Ann. Appl. Biol.* 59: 63-76.

Murant, A. F., Taylor, C. E., and Chambers, J. (1968). Properties, relationships and transmission of a strain of raspberry ringspot virus infecting raspberry cultivars immune to the common Scottish strain. *Ann. Appl. Biol.* 61: 175-186.

Murdock, D. J., Nelson, P. E., and Smith, S. H. (1976). Histopathological examination of Pelargonium infected with tomato ringspot virus. *Phytopathology* 66: 844-850.

Musil, M. (1970). Pea leaf rolling mosaic virus and its properties. Biologia, *Bratislava* 25: 379-392.

Narayanaswamy, P., and Jaganathan, T. (1975). Seed transmission of black gram leaf crinkle virus. *Phytopath. Z.* 82: 107-110.

Neergaard, P. (1977). Seed-borne viruses. Chapter 3 in "Seed Pathology, Vol. I." Macmillan Press, London and Madras. 839 pp.

Nelson, M. R., and Knuhtsen, H. K. (1969). Relation of seed transmission to the epidemiology of squash mosaic virus strains. *Phytopathology* 59: 1042.

Nelson, M. R., and Knuhtsen, H. K. (1973a). Squash mosaic virus variability: epidemiological consequences of differences in seed transmission frequency between strains. *Phytopathology* 63: 918-920.

Nelson, M. R., and Knuhtsen, H. K. (1973b). Squash mosaic virus variability: review and serological comparisons of six biotypes. *Phytopathology* 63: 920-926.

Nelson, R., and Down, E. E. (1933). Influence of pollen and ovule infection in the seed transmission of bean mosaic. *Phytopathology* 23: 25.

Nene, Y. L. (1972). A survey of viral diseases of pulse crops in Uttar Pradesh. *G. B. Pant Univ. Agr. Tech. Pantnagar, Res. Bull.* 4, 191 pp.

Newhall, A. G. (1923). Seed transmission of lettuce mosaic. *Phytopathology* 13: 104-106.

Nezhnev, Y. N. (1977). Bezrassadnaya kultura tomatov. (Cultivation of tomato without transplanting.) Tr. VNII. *Oroshaem. Ovoshchevod. i. Bakhchevod.* No. 6. pp. 70-74.

Nikitina, M. A. (1966). Rasprostranenia virusa Tabaka pri rassadnom i bezrassadnom sposobakh vyrashchivaniya Tomatov. *Izv. Akad. Nauk. Kazakh SSR. Ser. Biol.* 1966(3): 26-35.

Nitzany, F. E., and Gerechter, Z. K. (1962). Barley stripe mosaic virus host range and seed transmission tests among Gramineae in Israel. *Phytopathol. Mediterranea* 2: 11-19.

Noble, M., and Richardson, M. J. (1968). An annotated list of seed-borne diseases, 2nd edition. *Proc. Int. Seed Test. Ass.* 33: 1-191.

Noordam, D., Bijl, M., Overbeek, S. C., and Quiniones, S. S. (1965). Virussen uit *Campanula rapunculoides en Stellaria media* en hun relatie tot komkommermozaiek virus en tomaat-'aspermy'-virus. *Neth. J. Pl. Pathol.* 71: 61.

Nyland, G. (1962). Possible virus-induced genetic abnormalities in tree fruits. *Science* 137: 598-599.

O'Brien, M. J., and Raymer, W. B. (1964). Symptomless hosts of the potato spindle tuber virus. *Phytopathology* 54: 1045-1047.

Ogilvie, L., Mulligan, B. O., and Brian, P. W. (1935). Progress report on vegetable diseases. VI. *Annu. Rep. Agr. Hort. Res. Sta. Bristol* 1934: 175-190.

Ormerod, P. J. (1970). A virus associated with loganberry degeneration disease. *Annu. Rep. East Malling Res. Sta.* 1969: 165-167.

Owusu, G. K., Crowley, N. C., and Francki, R. I. B. (1968). Studies of the seed transmission of tobacco ringspot virus. *Ann. Appl. Biol.* 61: 195-202.

Pacini, E., and Cresti, M. (1977). Viral particles in developing pollen grains of Olea europaea. *Planta.* 137: 1-4.

Paguio, O. R., and Kuhn, C. W. (1974). Incidence and source of inoculum of peanut mottle virus and its effect on peanut. *Phytopathology* 64: 60-64.

Paludan, N. (1964). Virussygdomme hos *Asparagus officinalis.* (Virus diseases of *Asparagus officinalis*). *Manedsoversigt over Plantesygdomme* 407: 11-16.

Phatak, H. C. (1974). Seed-borne plant viruses—identification and diagnosis in seed health testing. *Seed. Sci. Technol.* 2: 3-155.

Phatak, H. C., and Summanwar, A. S. (1967). Detection of plant viruses in seeds and seed-stocks. *Proc. Int. Seed Test. Ass.* 32: 625-631.

Phatak, H. C., Diaz-Ruiz, J. R., and Hull, R. (1976). Cowpea ringspot virus: A seed transmitted cucumovirus. *Phytopath. Z.* 87: 132-142.

Pilowsky, M., and Cohen, S. (1974). Inheritance of resistance to tomato yellow leaf curl virus in tomatoes. *Phytopathology* 64: 632-635.

Porembskaya, N. B. (1964). Peredacha virusnykh boleznei lyupina cherez semena. *Trudy vses. Inst. Zashch. Rast.* 20: 54-55.

Pring, D. R. (1974). Barley stripe mosaic virus infection of corn and the "aberrant ratio" genetic effect. *Phytopathology* 64: 64-70.

Prochazkova, Z. (1977). Presumed role of mucilage of plantain seeds in spread of tobacco mosaic virus. *Biol. Plant.* 19: 259-263.

Provvidenti, R., and Braverman, S. W. (1976). Seed transmission of bean common mosaic virus in phasemy bean. *Phytopathology* 66: 1274-1275.

Provvidenti, R., and Cobb, E. D. (1975). Seed transmission of bean common mosaic virus in tepary bean. *Pl. Dis. Reptr.* 59: 966-969.

Purivirojkul, W., Sittiyos, P., Hsu, C. H., Poehlman, J. M., and Sehgal, O. P. (1978). Natural infection of mung bean (*Vigna radiata*) with cucumber mosaic virus. *Pl. Dis. Reptr.* 62: 530-534.

Quantz, L. (1953). Untersuchungen über ein samenübertragbares Mosaikvirus der Ackerbohne (*Vicia faba*). *Phytopath. Z.* 20: 421-448.

Quantz, L. (1954). Untersuchungen über die Viruskrankheiten der Acerkbohne, Mitt. Biol. Bund. Anst. Ld.-u. *Forstwe.* 80: 171-175.

Quantz, L. (1958). Untersuchungen zur Bestimmung masaikresistenter überempfindlicher Gartenbohnensorten (*Phaseolus vulgaris* L.) in Labortest. *Phytopath. Z.* 31: 319-330.

Rader, W. E., Fitzpatrick, H. F., and Hildebrand, E. M. (1947). A seed borne virus of muskmelon. *Phytopathology* 37: 809-816.

Ramaswamy, S., and Posnette, A. F. (1971). Properties of cherry ring mottle, a distinctive strain of prune dwarf virus. *Ann. Appl. Biol.* 68: 55-65.

Ramsdell, D. C., and Myers, R. L. (1978). Epidemiology of peach rosette mosaic virus in a concord grape vineyard. *Phytopathology* 68: 447-450.

Randles, J. W., and Duke, A. J. (1977). Three seed borne pathogens isolated from *Vicia faba* seed imported from the United Kingdom. *APPS Newsletter* 6: 37-38.

Randles, J. W., Harrison, B. D., and Roberts, I. M. (1976). *Nicotiana velutina* mosaic virus: purification, properties and affinities with other rod-shaped viruses. *Ann. Appl. Biol.* 84: 193-204.

Reddick, D., and Stewart, V. B. (1918). Varieties of beans susceptible to mosaic. *Phytopathology* 8: 530-534.

Reddick, D., and Stewart, V. B. (1919). Transmission of the virus of bean mosaic in seed and observations on thermal death point of seed and virus. *Phytopathology* 9: 445-450.

Reddy, K. R. C., and Nariani, T. K. (1963). Studies on mosaic diseases of vegetable Marrow (*Cucurbita pepo* L.) *Ind. Phytopathol.* 16: 260-267.

Reyes, T. T. (1961). Seed transmission of coffee ring spot by *Excelsa coffee (coffea excelsa) Pl. Dis. Reptr.* 45: 185.

Rohloff, I. (1962). Entwicklung einer Laboratoriumsmethode zur kurzfristigen Untersuchung von Salatsamen (*Lactuca sativa* L.) auf Befall mit Salatmosaik Virus (SMV). *Gartenbauwissenschaft* 27: 413-436.

Ross, J. P. (1963). Interaction of the soybean mosaic and bean pod mottle viruses infecting soybeans. *Phytopathology* 53: 887.

Ryder, E. J. (1964). Transmission of common lettuce mosaic virus through the gametes of the lettuce plant. *Pl. Dis. Reptr.* 48: 522-523.

Ryder, E. J., and Johnson, A. S. (1974). A method for indexing lettuce seeds for seed-borne lettuce mosaic virus by air stream separation of light from heavy seeds. *Pl. Dis. Reptr.* 58: 1037-1039.

Sachchidanada, J., Singh, S., Prakash, N., and Verma, V. S. (1973). Bean common mosaic virus on cowpea in India. *Z. PflKrankh. Pflschutz.* 80: 88-91.

Salazar, L. F., and Harrison, B. D. (1978). Host range, purification and properties of potato virus. *T. Ann. Appl. Biol.* 89: 223-235.

Salmon, E. S., and Ware, W. M. (1935). The chlorotic disease of the hop. IV. Transmission by seed. *Ann. Appl. Biol.* 22: 728-730.

Sander, E. (1959). Biological properties of red clover vein mosaic virus *Phytopathology* 49: 748-754.

Scarborough, B. A., and Smith, S. H. (1975). Seed transmission of tobacco- and tomato ring-spot viruses in Geraniums. *Phytopathology* 65: 835-836.

Schimanski, H. H., and Schade, C. (1974). Nachweis des nekrotischen Ringflecken-Virus der Kirsche im Pollen der Vogelkirsche (*Prunus avium* L.) und des chlorotischen Ring-flecken-Virus der Kirsche im Pollen der Steinweichsel (*Prunus mahaleb* L.) *Archiv. für Phytopathologie und Pflanzenschutz* 10: 3-6. VEG. Saatzucht-Baumschulen, Dresden (RPP 1975, 927).

Schippers, B. (1963). Transmission of bean common mosaic virus by seeds of *Phaseolus vulgaris* L. cultivar Beka. *Acta Bot. Neerl.* 12: 433-497.

Schmelzer, K. (1975). Untersuchungen über den Wirtspflanzenkreis des Tabakmauche Virus. *Phytoapth Z.* 30: 281-314.

Schmelzer, K. (1965). Hosts of cherry leaf roll virus among wild woody plants. *Zast. bilja.* 85-88: 485-489.

Schmelzer, K. (1966). Untersuchungen an Viren der Zier-und Wildgehölze. 5. Mitteilung: Virosen an Populus und Sambucus. *Phytopath Z.* 55: 317-351.

Schmelzer, K. (1969). Das Ulmenscheckungs-Virus. *Phytopath. Z.* 64: 39-67.

Sharma, S. R., and Varma, A. (1975). Three sap transmissible viruses from cowpea in India. *Ind. Phytopathol.* 28: 192-198.

Sharma, Y. R., and Chohan, J. S. (1973). Transmission of cucumis viruses 1 and 3 through seeds of cucurbits. *Ind. Phytopathol.* 26: 596-598.

Shepherd, R. J. (1964). Properties of a mosaic disease of cowpea and its relationship to bean pod mottle virus. *Phytopathology* 54: 466-473.

Shepherd, R. J. (1972). Transmission of viruses through seed and pollen. Chapter 10 in "Principles and Techniques in Plant Virology" (C. I. Kando and H. O. Agrawal, eds.), Van Nostrand Reinhold, New York.

Shepherd, R. J., and Fulton, R. W. (1962). Identity of a seed-borne virus of cowpea. *Phytopathology* 52: 489-493.

Shepherd, R. J., and Holdeman, Q. L. (1965). Seed transmission of the Johnson grass strain of the sugarcane mosaic virus in corn. *Pl. Dis. Reptr.* 49: 468-469.

Shoyinka, S. A., Bozarth, R. F., Reese, J., and Rossel, H. W. (1978). Cowpea mottle virus: a seed borne virus with distinctive properties infecting cowpeas in Nigeria. *Phytopathology* 68: 693-699.

Singh, R. P. (1970). Seed transmission of potato spindle tuber virus in tomato and potato. *Am. Potato J.* 47: 225-227.

Skotland, C. B., and Burke, D. W. (1961). A seed-borne bean virus of wide host range. *Phytopathology* 51: 565-568.

Slack, S. A., and Shepherd, R. J. (1975). Serological detection of seed-borne barley stripe mosaic virus by a simplified radial-diffusion technique. *Phytopathology* 65: 948-955.

Slykhuis, J. T. (1976). Virus and virus-like diseases of cereal crops. *Annu. Rev. Phytopathol.* 14: 189-210.

Smith, F. L., and Hewitt, W. B. (1938). Varietal susceptibility to common bean mosaic and transmission through seed. *Calif. Agr. Exp. Sta. Bull.* 621: 18 pp.

Smith, K. M. (1951). A latent virus in sugarbeets and marigolds. *Nature* 167: 1061.

Smith, K. M. (1972). A textbook of plant virus diseases. Longman, London. 684 pp.

Snyder, W. C. (1942). A seed borne mosaic of Asparagus bean, *Vigna sesquipedalis. Phytopathology* 32: 518-523.

Stelzner, G. (1942). Zur Frage der Virusübertragung durch Samen, insbesondere des X-Y- und Blattrollvirus der Kartoffel. *Züchter* 14: 225-234.

Stevenson, W. R., and Hagedorn, D. J. (1969). A new seed-borne virus of peas. *Phytopathology* 59: 1051.

Stevenson, W. R., and Hagedorn, D. J. (1973). Further studies on seed transmission of pea seed borne mosaic virus in *Pisum sativum. Pl. Dis. Reptr.* 57: 248-252.

Sutic, D. (1959). Die Rolle des Paprikasamens bei der Virusübertragung. *Phytopath. Z.* 36: 84-93.

Sutere, B. D., and Joshi, R. D. (1978). *Desmodium triflorum* mottle-a virus disease. *Geobios.* 5: 266-267.

Swenson, K. G., and Milbrath, J. A. (1964). Insect and mite transmission tests with the *Prunus* ringspot virus. *Phytopathology* 54: 399-404.

Takahashi, K., Tanaka, T., and Tsuda, Y. (1974). Soybean mild mosaic virus. *Ann. Phytopathol. Soc. Japan* 40: 103-105.

Taylor, C. E., and Thomas, P. R. (1968). The association of *Xiphinema diversicaudatum* (Micoletsky) with strawberry latent ringspot and arabis mosaic viruses in raspberry plantation. *Ann. Appl. Biol.* 62: 147-157.

Taylor, R. H., Grogan, R. G., and Kimble, K. A. (1961). Transmission of tobacco mosaic virus in tomato seed. *Phytopathology* 51: 837-842.

Thomas, W. (1973). Seed transmitted squash mosaic virus. *New Zealand J. Agr. Res.* 16: 561-567.

Thomas, W. D., Jr., and Graham, R. W. (1951). Seed transmission of red node virus in Pinto bean. *Phytopathology* 41: 959-962.

Thottappilly, G., and Schmutterer, H. (1968). Zur Kenntnis eines mechanisch, samen-pilz- und insektübertragbaren neuen Virus der Erbse. *Z. Pflkrankh. Pflschutz.* 75: 1-8.

Timian, R. G., and Sisler, W. W. (1955). Prevalence, sources of resistance, and inheritance of resistance to barley stripe mosaic virus (false stripe). *Pl. Dis. Reptr.* 39: 550-552.

Tomlinson, J. A. (1962). Control of lettuce mosaic by the use of healthy seeds. *Pl. Pathol.* 11: 61-64.

Tomlinson, J. A., and Carter, A. L. (1970a). Seed transmission of cucumber mosaic virus in chickweed. *Pl. Dis. Reptr.* 54: 150-151.

Tomlinson, J. A., and Carter, A. L. (1970b). Studies on the seed transmission of cucumber mosaic virus in chickweed (*Stellaria media*) in relation to the ecology of the virus. *Ann. Appl. Biol.* 66: 381-386.

Tomlinson, J. A., and Carter, A. L. (1971). Occurrence and seed transmission of tobacco ringspot virus in *Senecio vulgaris. Rep. Nat. Veg. Res. Sta.* 21: 109-110.

Tomlinson, J. A., and Walker, V. M. (1973). Further studies on seed-transmission in the ecology of some aphid-transmitted viruses. *Ann. Appl. Biol.* 73: 292-298.

Tomlinson, J. A., and Walkey, D. G. A. (1967). The isolation and identification of rhubarb viruses occurring in Britain. *Ann. Appl. Biol.* 59: 415-427.

Traylor, J. A., Williams, H. E., Weinberger, J. H., and Wagnon, H. K. (1963). Studies on the passage of prunus ringspot virus complex through plum seed. *Phytopathology* 53: 1143.

Troll, H. J. (1957). Zur Frage der Bräunevirus–Übertragung durch das Saatgut bei *Lupinus luteus. NachBl. dt. Pflschutzdienst, Berlin* 11: 218-222.

Troutman, J. L., Bailey, W. K., and Thomas, C. A. (1967). Seed transmission of peanut stunt virus. *Phytopathology* 57: 1280-1281.

Tsuchizaki, T., Yora, K., and Asuyama, H. (1970). Seed transmission of viruses in cowpea and Azuki bean plants. II. Relations between seed transmission and gamete infection. *Ann. Phytopathol. Soc. Japan* 36: 237-242.

Tu, J. C., and Ford, R. E. (1970). Maize dwarf mosaic virus infection in susceptible and resistant corn: virus multiplication, free amino acid concentration and symptom severity. *Phytopathology* 60: 1605-1608.

Tu, J. C., and Ford, R. E. (1971). Maize dwarf mosaic virus predisposes corn to root rot infection. *Phytopathology* 61: 800-803.

Tuite, J. (1960). The natural occurrence of tobacco ringspot virus. *Phytopathology* 50: 296-298.

Uyemoto, J. K., and Grogan, R. G. (1977). Southern bean mosaic virus: evidence for seed transmission in bean embryos. *Phytopathology* 67: 1190-1196.

Uyemoto, J. K., Gilmer, R. M., and Williams, E. (1971). Sap-transmissible viruses of elderberry in New York. *Pl. Dis. Reptr.* 53: 913-916.

Uyemoto, J. K., Taschenberg, E. F., and Hummer, D. K. (1977). Isolation and identification of a strain of grapevine Bulgarian latent virus in Concord grapevine in New York State. *Pl. Dis. Reptr.* 61: 949-953.

Valerie, C. C., and Walkey, D. G. A. (1978). Thermal inactivation of cherry leaf roll virus in tissue cultures of *Nicotiana rustica* raised from seeds and meristem tips. *Ann. Appl. Biol.* 88: 273-278.

Valleau, W. D. (1932a). Two seed-transmitted ringspot diseases of tobacco. *Phytopathology* 22: 28.

Valleau, W. D. (1932b). Seed transmission and sterility studies of two strains of tobacco ringspot. *Kentucky Agr. Exp. Sta. Bull.* 327: 43-80.

Valleau, W. D. (1939). Symptoms of yellow ringspot and longevity of the virus in tobacco seed. *Phytopathology* 29: 549-551.

Van Hoof, H. A. (1959). Seed transmission of lettuce mosaic virus in *Lactuca serriola Tijdschr. PlZiekt.* 65: 44-46.

Van Koot, Y. (1949). Enkele nieuwe gezichtspunten betreffende het virus van het tomatenmosaiek. *Tijdschr. PlZiekt.* 55: 152-166.

Van Velsen, R. J. (1961). Marginal chlorosis, a seedborne virus of *Arachis hypogea* variety 'Schwarz 21' in New Guinea. *Papua New Guin. Agr. J.* 14: 38-40.

Varma, P., and Gibbs, A. J. (1967). Preliminary studies on sap-transmissible viruses of red clover (*Trifolium pratense* L.) in England and Wales. *Ann. Appl. Biol.* 59: 23-30.

Vashisth, K. S., and Nagaich, B. B. (1965). A mosaic disease of runner bean. *Ind. Phytopathol.* 18: 311.

Voller, A., Bartlett, A., Bidwell, D. E., Clark, M. F., and Adams, A. N. (1976). The detection of viruses by enzyme-linked immunosorbent assay (ELISA). *J. Gen. Virol.* 33: 165-167.

Vorra-urai, S., and Cockbain, A. J. (1977). Further studies on seed transmission of broad bean stain virus and Echtes Ackerbohnenmosaik virus in field beans (*Vicia faba*). *Ann. Appl. Biol.* 87: 365-374.

Wagnon, H. K., Traylor, J. A., Williams, H. E., and Weinberger, J. H. (1960). Observations on the passage of peach necrotic leaf spot and peach ringspot viruses through peach and nectarine seeds and their effects on the resulting seedlings. *Pl. Dis. Reptr.* 44: 117-119.

Walkey, D. G. A. (1967). Chlorotic stunt of lettuce caused by arabis mosaic virus. *Pl. Pathol.* 16: 20-22.

Walkey, D. G. A., and Dance, M. C. (1979). High temperature inactivation of seed borne lettuce mosaic virus. *Pl. Dis. Reptr.* 63: 125-129.

Walkey, D. G. A., and Whittingham-Jones, S. G. (1970). Seed transmission of strawberry latent ringspot virus in celery (*Apium graveolens* var. dulce). *Pl. Dis. Reptr.* 54: 802-803.

Wallace, J. M. (1957). Virus-strain interference in relation to symptoms of psorosis disease of citrus. *Hilgardia* 27: 223-246.

Wallace, J. M., and Drake, R. J. (1953). Seed transmission of the avocado sun-blotch virus. *Citrus Leaves.* December 1952. Reprint. 2 pp.

Wallace, J. M., and Drake, R. J. (1962). A high rate of seed transmission of avocado sun-blotch virus from symptomless trees and the origin of such trees. *Phytopathology* 52: 237-241.

Watson, M. A., and Serjeant, E. P. (1962). Carrot motley dwarf. *Rep. Rothamsted Exp. Sta.*, 1961: 106-107.

Way, L. D., and Gilmer, R. M. (1958). Pollen transmission of necrotic ringspot virus in cherry, *Pl. Dis. Reptr*. 42: 1222-1224.

Wilkins, P. W., and Catherall, P. L. (1974). The effect of some isolates of ryegrass mosaic virus on different genotypes of *Lolium multiflorum. Ann. Appl. Biol.* 76: 209-216.

Williams, H. E., and Smith, S. H. (1967). Recovery of a virus from stored pear pollen. *Phytopathology* 57: 1011 (Abstr.).

Williams, L. E., Findley, W. R., Dollinger, E. J., and Ritter, R. M. (1968). Seed transmission studies of maize dwarf mosaic virus in corn. *Pl. Dis. Reptr.* 52: 863-864.

Wyatt, S. D., and Kuhn, C. W. (1979). Replication and properties of cowpea chlorotic mottle virus in resistant cowpeas. *Phytopathology* 69: 125-129.

Yakovleva, N. (1965). Borba s zelenoi mozaikoi Ogurtsov. (Control of green mosaic of cucumber.). *Zashch. Rast. Vredit. Bolez.* 10: 50-51.

Yamamoto, T. (1951). Studies on the sterility of barley. I. On the mechanism in occurrence of sterile grain. *Proc. Jap. Rep. Sci. Soc.* 20: 80-84.

Yang, A. F., and Hamilton, R. I. (1974). The mechanism of seed transmission of tobacco ringspot virus in soybean. *Virology* 62: 26-37.

Zink, F. W., Grogan, R. G., and Welch, J. E. (1956). The effect of the percentage of seed transmission upon subsequent spread of lettuce mosaic virus. *Phytopathology* 46: 662-664.

Zitter, T. A., and Murakishi, H. H. (1969). Nature of increased virulence in tobacco mosaic virus after passage in resistant tomato plants. *Phytopathology* 59: 1736-1739.

Zschau, K., and Janke, C. (1962). Samenübertragung des Luzernemosaik Virus an Luzerne. *NachBl. dt. Pfschutzdienst.* 16: 94-96.

Chapter 9

SEEDBORNE VIRUSES: VIRUS-HOST INTERACTIONS

Thomas W. Carroll

Department of Plant Pathology
Montana State University
Bozeman, Montana

9.1 INTRODUCTION

Seedborne plant viruses have been a curiosity in plant virology for over 45 yrs. How and why such viruses become seedborne is still somewhat of a mystery. It is known, however, that many factors associated with the viruses, their hosts and the environment, determine whether the viruses are, in fact, transmitted by seeds.

Some workers (Caldwell, 1934, 1952, 1962; Bennett, 1936, 1940, 1969; Crowley, 1957b, 1959; Baker and Smith, 1966) have characterized the seed transmission of plant viruses largely on the basis of virus-host interactions as they are expressed at the cytological and anatomical levels in the infected host plants. The concentration, distribution, and pathological effects of the viruses in floral, fruit, and seed parts were of utmost importance to their understanding of seed transmission. Considerable attention was also given to the avenues in host plants through which the viruses invaded the seeds or embryos of seeds.

According to the earlier workers, certain avenues were necessary for the *direct* invasion of embryos or seeds, and other avenues were necessary for the *indirect* invasion of embryos or seeds. Avenues for direct invasion included plasmodesmata that extended from the mother plants to developing seeds (Bennett, 1969), or to the embryos of developing seeds (Caldwell, 1934; Bennett, 1940, 1969; Crowley, 1957b). Avenues for direct invasion also included disintegrating tissues of the mother plants that surrounded the seeds (Bennett, 1969). Tissues such as these frequently come from fleshy or pulpy fruits. Thus, these tissues carry the viruses to the seeds, contaminating the seed surface.

Lack of plasmodesmata (Caldwell, 1934; Bennett, 1940, 1969; Crowley, 1957b), or plasmodesmatal breakdown (Caldwell, 1934) between embryos and surrounding seed and mother plant tissues were suggested as possible reasons for the inability of viruses to invade embryos directly. That plasmodesmata are probable avenues of virus movement in infected plants is suggested by the many studies in which virus or virus-like particles were seen in plasmodesmata (Esau *et al.*, 1967; Davidson, 1969; de Zoeten and Gaard, 1969; Kitajima and Lauritis, 1969; Lawson and Hearon, 1970; Roberts and Harrison, 1970; Kim and Fulton, 1971; Esau and Hoefert, 1972; Weintraub *et al.*, 1976).

Vascular connections between the mother plants and seeds or between the mother plants and embryos were thought to be unimportant avenues for direct invasion (Bennett, 1969), because viruses that were restricted to vascular tissues had not been shown to be seedborne.

Since indirect invasion of seeds or embryos occurs prior to initiation of the seeds (Bennett, 1940, 1969; Crowley, 1959; Baker and Smith, 1966), the avenues used consist of floral meristematic cells and tissues. These include gametophytic cells (Bennett, 1940, 1969; Crowley, 1959), gametes, and zygotes (Crowley, 1959; Bennett, 1969). Bennett (1940) and Fulton (1964) had observed that seed transmitted viruses are also pollen transmitted viruses. Proof that both male and female gametes transmit NEPO (Nematode transmitted viruses with polyhedral particles) viruses was given by Lister and Murant (1967).

In 1952, Caldwell demonstrated that tomato aspermy virus was not seed transmitted by tomato plants because it caused sterility of the pollen and ovules. This observation led him to propose that sterility was responsible for the rarity of seed transmission of other viruses. Later, however, Crowley (1957b) showed, that sterility could not account for the lack of seed transmission of five other viruses. Subsequent work on tomato aspermy virus (Crowley 1957b; Hollings and Stone, 1971), has verified that this virus is, indeed, not seed transmitted in tomato.

The earlier works (Caldwell, 1934, 1952, 1962; Bennett, 1936, 1940, 1969; Crowley, 1957b, 1959; Baker and Smith, 1966) on the cytological and anatomi-

cal aspects of seed transmission used mainly information from transmission experiments, and virus assays of floral, fruit, or seed parts. Additional facts came
from studies on the effects of seed transmitted viruses on mother plants and
seedlings. Electron microscopic evidence from ultrathin sections of seedborne viruses infecting reproductive cells, was not available when the earlier works were
conducted.

In 1974 Phatak listed 85 seedborne viruses. These viruses comprise a large
portion of the 200 or so plant viruses that have been described in some detail
in the literature.

This chapter will be concerned with the interactions between seedborne viruses and their host plants. The cytological and anatomical factors believed to influence or condition the seed transmission of plant viruses will be emphasized.
When known, the form, concentration, and distribution of the viruses in floral,
fruit, and seed tissues will be described. The few virus-host interactions for which
information is available, are discussed on the basis of what is observed in virus infected cells by means of the electron microscope. For those interactions, the visual aspects of the virus-host cell are presented. The chapter is also concerned
with the avenues of invasion of viruses from mother plants to seeds, and some of
the changes occurring in cells and tissues of the mother plants in response to infection. Special consideration is given to the morphological evidence that exists.
The chapter will also describe the pathological effects of the viruses in the hosts.
Some of these effects have been recently reviewed by Mathre (1978).

The reader is also referred to several other reviews that have appeared on various aspects of seedborne viruses and seed transmission (Fulton, 1964; Baker and
Smith, 1966; Bennett, 1969; Shepherd, 1972).

9.2 EXAMPLES OF VIRUS-HOST INTERACTIONS

9.2.1 Nonembryonic Transmission

Perhaps the only good example of a seedborne virus being transmitted by
nonembryonic tissues is tomato mosaic virus, also referred to as the tomato infecting form of tobacco mosaic virus (Hollings and Huttinga, 1976). To clarify
earlier contradictory findings on the importance of seed transmission in the origin of the virus, Taylor et al. (1961) studied the transmission of tomato mosaic
virus in tomato seeds. They were concerned with the location of the virus on or
in the seeds, the amount and mechanism of seed transmission of tomato mosaic
virus in tomato, and the effects of different methods of seed extraction and treatment on the elimination of the virus from tomato seeds and their effect on seed
transmission. Results of that study showed that based on infectivity assays, tomato mosaic virus was located in and on seed coats, and in a small percentage of
the endosperms, but not in embryos. Seed-coat virus could be eliminated by

some treatments, but not by others (Taylor *et al.*, 1961). Virus in endosperms was more difficult to eliminate by various treatments than that in seed coats. Endospermic virus was inactivated slowly during storage. The virus actually infested some of the tomato seeds, and when the seedlings developed, from these seeds, the contaminating virus infected the resulting seedlings, either through the roots or shoots.

Broadbent (1965) also reported that tomato mosaic virus was carried externally on tomato seeds in low concentrations. He also stated that the virus was carried internally in about 25% of the seed coats or endosperms. Sometimes high virus concentrations occurred in endosperms. Early infection versus late infection of mother plants resulted in an increase of abnormal and infected seeds, especially seeds with endosperm infections.

According to the information published by Ford (1966), pea streak virus also appears to be a seedborne virus that is transmitted by nonembryonic tissues. However, one important exception must be noted. Pea streak virus was only transmitted by immature seeds, not by mature seeds that were naturally dried and stored. Infectivity assays revealed the presence of pea streak virus in immature ovules and the seed coats of immature seeds of peas. No virus was detected in immature or mature embryos. Strangely enough, Bos (1973) reported no seed transmission for pea streak virus.

Electron microscopic observations of thin sections of floral, fruit, and seed parts from tomato plants infected with tomato mosaic virus have not appeared in the literature. Similarly, ultrastructural evidence has not been published for the interaction of pea streak virus in peas, relative to the seedborne nature of the virus.

Direct invasion of nonembryonic tissues appears to be the case for both tomato mosaic virus in tomato seeds and pea streak virus in pea seeds. Whereas tomato mosaic virus persists through the normal maturation of seeds and therefore infects emerging seedlings, pea streak virus becomes inactivated in seeds before they mature.

9.2.2 Nonembryonic or Embryonic Transmission?

Gilmer and Wilks (1967) discovered that tobacco mosaic virus was seedborne in apple and pear. The virus was isolated from inner seed coat and endosperm tissues of apple seeds. The virus was also found in embryos of dormant apple seeds, and in the radicle, hypocotyl, and cotyledons of germinating seeds. The tobacco mosaic virus isolates from apple resembled the type strain of the virus based on physical properties and most host reactions. Their findings suggest that nonembryonic or embryonic transmission of the virus may occur in apple seeds. Zaitlin (1975) reported that the wild type, common strain, or ordinary tobacco mosaic virus was not transmitted by seeds or pollen.

Studies on the seedborne nature of southern bean mosaic virus have produced confusing and contradictory results. In 1955 Cheo demonstrated that a bean strain of the virus was transmitted through immature, but not mature bean seeds. Furthermore, he detected southern bean mosaic virus in all flower and fruit parts of systemically infected bean cultivars. Both seed coats and embryos of immature seeds contained virus, but only seed coats of mature seeds had virus. From these results he concluded that the virus was inhibited as the seeds dried and matured. Crowley reported recovery of a bean strain of the virus from immature embryos, but not from seedlings germinated from 50 mature seeds. Crowley also stated that when flowers on healthy bean plants were pollinated from pollen from infected plants, some of the resulting seeds produced infected embryos, even though the leaf and pod tissues of the pollinated plants were virus-free. Crowley concluded that southern bean mosaic virus infected the embryos of bean seeds indirectly via pollen (presumably via eggs too) or, to some extent, directly by infecting the developing embryos. Seed transmission through mature embryos did not occur because the virus was inactivated as the embryos matured. Later, Shepherd and Fulton (1962) determined that a cowpea strain of southern bean mosaic virus was transmitted through 3-4% of the seeds produced by infected cowpea plants. In 1972, McDonald and Hamilton re-examined the distribution of the bean strain in bean seeds. They recovered the virus from immature embryos and seed coats. However, they found that when embryos and seed coats were surface decontaminated, the virus was eliminated from the embryos, but not from the seed coats. Infectious virus was also obtained from decontaminated mature seed coats. They speculated therefore that seed coat virus probably infected the embryo during germination of the bean seed. The cowpea strain of southern bean mosaic virus was once again studied in cowpeas by Gay (1973). He looked at the effect of host cultivar and infection age on the presence of the virus in floral parts and immature seeds. He detected infectious virus in sepals, petals, stamens, pistils, and pollen from two cultivars of cowpeas when these flower parts were assayed two days after inoculation. Virus was recovered from seed coats in two cultivars of cowpeas, and from cotyledons and embryos in one of the cultivars three days and in the other cultivar five days after inoculation of the plants. Finally, in 1977, Uyemoto and Grogan restudied southern bean mosaic virus in bean seeds. Supposedly, they worked with three bean strains of the virus. Results of their study showed that all three strains were seed transmitted through embryos of three bean cultivars. Immature and mature embryos were infected. Decontaminated, infected embryos or infected seeds produced infected seedlings. Southern bean mosaic virus was transmitted to a greater extent by immature seeds (57.0%) than by mature seeds (2.4%) when the type strain was tested in red kidney beans. Uyemoto and Grogan (1977) maintained that the virus was transmitted to seedlings through infected embryos, because they had taken proper precautions to surface decontaminate the embryos, thereby preventing any fortitous infection due to surface-borne virus. Shepherd (1971) stated that the virus may be seedborne in beans, but was definitely seed transmitted in cowpeas.

Ghanekar and Schwenk (1974), working with a Kansas strain of tobacco streak virus, discovered that the virus was transmitted through both immature and mature seeds of soybeans. Infectivity assays revealed that virus was present in flowers, immature seed coats and embryos, but not in mature seed coats and embryos. However, the authors stated that had more mature embryos been tested, mature embryo infections may have been detected. Earlier in 1971, Fulton mentioned that transmission of tobacco streak virus had been reported for bean, *Datura stramonium* and *Chenopodium quinoa*.

Electron micrographs depicting virus-host interactions at the cellular level for the above seedborne viruses have not appeared in the literature.

9.2.3 Embryonic Transmission

Bos *et al.* (1971) cite reports of seed transmission for alfalfa mosaic virus. Transmission through seed in certain cultivars of alfalfa is up to 6%, and in chili pepper it ranges from 1-5%. In 1974, Frosheiser described pollen and ovule transmission in alfalfa. When seeds produced from cross pollination between an alfalfa plant infected with a local lesion strain of the virus and a plant infected with a strain causing systemic symptoms in bean, were assayed it appeared that the virus could be transmitted through both pollen and ovules. Pollen transmission occurred more frequently (0.5-26.5%) than did ovule transmission (0-9.5%). Further work on the seedborne nature of alfalfa mosaic virus was reported by Wilcoxson *et al.* (1975). Electron microscopic examination of thin sections of infected alfalfa plants revealed virus particles in bud receptacles, anthers, pollen, ovary wall cells, and mature embryos. Wilcoxson *et al.* (1975) did not find virus particles in the sperm cells of pollen grains, nor in ovules. However, they stated that there was no apparent structural barrier to prevent the virus from invading the ovules because plasmodesmata were common between the ovary and ovule cells. Virus particles occurred mainly in rafts within the cytoplasms of the various cell types. Gallo and Ciampor (1977) also found alfalfa mosaic virus to be transmitted through seeds of *Nicandra physaloides*. For one strain of the virus, the average seed transmission rate was 23%. Electron microscopy detected aggregates of virus particles of one strain of the virus in embryo cotyledons.

Medina and Grogan (1961) studied the seed transmission of bean common mosaic virus in beans. Seeds obtained from cross pollination experiments, in which pollen from infected plants was applied to healthy plants, and vice versa, showed that the virus was transmitted by both pollen and ovules. Pollen usually transmitted virus to more progeny plants than ovules. Infectivity assays determined infectious virus was present in flower petals, pistils, anthers, and mature embryos. In 1963, a detailed report by Schippers on the transmission of bean common mosaic virus by bean seed appeared. The report emphasized anatomical and cytological aspects of the virus-host interaction. One interesting fact that came to light was that to obtain infected bean seeds, inoculation of the mother plants had to occur before the flower buds had passed a critical period. Even under optimal conditions of host plant age and environment, seed transmission only averaged 15%. By infectivity assays and the dipping method for electron microscopy, virus was detected in ovaries, ovules, and seeds. Infectivity assays alone showed virus present in immature and mature embryos. Cross pollination tests demonstrated both pollen and ovule transmission. Electron microscopical examination of thin sections of the embryo sac and surrounding tissues revealed that no plasmodesmata occurred in the cell walls of the inner integument bordering the embryo sac. Microscopy also revealed that just prior to fertilization the nucellar tissue at the chalazal end of the embryo sac had begun to disintegrate. The absence of plasmodesmata between the inner integument and the embryo sac, and the presence of disintegrating nucellar tissue, may prevent virus invasion of the embryo sac after the ovule has reached that critical period in its development, according to Schippers (1963). The occurrence of virus particles in ovules or embryo sacs was not reported. In 1971 Bos cited many references on the seed transmission of bean common mosaic virus. Shortly thereafter, Ekpo and Saettler (1974) investigated the distribution of bean common mosaic virus in developing bean seeds by infectivity assay with a new indicator host plant, because they believed that little information was available on the mechanism of seed infection and transmission. Their results showed that opened and unopened blossoms, as well as young pods contained infectious virus. Bean common mosaic virus was also detected in cotyledons, and embryos, but not in seed coats of seeds developing on infected plants. The distribution of the virus in cotyledons and embryos was not affected appreciably by seed drying during maturation. Evidence for the ultrastructural localization of bean common mosaic virus in dormant and germinating bean seeds was provided in 1978 by Hoch and Provvidenti. Their electron micrographs depicted virus particles, virus inclusions, and virus-related inclusion materials in mature embryos from dehydrated seeds. Virus-related inclusion materials, and pinwheel and scroll inclusion bodies were visible in mature embryos from rehydrated seeds.

Crowley (1959) conducted two experiments on the time of infection of soybean embryos by tobacco ringspot virus. Results of those experiments showed that a low level of infection occurred in immature embryos. This work was followed in 1968 by further studies on the seed transmission of tobacco ringspot virus in soybean (Owusu et al., 1968). They found that the age of the mother plants at the time they became infected was the most important factor determining the amount of infected seed produced. Infectivity assays showed that tobacco ringspot virus was consistently present in embryos and perisperm tissues of infected mature seeds but not in seed coats. Owusu et al. (1968) suggested that the inability of the virus to infect maturing embryos was not due to the slowness of virus movement in the mother plants, but may be due to some physical barrier between the embryos and the mother plants. According to Stace-Smith (1970), tobacco ringspot virus transmission through seed is common in soybean, petunia, *Nicotiana glutinosa*, *Gomphrena globosa* and *Taraxacum officinale*; rare in tobacco, cantaloupe, cucumber, muskmelon and lettuce. In 1974 Yang and Hamilton investigated the mechanism of seed transmission of tobacco ringspot virus in soybean. Using the thin sectioning technique for electron microscopy, they detected virus like particles in the intine of the pollen wall, the wall and cytoplasm of the cell, the integuments, nucellus, embryo sac wall and in the megagametophytic cells. Anthers from infected plants produced less pollen than those from healthy plants. Cross-pollination experiments provided evidence that suggested that infection of megagametophytes was the main factor contributing to the seed transmission of the virus in soybean.

In 1962, Bos briefly mentioned reports of soybean mosaic virus transmission through soybean seeds. As much as 30% seed transmission has occurred. Only plants infected before and during flowering produce infected seeds. Virus has been found in seed coats, embryos, immature and mature seeds. Immature seeds contain more virus than mature seeds. Some years later, Tu (1975) detected virus particles and aggregates resembling soybean mosaic virus in thin sections of the embryo and endosperm, but not the seed coat of mature, dried seeds of soybean. Infectious virus was present in the embryo and endosperm, but not in the seed coat.

Bock and Kuhn (1975) indicated that seed transmission of peanut mottle virus occurred in groundnut at low percentages (0.02-2.0%). The virus, however, was not seed transmitted in cowpea, soybean, *Pisum sativum*, or *Cassia obtusifolia*. A trace level of seed transmission has also been reported in bean. Later, in 1977, Adams and Kuhn studied the location of soybean mosaic virus in infected peanut seeds and the factors that affected seed transmission. They found that seed transmission was the result of embryo infection. Embryos contained virus, but not seed coats or cotyledons. Two virus isolates were detected in flowers, pegs, pods, and immature seeds of infected plants of two peanut cultivars.

The seedborne nature of tomato black ring virus in different plant hosts, including soybean, has been examined by Lister and Murant (1967). They determined that the virus was present in embryos, but not in seed coats of soybean seeds. In 1970, Murant cited examples of seed transmission in fifteen botanical families by tomato black ring virus.

9.3 BARLEY STRIPE MOSAIC VIRUS IN BARLEY

9.3.1 Seedborne Enigma?

The seedborne nature of barley stripe mosaic virus has been difficult to characterize. Previous work has shown that the virus could be seedborne (Gold *et al.*, 1954; Hagborg, 1954; Eslick and Afanasiev, 1955; Crowley, 1959; Singh *et al.*, 1960; Inouye, 1962; Hamilton, 1965; McKinney and Greeley, 1965; Timian, 1967), but it also could be nonseed borne (Hamilton, 1965; McKinney and

Greeley, 1965; Shivanathan, 1970). Furthermore, both direct invasion of developing embryos (Eslick and Afanasiev, 1955; Crowley, 1959; Singh *et al.*, 1960), and indirect invasion of embryos (Hagborg, 1954; Inouye, 1962; Timian, 1967) have been suggested as means by which the virus becomes seedborne, in those cases where it, in fact, was seedborne. Why has there been the disparity in the results of seed transmission experiments? Did anomolies actually exist for the seedborne nature of barley stripe mosaic virus? This mystery surrounding the virus was one of the reasons I felt compelled to study its relations with seeds.

A careful study of the literature indicated to me that most, if not all, of the discrepancies pertaining to the seed transmission of barley stripe mosaic virus could be attributed to the different virus strain—host cultivar—environment interactions used in previous studies. Therefore, to partly overcome this difficulty, I also chose to work, in part, with the two virus-host combinations that had been used earlier to good advantage by Hamilton (1965) and Shivanathan (1970). These combinations consisted of single infections of the type strain, which is seedborne, and the nonseed passage strain of barley stripe mosaic virus in the Atlas cultivar of barley. These two combinations permitted the comparison of two different virus strains in the same host plant, under identical environments. These two strains, along with other seedborne strains of the virus in the same or additional barley cultivars, were used to further define the seedborne nature of barley stripe mosaic virus in barley.

9.3.2 Seedborne Strains: Virus-host Interactions

Evidence to support the hypothesis that some seedborne strains of barley stripe mosaic virus invade embryos directly comes from several studies on the influence of host age at the time of inoculation. Eslick and Afanasiev (1955) found that when the virus infected mother plants after flowering, some seed transmission resulted. Their findings were later confirmed by Crowley (1959) and Singh *et al.* (1960). In the case of Eslick and Afanasiev (1955) and Singh *et al.* (1960), the possibility existed that some of the infected seeds produced by mother plants actually came from developing flowers on late tillers or stems. In the study of Crowley (1959), this was not the case, for he inoculated Mars and Compana barleys and harvested them 14 days later and he got 1% seed transmission of the virus. Proof of the direct invasion of developing embryos *in situ* by the type strain of barley stripe mosaic virus was not obtained by Carroll (1972). When naturally ripened seeds were harvested from mother plants that had developing embryos, unfertilized ovules, and immature florets at the time of infection, only embryos from unfertilized ovules and immature florets were infected. Serological tests of the embryos of seeds from the developing embryos showed that 0 of 5502 (0%) had virus antigen present. By contrast, in seeds from unfertilized ovules, 49 of 386 (12.7%) of the embryos were infected, and 240 of the 1665 (14.4%) of the embryos from immature florets contained virus.

Light and electron microscopy showed that to invade the *in situ* embryos directly, type strain virus would have to come from mother tissues and penetrate the embryos. Developing embryos are bounded by endosperm and nucellar tissues. Mature embryos are surrounded by endosperm and caryopsis coat tissues (Carroll, 1972). Ultrathin sections of developing embryos revealed that no plasmodesmata connected the embryos with surrounding tissues (Carroll, 1972). Thus, the implication is that to invade embryos, the virus must enter embryo cells, by avenues other than plasmodesmata. This evidence supports the suggestion of Caldwell (1934) and Bennett (1940) that virus infection of embryos through plasmodesmata was improbable.

Considerable work has been done on the distribution in the host of seedborne strains of barley stripe mosaic virus in barley. Infectivity assays, serological tests, electron microscopical observations, ultracentrifugation analyses, seed transmission tests or germination tests with mature seeds and cross-pollination experiments were used to determine the presence of virus and occassionally the titer of virus, in various floral, fruit and seed tissues. Virus was detected by one or more of the methods mentioned above in glumes (Inouye, 1962; Shivanathan, 1970; Carroll, 1972), anthers (Gold *et al.*, 1954; Inouye, 1962; Shivanathan, 1970; Carroll, 1972), pollen (Gold *et al.*, 1954; Gardner, 1967; Carroll, 1974), pistils (Gold *et al.*, 1954; Carroll, 1972), ovaries (Inouye, 1962; Shivanathan, 1970), ovules (Shivanathan, 1970, Mayhew and Carroll, 1974b), immature seeds (Shivanathan, 1970, Carroll, 1972), zygotes (Brlansky and Carroll, 1978), developing embryos (Inouye, 1962; Shivanathan, 1970; Carroll, 1972), nonembryonic tissues of developing seeds (Inouye, 1962; Shivanathan, 1970; Carroll, 1972), mature embryos (Gold *et al.*, 1954; Crowley, 1959; Hamilton, 1965; Inouye, 1966; Carroll, 1969, 1972), and mature endosperms (Gold *et al.*, 1954; Crowley, 1959; Inouye, 1966; Carroll, 1972). In summary, it appears that seedborne strains of barley stripe mosaic virus infect all floral, fruit, and seed tissues. Their presence as infectious virus in naturally dried seeds suggests that they have some tolerance to the normal maturation and drying processes of seeds.

Information obtained on the influence of developmental stage of the mother plants at the time of infection on the infection of embryos by seedborne strains of barley stripe mosaic virus lends support to the hypothesis that some strains only invade embryos indirectly. Hagborg (1954), Inouye (1962), and Timian (1967) found that if barley was infected after fertilization (pollination), the resultant seeds would not give rise to transmission of barley stripe mosaic virus. These results were later confirmed by Carroll (1972), who showed by serological testing that virus was present only in embryos of seeds that had developed from unfertilized pistils or immature florets at the time the mother plants became infected.

Ovule and pollen transmission experiments using the cross pollination method, and electron microscopic observations of virus in gametophytes and gametes also strengthen the hypothesis that some strains of barley stripe mosaic virus only infect embryos indirectly.

On the basis of circumstantial evidence, Crowley (1959), and Bennett (1969) suggested that early infection of the embryo sac in the developing ovule was one essential way for achieving successful seed infection. Results of ovule transmission experiments, in which infected mother plants are emasculated, and then pollinated with pollen from virus-free plants, substantiate their earlier suggestion that early ovule infection leads to seed transmission. Inouye (1962), and Carroll and Mayhew (1976a, 1976b), showed that ovule transmissions produced higher percentages of seed transmission than did pollen transmissions of the reciprocal crosses where infected pollen was applied to virus-free ovules. For field grown plants, the transmission by ovules (Carroll and Mayhew, 1976b) and pollen (Carroll and Mayhew, 1976a) was 59% and 32%, respectively.

Direct experimental evidence for the presence of virus particles of a seedborne strain of barley stripe mosaic virus in the female gametophyte or tissues adjoining it was presented by Mayhew and Carroll (1974b) and Carroll and Mayhew (1976b). The particles were seen in thin sections of developing ovules by electron microscopy. They occurred in sporogenous cells of young floral meristems, megaspore mother cells, megaspores (Carroll and Mayhew, 1976b), and egg cells (Mayhew and Carroll, 1974b; Carroll and Mayhew, 1976b). The particles were mainly in the cytoplasm of the different cell types, and quite frequently they were associated with wall, cytoplasmic, or spindle microtubules (Mayhew and Carroll, 1976b).

The presence of virus particles in the female sporogenous or archesporial cells appeared to be critical for infection of egg cells and hence transmission by ovules. Sporogenous cells give rise ultimately to embryo sacs, including egg cells. Sporogenous or archesporial cells are found beneath the protoderm at the apex of the ovule primordiums. They are bounded by primary cellulosic walls and during early premeiosis they appeared to possess plasmodesmata connecting them to adjacent cells. Later, during premeiosis, the sporogenous cells enlarged to form megaspore mother cells (otherwise known as embryo sac mother cells or macrospore mother cells).

The single megaspore mother cell seen in each young ovule, assumed a roughly rectangular shape by the end of the premeiotic stage of ovule and embryo sac development (Fig. 1). At this stage, the megaspore mother cell became completely surrounded by a layer of callose. No plasmodesmata were seen extending through this callose layer to the adjacent cells. The callose layer persisted during meiosis of the megaspore mother cell, enveloping the one functional and three degenerating megaspores that were produced (Carroll and Mayhew, 1976b). Subsequently, the functional megaspore divided mitotically giving rise to the embryo sac, including the egg cell.

The egg is a large cell with a spherical nucleus (Fig. 2A). Its cytoplasm contains numerous vacuoles and starch grains. Virus particles in the egg cell cytoplasm are depicted in figure 2B.

To accommodate the rapidly expanding size of the developing embryo sac, nucellar cells adjacent to the sac die and collapse (Carroll and Mayhew, 1976b).

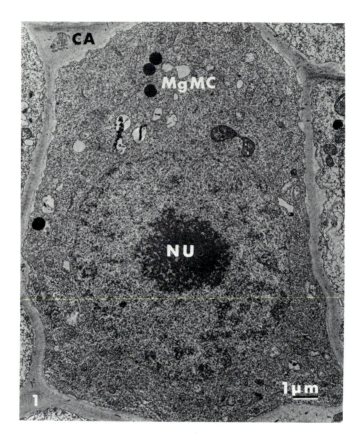

FIG. 1. Elongate megaspore mother cell (MgMC) with large spherical nucleus and single prominent nucleolus (NU). The cell is surrounded by a special callose (CA) layer that lacks plasmodesmata. × 6000. Bar represents 1 μm.

The embryo sac lacks plasmodesmatal connections with the surrounding nucellar cells.

Abnormal maternal development has been reported for CM67 barley infected by a selected strain of barley stripe mosaic virus (Slack *et al.*, 1975).

Many seed transmitted viruses are also pollen transmitted (Fulton, 1964; Bennett, 1969; Shepherd, 1972). This has been verified for barley stripe mosaic virus in barley. Gold *et al.* (1954) first reported pollen transmission of the virus. They observed that about 10% of the seedlings from seeds of healthy plants pollinated from infected plants showed symptoms of barley stripe mosaic. Later, Inouye (1962) reported that for one barley variety, 30% seed transmission was demonstrated when healthy pistils were pollinated with pollen from diseased plants. Timian (1967), and Carroll and Mayhew (1976a, 1976b), showed that seed transmission was lowest when only the pollen parent was infected.

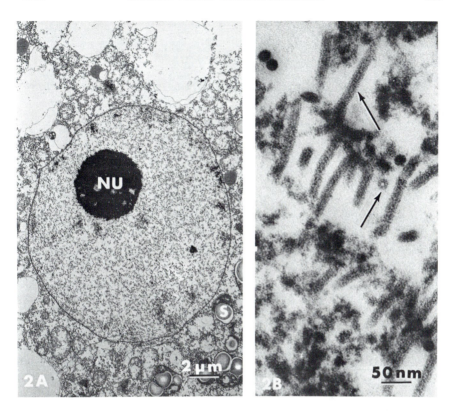

FIG. 2A.   Portion of a barley egg. A vacuolate nucleolus (NU) is evident in the large spherical nucleus. Also seen is the highly vacuolate cytoplasm that contains starch grains (S). ×
4000. Bar represents 2 μm.

FIG. 2B.   Virus particles (arrows) in the cytoplasm of an egg. × 150,000. Bar represents 50
nm.

Gold *et al.* (1954) found virus particles of barley stripe mosaic virus with the
electron microscope in triturated pollen that had originated from infected plants.
Later, Gardner (1967) detected intracellular virus particles in the cytoplasm of
the vegetative cell of barley pollen. In 1972 Carroll not only found virus particles
in vegetative cells, confirming earlier work, but he also found particles in sperm
cells (Carroll, 1974). These studies provided important cytological evidence for
the infection of the male gametophytes and gametes by some seed-transmitting
strains of the virus, but they did not indicate the manner by which pollen and
sperms had become infected. So, in 1976 Carroll and Mayhew studied anther
and pollen infection as they pertained to pollen transmission. Examination of
sectioned anthers and pollen revealed that the seed- and pollen-transmitted strain
of barley stripe mosaic virus invaded the male floral meristem early and subsequently infected pollen mother cells and sperms. During the premeiotic and mei-

otic stages of anther and pollen development, most virus particles were seen attached to microtubules, including those of the spindle (Mayhew and Carroll, 1974a, 1976a). In later stages, the association of virus particles with microtubules diminished.

In young anthers during premeiosis, many sporogenous or archesporial cells were seen in the anther locules. These cells appeared to be bounded by primary cellulosic walls. Plasmodesmata connecting these cells with themselves and surrounding cells were visible within the primary walls. During premeiosis the archesporial cells produced pollen mother cells and later during meiosis the pollen mother cells produced microspores. All meiotic stages of the pollen mother cells were enveloped in callose. A pollen mother cell in metaphase II of meiosis is seen with its callose layer in figure 3. The callose layer separated the pollen mother cells and microspores from the surrounding tapetal or anther cells. The callose layer was devoid of plasmodesmata (Carroll and Mayhew, 1976a).

Following meiosis, the microspores divided mitotically two times and then matured into pollen grains. An exine or outer wall formed around each pollen grain. No plasmodesmata were seen in the exine. A mature pollen grain contained two sperms, a vegetative cell nucleus and numerous starch grains (Fig. 4A). Some sperms were full of virus particles (Fig. 4B).

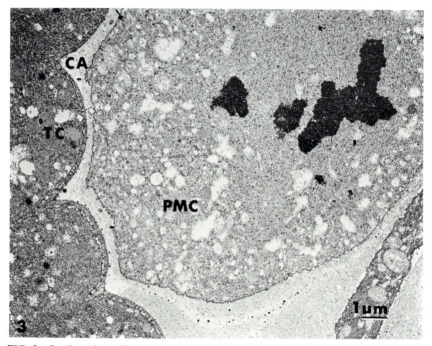

FIG. 3. Portion of a pollen mother cell (PMC) in metaphase II. A callose layer (CA) separates the pollen mother cell from surrounding tapetal cells (TC) in the anther. × 8000. Bar represents 1 μm.

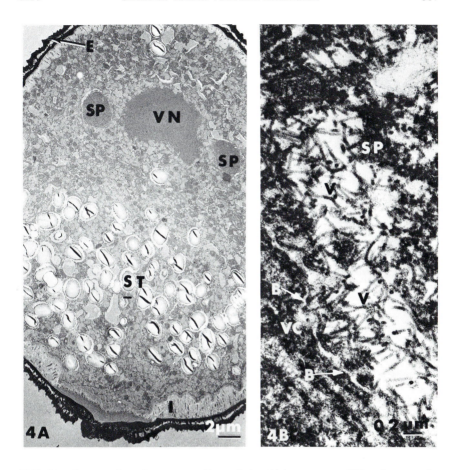

FIG. 4A. Cross section of a mature pollen grain in which two sperms (SP) and a vegetative nucleus (VN) are apparent. Also seen are starch grains (ST), intine (I), and exine (E). × 3000. Bar represents 2 μm.

FIG. 4B. Scattered virus particles (V) in the cytoplasm of a sperm (SP) cell. A special boundary (B) separates the sperm cell from the vegetative cytoplasm (VC). × 50,000. Bar represents 2 μm.

The finding that more virus particles (39.2%) were present in the cytoplasm rather than in the nucleus (14.4%) of sperm cells (Carroll, 1974), and determinations of the level of pollen transmission in the field (32%) (Carroll and Mayhew, 1976a), suggest that some or all of the cytoplasmic contents, in addition to the nucleus of the sperm, become absorbed into the egg cytoplasm during fertilization.

It has been shown previously (Yamamoto, 1951; Inouye, 1962) that most sterility caused by barley stripe mosaic virus in barley is due to a disorder of the

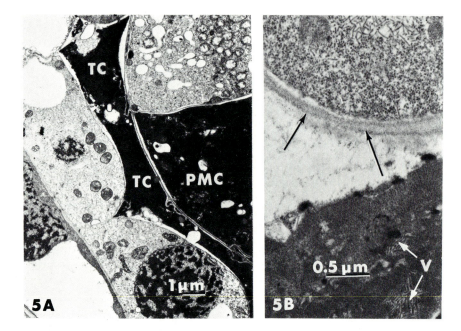

FIG. 5A. Degenerated pollen mother cell (PMC) and tapetal cells (TC). × 5000. Bar represents 1 μm.

FIG. 5B. Virus particles (V) in the cytoplasm of a disintegrating pollen mother cell. Part of a persisting pollen mother cell (arrows) with its included virus particles is also evident. × 24,000. Bar represents 0.5 μm.

anther and pollen. Virulent strains of the virus in highly susceptible cultivars of barley are responsible for underdeveloped anthers, abnormal anther dehiscence, and the reduction of normal pollen. In Japan, workers have stated that this disorder gives rise to semisterility known as 'Chochin-bo' or 'Lantern head' (Inouye, 1962). The Japanese referred to it as semisterility because the pistils were mostly fertile. Results of the 1976 study (Carroll and Mayhew) also indicated that virus infection by a seed transmitted strain of barley stripe mosaic virus enhanced abnormal anther and pollen development. A degenerating pollen mother cell and abnormal tapetal cells are depicted in figure 5A. Virus particles have been found in many of the disintegrating pollen mother cells (Fig. 5B). For fieldgrown plants, the percentage average sterility for the infected plants was 37.1% and 4.6% for the virus-free plants (Carroll and Mayhew, 1976a). This much sterility did not eliminate pollen transmission of the virus (Carroll and Mayhew, 1976a).

The presence and importance of seedborne strains of barley stripe mosaic virus in barley embryos are well documented. Gold *et al*. (1954) found virus particles in embryo extracts using the electron microscope. Crowley (1959) and Inouye (1962) detected virus in barley embryos by infectivity assays. Hamilton

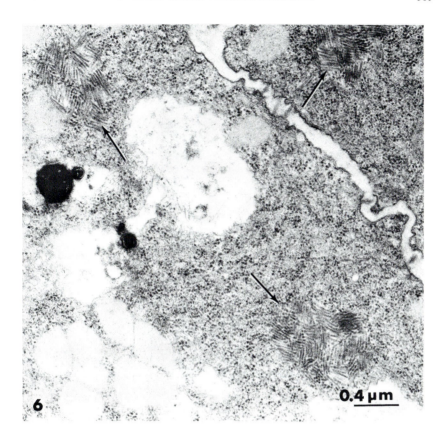

FIG. 6.   Aggregates of virus particles (arrows) in an immature embryo. × 30,000. Bar represents 0.4 μm.

(1965) showed by serological analysis that a seed-transmitted strain of the virus occurred in 25-30% of the embryos and seedlings obtained from infected Atlas barley seeds. In 1966, Inouye studied the relationship of embryo infection and seed transmission, by an embryo culture method and electron microscopic observation of embryo extracts. He found that infected seedlings arose from infected embryos. Later, barley stripe mosaic virus was demonstrated in thin sections of fully developed (Carroll, 1969), and immature (Carroll, 1972) barley embryos with the electron microscope. Virus particles are seen in an immature embryo in figure 6.

9.3.3   A Nonseedborne Strain: Virus-host Interaction

Hamilton (1965) found that the nonseed passage strain of barley stripe mosaic virus was not seed transmitted in Atlas barley. When the embryos of 150 seeds from mother plants that had become infected via mechanical inoculation were tested serologically, no virus antigen was detected. Furthermore, no virus was detected in leaves of seedlings produced from other seeds that had come from the mechanically inoculated mother plants. In 1970, Shivanathan extended the earlier work of Hamilton and he studied the influence of the nonseed passage strain and the environment on seed transmission of barley stripe mosaic virus in Atlas barley. Shivanathan (1970) demonstrated by infectivity tests and serological assays that the nonseed passage strain was absent from anthers, pollen, ovules, or seeds. This finding, along with the results of experiments on the effects of temperature on seed transmission and information obtained on the virological properties of the strain itself, formed the basis for a hypothesis to explain the lack of seed transmission by the nonseed passage strain in Atlas barley. Shivanathan (1970) hypothesized that the nonseed transmissibility of the strain was due to the fact that the virus was defective. As a consequence, he maintained, it was unstable in plant tissues and that probably prevented it from occurring in certain floral tissues and in seeds.

In 1972 (Carroll) floral parts, seeds, and embryos of plants which were infected in the seedling stage with the nonseed passage strain were studied at anatomical and cytological levels using serological and microscopical methods to determine the presence or absence of virus. Results of that study showed that the nonseed passage strain was present in 1.8% of the anthers, 3.2% of the unfertilized pistils, less than 5% of the glumes, 9.2% of immature seeds about 4-10 days old, and in 0.9% of immature seeds, about 11-21 days old. The strain was absent in pollen, immature and mature embryos, and in endosperm tissue of mature seeds. When the virus occurred infrequently in unfertilized pistils and in immature seeds it was confined to the ovary walls. Virus particles of the nonseed passage strain of barley stripe mosaic virus are shown in an ovary wall cell in figure 7A. Direct and circumstantial evidence suggested that the virus was not seed transmitted because it was excluded from embryos, ovules, and pollen and that when it rarely occurred in the outer tissues of immature seeds it became eliminated from them as they matured. It is interesting to note that this suggestion given for the nonseed transmissibility of the nonseed passage strain is just the opposite of the two mechanisms proposed to explain the prevention of seed transmission by Fulton (1964); elimination of the virus from the embryo of maturing seed and exclusion of the virus from developing seed.

In light of the earlier discussions on the importance of ovule and pollen infection to the seed transmission of plant viruses (Nelson, 1932; Fulton, 1964; Bennett, 1969), and the results of the studies on the nonseed passage strain (Shivanathan, 1970; Carroll, 1972), two more investigations were conducted on the lack of transmission by the virus. One study was to determine if virus particles were present in anthers and pollen (Carroll and Mayhew, 1976a), and the other study was to determine if particles were in ovules and embryo sacs (Carroll and Mayhew, 1976b). Electron microscopic examination of sectioned anthers and pollen revealed the presence of the nonseed passage strain in wall cells of only a single

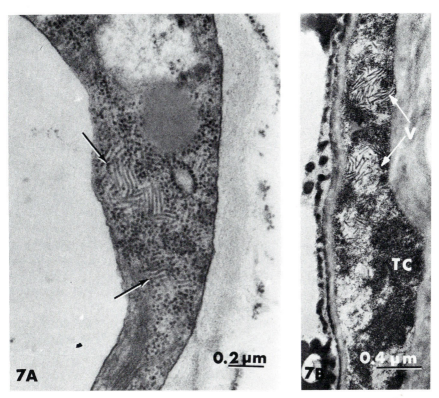

FIG. 7A. Virus particles (arrows) of the non seed passage strain in the cytoplasm of an ovary wall cell. × 45,000. Bar represents 0.2 μm.

FIG. 7B. Clusters of virus particles (V) of the non seed passage strain in a collapsed tapetal cell (TC). × 30,000. Bar represents 0.4 μm.

mature anther. Small groups of virus particles were visible in collapsed tapetal cells (Fig. 7B) or epidermal cells. No particles were seen in young anthers or pollen. It was also shown that the virus adversely affected anther and pollen development, but to a lesser extent than did a seed transmitted strain (Carroll and Mayhew, 1976a). Virus particles of the nonseed passage strain were not detected by electron microscopy in thin sections of developing ovules or embryo sacs (Carroll and Mayhew, 1976b). The combined results of the pollen (Carroll and Mayhew, 1976a) and ovule (Carroll and Mayhew, 1976b) studies supported Bennett's (1969) hypothesis pertaining to the seed transmission of plant viruses. The seed-transmitted viruses in Bennett's opinion, were those viruses which invaded the primary meristem of the host early in its development. This permitted the infection of both male and female gametophytes and gametes. Failure to invade the meristem early, Bennett (1969) maintained, allowed escape of the pollen mother cells and megaspore mother cells from virus invasion, resulting in the production of virus-free pollen and embryo sacs. Evidence from the latter two studies (Carroll and Mayhew, 1976a, 1976b) suggested that the nonseed passage strain of barley stripe mosaic virus did not invade male or female meristems early enough to initiate pollen or ovule infection. Thus, the male and female gametophytes and gametes escaped infection. Lack of gamete infection was probably why no seed transmission of the non seed passage strain occurred in Atlas barley.

9.4   CLOSING REMARKS

During the last 10 yrs., cytological and anatomical studies have given us important clues on how viruses become seedborne. These studies have revealed the presence and pathological effects of seedborne viruses in floral, fruit, and seed parts. In addition, they have given us a better indication of what avenues the viruses take to invade seeds. These studies have not told us, however, why viruses become seedborne. This will undoubtedly require unraveling the mystery of the physiological or biochemical factors that influence or condition seed transmission.

Clearly, there is a need to examine virus-host interactions for more of the seedborne viruses by the ultrathin sectioning technique for electron microscopy. Primary emphasis should be on what direct and indirect avenues of invasion seedborne viruses take to infect or contaminate seeds. Information gathered in the future on these avenues may or may not support the suggestions of earlier investigators. To date, however, it appears that the earlier investigators were mostly correct in their assessment of what avenues viruses used in the mother plants to become seedborne. For example, Bennett (1969) suggested that direct invasion of developing seeds occurred through plasmodesmata that extended from tissues of the mother plants to the seeds. Studies of ovaries and ovules have confirmed that, indeed, plasmodesmata are present in the cells that connect ovaries and ovules. Thus, it is probable that viruses move in plasmodesmata from placental cells of the ovary through the funiculus attaching the ovule to the ovary, and into cells of the integuments and nucellus. This would result in the direct invasion of a developing seed, wherein nonembryonic tissues could become infected. Bennett (1969) also suggested that viruses could directly invade seeds through the disintegrating tissues of the mother plants. This suggestion came from convincing evidence obtained on the seed transmission of tomato mosaic virus by tomato seeds (Taylor et al., 1961; Broadbent, 1965). Virus in disintegrating ovarian or fruit tissues could contaminate the seed surface.

Although plasmodesmata have been implicated as avenues through which viruses invaded the embryos of developing seeds (Caldwell, 1934; Bennett, 1940, 1969; Crowley, 1957b), cytological proof for this is lacking. So far, developing embryos have not revealed the presence of plasmodesmata connecting them with surrounding tissues. This means, then, that avenues other than plasmodesmata would be required for virus invasion. It is not known if intact macromolecules of viral nucleoprotein or ribonucleic acid (RNA) move directly through cell walls.

In nearly all cases where indirect invasion of seeds has been suggested and later documented, infected embryos are involved. For indirect invasion of the seeds to occur, the viruses must attack the floral meristematic tissues early in their development, as was previously hypothesized by Bennett (1940, 1969), Crowley (1959), and Baker and Smith (1966). Gametophytic cells (Bennett, 1940, 1969; Crowley, 1959), gametes, and zygotes (Crowley, 1959; Bennett, 1969) were suggested as the avenues taken by the viruses to infect the embryos. Past work on pollen and ovule transmission using the cross pollination technique and infectivity assays of pollen and ovules, and more recent studies on the electron microscopy of embryo infecting viruses support the suggestion that the viruses passed through reproductive cells. Thin sections have revealed the presence of virus particles in young floral meristems (Wilcoxon, *et al.*, 1975; Carroll and Mayhew, 1976a, 1976b), male gametophytes (Gardner, 1967; Carroll, 1972, 1974; Yang and Hamilton, 1974; Wilcoxson *et al.*, 1975; Carroll and Mayhew, 1976a), female gametophytes (Mayhew and Carroll, 1974b; Yang and Hamilton, 1974; Carroll and Mayhew, 1976b), male gametes (Carroll, 1974; Carroll and Mayhew, 1976a), and female gametes (Mayhew and Carroll, 1974b; Carroll and Mayhew, 1976b). The finding that both pollen mother cells (Carroll and Mayhew, 1976a), and megaspore mother cells (Carroll and Mayhew, 1976b) were surrounded by a callose layer that lacked plasmodesmata, suggested that invasion of the precursor archesporial or sporogenous cells was essential for the infection of the mother cells. Electron microscopical details of virus transfer from infected sperms to eggs and endosperm mother cells during double fertilization are not presently known.

It is difficult to understand how endosperms, but not embryos, of tomato seeds become infected by tomato mosaic virus. Broadbent (1965) stated that there was a possibility that the pollen from infected tomato plants may be infected.

Seedborne plant viruses are undeniably special, and their transmission by seeds is, indeed, an interesting phenomenon.

9.5 ACKNOWLEDGMENTS

Reports on barley stripe mosaic virus by the author of this chapter and his colleagues resulted from investigations of the Agricultural Experiment Station at Montana State University. The work was supported in part by NSF grants GB-8082 and GB-35323. I am grateful to Sherri L. Johnson for her assistance with the preparation of the photographs. I also wish to thank D. E. Mathre who made helpful suggestions about the manuscript.

9.6 REFERENCES

Adams, D. B., and Kuhn, C. W. (1977). Seed transmission of peanut mottle virus in peanuts. *Phytopathology* 67: 1126-1129.

Baker, K. F., and Smith, S. H. (1966). Dynamics of seed transmission in plant pathogens. *Ann. Rev. Phytopathol.* 4: 311-334.

Bennett, C. W. (1936). Further studies on the relation of the curly top virus to plant tissues. *J. Agric. Res.* 53: 595-620.

Bennett, C. W. (1940). The relation of viruses to plant tissues. *Botan. Rev.* 6: 427-473.

Bennett, C. W. (1969). Seed transmission of plant viruses. *Adv. Virus Res.* 14: 221-261.

Bock, K. R., and Kuhn, C. W. (1975). Peanut mottle virus. *CMI/AAB Descriptions of Plant Viruses*, No. 141. Commonw. Mycol. Inst., Assn. Appl. Biol., Kew, Surrey, England. 4 pp.

Bos, L. (1971). Bean common mosaic virus. *CMI/AAB Descriptions of Plant Viruses*, No. 73. Commonw. Mycol. Inst., Assn. Appl. Biol., Kew, Surrey, England. 4 pp.

Bos, L. (1972). Soybean mosaic virus. *CMI/AAB Descriptions of Plant Viruses*, No. 93. Commonw. Mycol. Inst., Assn. Appl. Biol., Kew, Surrey, England. 4 pp.

Bos, L., and Jaspars, E. M. J. (1971). Alfalfa mosaic virus. *CMI/AAB Descriptions of Plant Viruses*, No. 46. Commonw. Mycol. Inst., Assn. Appl. Biol., Kew, Surrey, England. 4 pp.

Bos, L. (1973). Pea streak virus. *CMI/AAB Descriptions of Plant Viruses*, No. 112. Commonw. Mycol. Inst., Assn. Appl. Biol., Kew, Surrey, England. 4 pp.

Brlansky, R. H., and Carroll, T. W. (1978). Transmission of barley stripe mosaic virus by sperm and egg cells in barley. Abst. No. 341, p. 200, *70th Ann. Meeting APS*, Tucson, AZ, USA.

Broadbent, L. (1965). The epidemiology of tomato mosaic. *Ann. Appl. Biol.* 56: 177-205.

Caldwell, J. (1934). The physiology of virus diseases in plants. V. The movement of the virus agent in tobacco and tomato. *Ann. Appl. Biol.* 21: 191-205.

Caldwell, J. (1952). Some effects of plant viruses on nuclear divisions. *Ann. Appl. Biol.* 39: 98-102.

Caldwell, J. (1962). Seed-transmission of viruses. *Nature* 193: 457-459.

Carroll, T. W. (1969). Electron miscroscopic evidence for the presence of barley stripe mosaic virus in cells of barley embryos. *Virology* 37: 649-657.

Carroll, T. W. (1972). Seed transmissibility of two strains of barley stripe mosaic virus. *Virology* 48: 323-336.

Carroll, T. W. (1974). Barley stripe mosaic virus in sperm and vegetative cells of barley pollen. *Virology* 60: 21-28.

Carroll, T. W., and Mayhew, D. E. (1976a). Anther and pollen infection in relation to the pollen and seed transmissibility of two strains of barley stripe mosaic virus in barley. *Can. J. Bot.* 54: 1604-1621.

Carroll, T. W., and Mayhew, D. E. (1976b). Occurrence of virions in developing ovules and embryo sacs of barley in relation to the seed transmissibility of barley stripe mosaic virus. *Can J. Bot.* 54: 2497-2512.

Cheo, P. C. (1955). Effect of seed maturation on inhibition of southern bean mosaic virus in bean. *Phytopathology* 45: 17-21.

Crispin Medina, A., and Grogan, R. G. (1961). Seed transmission of bean mosaic viruses. *Phytopathology* 51: 452-456.

Crowley, N. C. (1957a). The effect of developing embryos on plant viruses. *Australian J. Biol. Sci.* 10: 443-448.

Crowley, N. C. (1957b). Studies on the seed transmission of plant virus diseases. *Australian J. Biol. Sci.* 10: 449-464.

Crowley, N. C. (1959). Studies on the time of embryo infection by seed-transmitted viruses. *Virology* 8: 116-123.

Davidson, E. M. (1969). Cell to cell movement of tobacco ringspot virus. *Virology* 37: 694-695.

de Zoeten, G. A., and Gaard, G. (1969). Possibilities for inter- and intra-cellular translocation of some icosahedral plant viruses. *J. Cell Biol.* 40: 814-823.

Ekpo, E. J. A., and Saettler, A. W. (1974). Distribution pattern of bean common mosaic virus in developing bean seed. *Phytopathology* 64: 269-270.

Esau, K., Cronshaw, J., and Hoefert, L. L. (1967). Relation of beet yellows virus to the phloem and to movement in one sieve tube. *J. Cell Biol.* 32: 71-87.

Esau, K., and Hoefert, L. L. (1972). Ultrastructure of sugarbeet leaves infected with beet western yellows virus. *J. Ultrastruct. Res.* 40: 556-571.

Eslick, R. F., and Afanasiev, M. M. (1955). Influence of time of infection with BSMV on symptoms, plant yield and seed infection of barley. *Plant Dis. Reptr.* 39: 722-724.

Ford, R. E. (1966). Recovery of pea streak virus from pea seed parts and its transmission by immature seed. *Phytopathology* 56: 858-859.

Frosheiser, F. I. (1974). Alfalfa mosaic virus transmission to seed through alfalfa gametes and longevity in alfalfa seed. *Phytopathology* 64: 102-105.

Fulton, R. W. (1964). Transmission of plant viruses by grafting, dodder, seed, and mechanical inoculation. *In* "Plant Virology" (Corbett, M. K. and Sisler, H. D., eds.), pp. 39-67. Univ. of Florida Press, Gainesville, Florida. 527 pp.

Fulton, R. W. (1971). Tobacco streak virus. *CMI/AAB Descriptions of Plant Viruses*, No. 44. Commonw. Mycol. Inst., Assn. Appl. Biol., Kew, Surrey, England, 4 pp.

Gardner, W. S. (1967). Electron microscopy of barley stripe mosaic virus: Comparative cytology of tissues infected during different stages of maturity. *Phytopathology* 57: 1315-1326.

Gallo, J., and Ciampor, F. (1977). Transmission of alfalfa mosaic virus through *Nicandra physaloides* seeds and its localization in embryo cotyledons. *Acta Virol.* 21: 344-346.

Gay, J. D. (1973). Effect of plant variety and infection age on the presence of southern bean mosaic virus in floral parts and unripe seeds of *Vigna sinensis*. *Plant Dis. Reptr.* 57: 13-15.

Ghanekar, A. M., and Schwenk, F. W. (1974). Seed transmission and distribution of tobacco streak virus in six cultivars of soybeans. *Phytopathology* 64: 112-114.

Gilmer, R. M., and Wilks, J. M. (1967). Seed transmission of tobacco mosaic virus in apple and pear. *Phytopathology* 57: 214-217.

Gold, A. H., Suneson, C. A., Houston, B. R., and Oswald, J. W. (1954). Electron microscopy and seed and pollen transmission of rod-shaped particles associated with the false stripe virus disease of barley. *Phytopathology* 44: 115-117.

Hagborg, W. A. F. (1954). Dwarfing of wheat and barley by the barley stripe-mosaic (false stripe) virus. *Can. J. Bot.* 32: 24-37.

Hamilton, R. I. (1965). An embryo test for detecting seed-borne barley stripe mosaic virus in barley. *Phytopathology* 55: 798-799.

Hoch, H. C., and Provvidenti, R. (1978). Ultrastructural localization of bean common mosaic virus in dormant and germinating seeds of *Phaseolus vulgaris*. *Phytopathology* 68: 327-330.

Hollings, M., and Huttinga, H. (1976). Tomato mosaic virus. *CMI/AAB Descriptions of Plant Viruses*, No. 156. Commonw. Mycol. Inst., Assn. Appl. Biol., Kew, Surrey, England. 6 pp.

Hollings, M., and Stone, O. M. (1971). Tomato aspermy virus. *CMI/AAB Descriptions of Plant Viruses*, No. 79. Commonw. Mycol. Inst., Assn. Appl. Biol., Kew, Surrey, England. 4 pp.

Inouye, T. (1962). Studies on barley stripe mosaic in Japan. *Ohara* (Japan) *Inst. Landw. Biol. Ber. Univ.* 11: 413-496.

Inouye, T. (1966). Some experiments on the seed transmission of barley stripe mosaic virus in barley with electron miscroscopy. *Ohara* (Japan) *Inst. Landw. Biol. Ber. Univ.* 13: 111-122.

Kim, K. S., and Fulton, J. P. (1971). Tubules with virus-like particles in leaf cells infected with bean pod mottle virus. *Virology* 43: 329-337.

Kitajima, E. W., and Lauritis, J. A. (1969). Plant virions in plasmodesmata. *Virology* 37: 681-684.

Lawson, R. H., and Hearon, S. (1970). Subcellular localization of chrysanthemum tissue. *Virology* 41: 30-37.

Lister, R. M., and Murant, A. F. (1967). Seed-transmission of nematode-borne viruses. *Ann. Appl. Biol.* 59: 49-62.

Mathre, D. E. (1978). Disrupted reproduction. *In* "Plant Disease, An Advanced Treatise," Vol. III, (J. G. Horsfall and E. B. Cowling, eds.), pp. 257-278. Academic Press, New York. 488 pp.

Mayhew, D. E., and Carroll, T. W. (1974a). Barley stripe mosaic virions associated with spindle microtubules. *Science* 185: 957-958.

Mayhew, D. E., and Carroll, T. W. (1974b). Barley stripe mosaic virus in the egg cell and egg sac of infected barley. *Virology* 58: 561-567.

Murant, A. F. (1970). Tomato black ring virus. *CMI/AAB Descriptions of Plant Viruses,* No. 38. Commonw. Mycol. Inst., Assn. Appl. Biol., Kew, Surrey, England. 4 pp.

McDonald, J. G., and Hamilton, R. I. (1972). Distribution of southern bean mosaic virus in the seed of *Phaseolus vulgaris. Phytopathology* 62: 387-389.

McKinney, H. H., and Greeley, L. W. (1965). Biological characteristics of barley stripe-mosaic virus strains and their evolution. *U.S. Dept. Agr. Tech. Bull.* No. 1324. 84 pp.

Nelson, R. (1932). Investigations in the mosaic disease of bean. *Michigan State College Agr. Expt. Sta. Tech. Bull.* No. 118. pp. 1-71.

Owusu, G. K., Crowley, N. C., and Francki, R. I. B. (1968). Studies of the seed-transmission of tobacco ringspot virus. *Ann. Appl. Biol.* 61: 195-202.

Phatak, H. C. (1974). Seed-borne plant viruses—identification and diagnosis in seed health testing. *Seed Sci. & Technol.* 2: 3-155.

Roberts, I. M., and Harrison, B. D. (1970). Inclusion bodies and tubular structures in *Chenopodium amaranticolor* plants infected with strawberry latent ringspot virus. *J. Gen. Virol.* 7: 47-54.

Schippers, B. (1963). Transmission of bean common mosaic virus by seed of *Phaseolus vulgaris* L. Cultivar Beka. *Acta Botanica Neerlandica* 12: 433-497.

Shepherd, R. J. (1971). Southern bean mosaic virus. *CMI/AAB Descriptions of Plant Viruses,* No. 57. Commonw. Mycol. Inst., Assn. Appl. Biol., Kew, Surrey, England. 4 pp.

Shepherd, R. J. (1972). Transmission of viruses through seed and pollen. *In* "Principles and Techniques in Plant Virology" (C. I. Kado and H. O. Agrawal, eds.), pp. 267-292. Van Nostrand-Reinhold, Princeton, New Jersey. 688 pp.

Shepherd, R. J., and Fulton, R. W. (1962). Identity of a seed-borne virus of cowpea. *Phytopathology* 52: 489-493.

Shivanathan, P. (1970). Studies on the non-seed transmissibility of the NSP strain of barley stripe mosaic virus. PhD. Thesis, Univ. of McGill, Montreal, Canada. 144 pp.

Singh, G. P., Arny, D. C., and Pound, G. S. (1960). Studies on the stripe mosaic of barley, including effects of temperature and age of host on disease development and seed infection. *Phytopathology* 50: 290-296.

Slack, S. A., Shepherd, R. J., and Hall, D. H. (1975). Spread of seedborne barley stripe mosaic virus and effects of the virus on barley in California. *Phytopathology* 65: 1218-1223.

Stace-Smith, R. (1970). Tobacco ringspot virus. *CMI/AAB Descriptions of Plant Viruses,* No. 17. Commonw. Mycol. Inst., Assn. Appl. Biol., Kew, Surrey, England. 4 pp.

Taylor, R. H., Grogan, R. G., and Kimble, K. A. (1961). Transmission of tobacco mosaic virus in tomato seed. *Phytopathology* 51: 837-842.

Timian, R. G. (1967). Barley stripe mosaic virus seed transmission and barley yield as influenced by time of infection. *Phytopathology* 57: 1375-1377.

Tu, J. C. (1975). Localization of infectious soybean mosaic virus in mottled soybean seeds. *Microbios* 14: 151-156.

Uyemoto, J. K., and Grogan, R. G. (1977). Southern bean mosaic virus: evidence for seed transmission in bean embryos. *Phytopathology* 67: 1190-1196.

Weintraub, M., Ragetli, H. W. J., and Leung, E. (1976). Elongated virus particles in plasmodesmata. *J. Ultrastruct. Res.* 56: 351-364.

Wilcoxson, R. D., Johnson, L. E. B., and Frosheiser, F. I. (1975). Variation in the aggregation forms of alfalfa mosaic virus strains in different alfalfa organs. *Phytopathology* 65: 1249-1254.

Yamamoto, T. (1951). Studies on the sterility of barley. I. On the mechanism in occurrence of sterile grain. (Japanese with English summary). *Proc. Japan Crop Sci. Soc.* 20: 80-84.

Yang, A. F., and Hamilton, R. I. (1974). The mechanism of seed transmission of tobacco ringspot virus in soybean. *Virology* 62: 26-37.

Zaitlin, M., and Israel, H. W. (1975). Tobacco mosaic virus (type strain). *CMI/AAB Descriptions of Plant Viruses*, No. 151. Commonw. Mycol. Inst., Assn. Appl. Biol., Kew, Surrey, England. 5 pp.

Chapter 10

MAN-MADE EPIDEMIOLOGICAL HAZARDS IN MAJOR CROPS OF DEVELOPING COUNTRIES[1]

L. Chiarappa

Food and Agricultural Organization of the United Nations
Rome, Italy

[1] The contents of this paper reflect the ideas of the author and not necessarily those of the Organization he represents.

10.1 INTRODUCTION

In 1973 Commoner stated: "In seeking a remedy for the pervasive errors in international development I suggest that we reverse the order of relationships which now connect economic need, technology, engineering and the biology of the natural world. What is called for, I believe, is a new technology, designed to meet the needs of the human condition and of the natural environment rather than of "the new industrial state."

As far as I am aware, no comprehensive review has been made in the field of plant pathology to see if and which "pervasive errors in international development" have been made during this last quarter century and if there is need to "reverse the order of relationships" that have influenced the application of this field of science to world agriculture. The purpose of this paper is to make a brief analysis of past or existing trends in disease control in developing countries and to suggest possible departures for the future. This analysis will be achieved by examining first some of the disease hazards that have been introduced into the farming systems of developing countries during the process of recent agricultural modernization and by suggesting possible ways for overcoming these problems.

10.2 GENETIC VULNERABILITY OF CROPS

Much has been written, but little has actually been done, on the problem of genetic vulnerability in crops following the epidemic of corn leaf blight in North America that nearly destroyed much of the 1969 corn harvest (Anon., 1972; Day, 1977; Harlan, 1976). Whereas this epidemic had relatively few socio-economic consequences in the United States, due to the vast resources of that country, its perfect organization and capability of promptly applying the necessary remedial action, a similar situation in a developing country could have had disastrous consequences.

The basis of the problem which brought about the corn situation in the United States is the general trend of modern plant breeders to produce crop varieties adapted to large-scale monotypic monocultures with the broadest range of adaptation and the highest yielding capacities. In pursuing this scope, only a few, genetically homogenous cultivars, that have no or minimal resistance to plant parasites, are being used.

Furthermore, when problems of disease resistance are considered, they are usually resolved in the easiest and quickest possible way: by the use of single resistance genes that protect against only one or a few of the most important parasites. This is often done in full ignorance of where the new cultivar is going to be grown and which other parasites it may eventually encounter. This type of resistance, being fully operative in the plant breeder's plot (and in the farmer's field when first used), is one of the major contributors to the epidemiological hazard of most modern cultivars once these are widely grown and changes occur in the virulence of the parasite population. The resistance is then known to breakdown and replacement varieties must then be hurriedly bred and distributed to the growers.

In spite of the recognition of this problem by plant breeders and plant pathologists, the tendency over recent years has been to continue or even to expand activities following these directives. This tendency somewhat reflects the philosophy and approach of the industrial sector in technologically advanced countries where a new mechanical device, for example a new car, is built to last only a few years to be soon replaced by a newer model. As in the case of cars, the new crop cultivar is often produced in far away "factories" from which it is distributed throughout the world either fully assembled or in sections for *in loco* assembling. While recognizing the efficiency and commercial soundness of this approach, it is hard to believe that the same philosophy can be successfully applied to agriculture and especially to conditions existing in developing countries.

A few examples of man-made epidemiological hazards due to genetic vulnerability are given.

10.2.1 Wheat

Over the past nine years major changes have occurred in wheat growing throughout the world and particularly in the Indo-Pakistan subcontinent, north Africa and the Near East. This is what has been called the "green revolution." The key to this "revolution" has been the use of high-yielding dwarf or semi-dwarf varieties with broad adaptation capacity and with good capability to respond to the use of fertilizers, irrigation and improved cultural practices. The popularity of these new high yielding wheats has been so great that in only a few years they have been planted in millions of hectars. The outstanding results obtained in raising yield deserve the greatest recognition.

However, while varieties involved in the program at time of release were resistant to stem rust, yellow rust, loose smut, bunt and powdery mildew (Saari *et al.*, 1974), all this resistance was vertical resistance (Van der Plank, 1963). This, to operate, depends on only a few genes and is likely to be unstable. This instability has been the Achille's heel of the "green revolution," and, in fact, it has considerably contributed to the epidemiological hazard of individual countries or of entire regions. This has already been experienced in the Septoria epidemic in Morocco and Tunisia in 1968-1969; in the leaf rust epidemic in the Indo-Pakistan subcontinent in 1971-1972 and 1972-1973; in the outbreak of *Puccinia recondita* in the Yaqui Valley of Mexico in 1977; and in the Karnal bunt situation in Punjab in 1979.

10.2.2 Coconut

Coconut (*Cocos nucifera* L.) is a crop that until recently has undergone little genetic improvement. The resulting effect has been that most of the coconuts extensively grown in Asia and the Pacific have only been selected under cultivation. This has given rise to a great variety of coconut lines which have been grown for centuries in good balance with their agro-ecosystems and have experienced only a relatively few parasite problems. In addition to the coconuts selected under cultivation, which, according to Harries (1978) belong to the "Niu Vai" group, there is another group—the so called "Niu Kafa"—which includes varieties that have instead evolved naturally. The differentiation between these two groups is possible on the basis of certain morphologic and physiologic characters, on fruit component analysis and, quite important, on their susceptibility to one of the most serious diseases of coconut, the so called lethal yellowing. This disease has almost entirely destroyed the susceptible Jamaica Tall coconut in Jamaica and Florida as well as the susceptible West African Tall in Ghana and Togo. Both of these coconuts belong to the Niu Kafa group, while all coconuts derived from Malaysia, Indonesia, Sarawak, Cambodia, Thailand, Peru and Panama are included in the Niu Vai group and show certain levels of resistance to the disease (Harries, 1978). This and other considerations led to the suggestion that lethal yellowing probably originated in South East Asia where the disease is still endemic causing little damage in the moderately resistant populations that prevail there (Chiarappa, 1979).

In recent years the discovery that hybrid coconuts could produce earlier and give higher yields per unit area than traditional tall varieties gave rise to the large scale production of such hybrids. The greatest impetus to this work was given in Ivory Coast with the production of the high yielding cross West African Tall x Malayan Dwarf. Large quantities of seed of this hybrid are being introduced into Indonesia, the Philippines, and Malaysia. Plans are already under way for large scale replacements of senile local varieties with these new hybrids. In the Philippines alone these plans call for the massive planting of 60,000 ha/year over the next 40 years.

The epidemiological hazard of this undertaking can be easily visualized. In fact, either one of the following events is likely to take place: if lethal yellowing disease is already present in S.E. Asia—then the replacement of the locally resistant coconuts with the susceptible exotics will determine epidemiological conditions similar to those existing in West Africa or Jamaica. Or, if lethal yellowing is not yet present, it could be accidentally introduced in the future and cause the same type of destruction.

10.2.3 Rice

Changes similar to those due to the "green revolution" in wheat have taken place with rice cultivation in Asia. High yielding varieties produced in rapid succession by national and international research centers have been planted over vast areas. As in wheat, the second generation problems with rice were the unexpected outbreaks of previously unimportant pests and diseases. Probably the most difficult of such parasite problems has been the brown planthopper, *Nilaparvata lugens*, in South East Asia, and particularly in Indonesia. Breeding for resistance against this pest started in the Philippines as early as 1968. From 1973 on, five resistant varieties were released in rapid succession, all having a single dominant resistant gene against biotype 1 of the insect. By the 1975-1976 wet seasons more than 400,000 ha of these resistant varieties were planted in Indonesia. This acreage tripled by the 1976-1977 dry season, covering 25% of the total rice crop (Oka, 1978). Resistance to biotype 1, however, only lasted 4 to 5 rice cropping seasons. In 1977 the occurrence of biotype 2 of the brown planthopper forced a change to another three varieties also carrying single recessive resistance genes. Following this change, a new series of varieties resistant to biotype 1 and 3 were also released. In the meantime, losses caused by this insect together with those due to two insect-transmitted viruses (grassy stunt and ragged stunt) caused great concern to Indonesia.

A few statistics could usefully illustrate this point. The population of Indonesia, estimated at 135.2 million will rise to 243 million by the turn of the century. In Java and Bali there will be an estimated population density of 1,100 habitants per square km. (Sumitro, 1977). This means that to maintain the present rate of food consumption, supply has to be doubled. Instead, during the period 1952-1972 food production has increased at only a yearly average of 2% as aginst a 2.6% increase in demand. In Indonesia, rice accounts for two-thirds of the total energy intake. Losses due to the brown planthopper and to the two vectored viruses in 1976-1979 corresponded to 364,500 tones of milled rice valued at more than $100 million, sufficient to feed three million Indonesians for one full year (Oka, 1979). It is clear that, if a solution to this problem is not readily found, Indonesia will be facing a very serious food shortage in the future.

10.2.4 Cacao

Emphasis on high yields resulted in large introductions into Ecuador toward the end of last century of cacao types such as Trinitario and other high yielding foreign types. Up to this time Ecuador had been the world's leading cacao producer, most famous for its high quality "cacao nacional." Large estates had been successfully planted with this indigenous material because the "cacao nacional" required very little attention and appeared fully adapted to the special climate of the coastal plain of Ecuador where it originated or had been domesticated many centuries before. Hybridization between the introduced high yielding exotics and the "cacao nacional" took place in a few years with the resulting formation of new local hybrids. In the 1920's two "new" cacao pathogens were identified:

Monilia roreri, causing a watery decay of pods and their mummification, and *Crinipellis perniciosa* (so called witches' broom) causing an abnormal proliferation of buds and the hypertrophy and formation of "brooms."

These diseases soon became the major production constraint causing dramatic reduction in yield and forcing the abandonment of many cacao estates. Even today the combined effect of these two diseases is estimated to cause a loss varying from 27 to 68% (Evans *et al.*, 1977). As a result, Ecuador is no longer the world leading producer of cacao and banana growing has largely replaced this crop.

10.2.5 Banana

Between 1900 and 1960 Panama disease, caused by the fungus *Fusarium oxysporum* f. sp. *cubense*, was the most destructive disease of commercial bananas in Central and South America, where nearly 80% of the fruit entering world markets was produced. Over this period nearly 100,000 acres of bananas were either destroyed or abandoned. Panama disease forced the substitution of the susceptible Gros Michel variety with the resistant Cavendish varieties. These high yielding bananas, producing twice as much as the Gros Michel, were planted over vast acreages in spite of their known susceptibility to nematode root rots, to pitting disease (caused by the fungus *Piricularia grisea*) and, most important, to the common sigatoka disease (caused by the fungus *Mycosphaerella fijensis* var. *musicola*). This preference was mainly due to the shorter stature and greater resistance of the Cavendish to losses from windstorms, and to the possibility of chemical protection against sigatoka, which was at that time effective and economically feasible.

Recently, however, banana growers in Mexico and Central America have been facing another serious problem: the appearance and spread of black sigatoka disease. This malady is caused by the fungus *Mycosphaerella fijensis* var. *difformis* which, according to Stover (1974, 1976), is a possible mutant of *M. fijensis* var. *musicola*, the cause of common sigatoka. The problem with this mutant is that it is more virulent than the normal strain and that it readily attacks the Cavendish varieties. This, coupled with the genetic uniformity of the crop and the continuity in space and time of susceptible leaf tissue, has created an entirely new epidemiological condition. To cope with this, a greater concentration of fungicides is required with an increased number of applications during the entire growing cycle. Furthermore, the new strain also attacks the cooking bananas (plantains), previously not susceptible to the old strain of the pathogen. This, in turn, creates a new socio-economic problem because plantains are an important source of food for the small farmers who cannot afford the cost of fungicide protection. This means that plantains in this part of the world may disappear as a crop if no remedial action is sought.

10.3 THE PESTICIDE HAZARD IN CROP PRODUCTION

Over the last thirty years great advances have been made in chemical control of plant diseases. These have been most rapid and with the greatest significance in the industrialized countries, particularly in the area of fungal disease control. However, less spectacular results have been obtained in the control of diseases caused by bacteria, viruses or mycoplasmas.

In developing countries, instead, one of the trends which have characterized this period has been the widespread application of fungicides and other pesticides to food crops such as wheat or rice—typical of extensive agricultures and previously considered outside the range of intensive chemical protection. This trend, although justified by the pressure of feeding a rapidly growing world population, has generated a series of new problems, mostly due to the lack of appropriate technology, of suitable infrastructures and of financial resources.

What is now known as the "pesticide hazard" is a problem which is taking different and varied forms including misuse of pesticides and environmental pollution, dependency on pesticide use, uncertain pesticide supplies, parasite resistance and resurgence problems, unwanted side effects and outright economic losses. Some of these problems will be briefly reviewed within the context of the crops mentioned earlier.

10.3.1 Wheat

Large-scale human poisoning due to the misuse of pesticides in wheat is sufficiently well documented. In late 1971 and early 1972, several thousands of people were poisoned in Iraq due to the ingestion of seed treated with alkylmercury compounds or of meat from domestic animals that had been feed with fungicide-treated grains. Mortality was as high as 11% with many people remaining permanently crippled or demented (Anon., 1974).

In Turkey, between the years 1955 and 1959 hundreds of people were also poisoned due to the use of hexachlorobenzene.

Newly acquired dependency on pesticide use in wheat is well documented in Brazil and Paraguay. In the first country, following the demonstration that foliar diseases contributed to a loss of 69 per cent of the wheat yield, chemical control increased from treatment of 500 ha in 1973 to 1,000,000 ha in 1976 at a yearly cost estimated around US$ 30 million. In spite of the newly acquired chemical protection, in 1977 there were serious losses in yield. These ranged from nearly 21% in the Parana district (1,440,000 ha of wheat) to as much as 63% in Rio Grande do Sul (1,573,000 ha). It was calculated then that three applications of insecticides (required to control the aphid vectors of barley yellow dwarf virus), and two to three applications of foliar fungicides would cost the farmer around

$55/ha. This was the equivalent of 312 kg of wheat in a country where the average production (1962-1977) is of only 926 kg/ha. It should be noted that a much higher wheat production is obtained in Argentina (1,700 kg/ha) or Italy (2,700 kg/ha) practically without the use of pesticides.

A comparable situation of dependency on pesticides has developed in Paraguay where large scale wheat growing practically started in 1967. From 1968 to 1971 there was a steady increase in acreage (from 8.5 to 51.5 thousand ha) and in total production (from 9.1 thousand to 54.8 thousand tons). In 1972 a serious epiphytotic caused a severe reduction in yield (48%) with a consequent drop of total national production (68%). This forced the adoption of insecticides and fungicides (three applications per season of each). Although this helped in stabilizing production and reducing risk, it forced importation into the country of large quantities of chemicals and it reduced the farmer's already narrow profit margin.

Uncertain pesticide supplies is the other risk that most of the developing countries are likely to face. This was quite evident in 1974 when a sudden shortage of plant protectant chemicals resulted in serious wheat losses in India and Pakistan.

10.3.2. Coconut

So far there are no records of widespread use of pesticides for the control of coconut parasites in developing countries. Even where pesticides have been used for the control of such diseases as red ring and hartrot, their application has been rather limited. As indicated before, relatively few parasites are known to attack the coconut palm. Furthermore, coconut is mostly a subsistence crop requiring a minimum of inputs.

However, an example of the potentials of pesticide hazards could be drawn from what has occurred in Florida for the control of Lethal Yellowing.

The disease was first reported in 1955 in Key West and in 1971 on the mainland of Florida where coconut is widely used as an ornamental tree. The antibiotic oxytetracycline was shown to be effective in controlling the disease if injected into the trunks prior to yellowing of the foliage. Injections, to be effective, were to be repeated at intervals of four months.

By 1977 the Florida Division of Plant Industry had distributed 3,000 kg of oxytetracycline for injections purpose. This did not include the amounts of chemicals distributed by private individuals (Gwin, 1977). This quantity was sufficient to inject 3,000,000 trees. In spite of this intensive campaign, Lethal Yellowing caused an estimated loss of $1,200,000 plus an additional cost of $3,000,000 in tree removal (Soowal, 1979). The main risk involved here was the widespread use of an antibiotic such as oxytetracycline which is used in human and animal medicine and which is also known to induce resistant strains of other microorganisms. This is a very serious risk, which certainly should not be taken by any country.

10.3.3   Rice

In 1950 phenylmercury compounds were found to be very effective for the control of rice blast (*Piricularia oryzae*) in Japan. These compounds were easily absorbed on the rice plant tissues where they remained active for a prolonged period of time, preventing infection and sporulation of the blast fungus. As a result, phenylmercury fungicides were widely adopted in rice growing throughout the country. Disease control was so good that farmers were able to plant more susceptible but higher-yielding varieties, they could apply higher nitrogen levels than ever used before and they could grow rice in areas where this crop could not be grown earlier because of severe losses caused by the disease. Most important, Japanese farmers, with the use of this fungicide and other pesticides, were able to stabilize rice production irrespective of climatic conditions (Okamoto, 1965).

The remarkable achievements obtained in Japan through crop protection in rice are most easily summarized in the graph shown in Fig. 1. Here the average yields of brown rice are plotted over a 50-year period (1925-1975) together with the average losses per hectare due to parasite attack. Losses are given only for the 1950-1975 period.

It can be seen that until 1954 the rice yield in Japan remained under 3.5 tons/ha and was characterized by considerable seasonal fluctuations. Losses due to parasites in the late 40's were over 600 kg/ha. Beginning with the widespread use of phenylmercury and other fungicides in 1954, losses were drastically reduced to around 200 kg/ha and have remained pretty much at this level for the subsequent twenty five years, in spite of the increasing production levels reached during this period. Also outstanding is the reduction achieved in seasonal yield variation.

The Japanese experience not only demonstrates the important role of plant protection in reducing yield losses, but also how this can function as a stabilizing factor in agricultural production. There is one point, however, that should not be overlooked: the very rapid increase of dependency on pesticide use from 1950 onward. The value of pesticides used during this period appears to grow exponentially and with extreme rapidity. Even taking into account inflationary effects, it is evident that the whole rice production system in Japan is highly dependent on energy-related pesticides, which absorb a manufacturing potential that could probably be more effectively utilized elsewhere. This results in two innate risks: the vulnerability of the system in the event of an energy crisis, and the related toxicological hazard.

As for the latter, when it was discovered in 1956 that phenylmercury acetate

YIELD OF BROWN RICE (tons/ha)

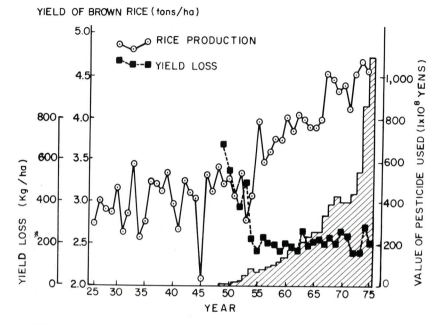

FIG. 1. Rice production in Japan from 1925 to 1975 as related to yield losses and value of pesticides used.

applied to rice foliage was easily translocated to the grains, as much as 300 to 400 g of metallic mercury were being applied per hectare in Japan. It took until 1968 to ban this fungicide from use for evident toxicological reasons.

Also the point should be made that the above results cannot easily be transferred to most of the developing countries because they still lack the infrastructures, industry and organization of Japan.

In Indonesia several attempts were made to control the brown plant hopper and other parasites of the rice crop through the intensive and widespread use of pesticides. This was mainly done through the so called BIMAS scheme, a governmental intensification programme started in the 60's and aiming at doubling rice production. Blanket treatment of pesticides were applied over vast areas of the country with the result that many of the known side effects of pesticides use were observed in the country. It is not unlikely that the new pest status of the brown plant hopper observed for the first time in 1969 was the result of these blanket treatments. This was further aggravated by the use of susceptible "high yielding" rice varieties, continuous and staggered rice planting and increased use of nitrogenous fertilizers. During outbreak situations, pesticides were applied every week or even every 2-3 days (Oka, 1979) for as many as 12 applications (Mochida, 1979). An indication of the dependency on pesticides for the growing of rice in Indonesia is given in Table I.

TABLE I. Pesticide use in Rice Intensive Production Schemes in Indonesia[1]

Years	1973/74	1974/75	1975/76	1976/77
Kg active ingredient	1,323,934	2,014,513	3,309,022	3,317,334

[1] Source: Anon. (1977)

10.3.4 Cacao

São Tome and Principe are two African territories that faced a serious plant disease problem soon after acquiring independence: pod rot disease, caused by *Phytophthora palmivora*. Cocoa production is the main source of income in both territories. However, to produce cocoa, the disease must be kept under control.

Following independence, deterioration of spray equipment, lack of trained personnel and unavailability of fungicides resulted in near abandonment of disease control practices. Consequently, cocoa production dropped to nearly half that obtained in preindependence years. The pesticide risk, in this case, took the form of deterioration of facilities and infrastructures previously available, coupled with temporarily inaccessible crop protection means.

10.3.5 Banana

Sigatoka disease was not known in the Western Hemisphere prior to 1933 and no fungicides were used to grow the Gros Michel banana.

As a result of the first epidemic of common sigatoka in the Ulua Valley of Honduras in 1935-1936, a very elaborate system was developed to apply bordeaux mixture sprays on a routine basis during the entire growing cycle.

The bordeaux era lasted from 1935 to 1958. In 1953 it was discovered that petroleum oils controlled sigatoka with or without the addition of other fungicides (Guyot, 1953). By 1958, oil-in-water emulsions with dithiocarbamate fungicides became very popular in Central America and this type of control lasted until about 1973. Fourteen to 18 sprays were applied each year at intervals of 14 to 28 days. Following the 1973-1974 epidemic of black sigatoka in the Ulua Valley of Honduras, it was necessary to use from 22 to 28 sprays a year to control this new disease (Stover, 1974). It was at this time that the systemic fungicide benomyl began to be used in oil-in-water emulsions. Under normal conditions, 24 to 36 applications were made to secure control. In Honduras, however, due to the development of strains of the pathogen resistant to benomyl (after only one year of use), 36 to 45 treatments per year have become necessary, with the use of alternative fungicides (such as Bravo [daconil]). These materials cost more to the farmers because of the higher concentrations and the more sophisticated application technology required.

In economic terms this escalation in fungicide usage to control leaf spot disease in Central America has raised the cost four to five times in only six years. It is now estimated that the control of black sigatoka in Costa Rica (where the disease was found in October 1979), costs around $1,000/ha. It is quite evident that at such level it will be very difficult for the small banana growers to be able to produce a crop economically, and it will be impossible for the plantain growers to raise cooking bananas.

10.4 THE QUARANTINE RISK

The drive for increasing agricultural production over the past years, has accelerated on a global scale the transfer of plant propagation material. This has been particularly true in developing countries, and especially in those that have become independent and have directed their efforts to the diversification of agriculture.

Such transfer has been encouraged not only by most technical assistance programs but also by the various international research centers now operating in several continents. In fact, the working practice of these centers has been essentially directed to the establishment of centralized breeding programs coupled with multilocation testing in individual countries. This has generated a series of difficulties particularly where bulk transfer of propagating material was involved, where this consisted of vegetative material and where quarantine infrastructures at the receiving end were either limited or malfunctioning.

These difficulties have essentially arisen from the unknown health status of the plant material being transferred, from the ignorance of the epidemiological competency of foreign parasites and from the inconsistencies or outright violation of plant quarantine regulations.

A few examples of the quarantine hazards recently experienced can give an idea of the problem.

10.4.1 Bananas and Moko Disease

Bacterial wilt of bananas, caused by *Pseudomonas solanacearum*, is a destructive disease present in Trinidad, Grenada, Central and South America, where it is known as "moko disease." Race 2 of this bacterium can attack indigenous *Heliconia* species, the plantain and the cultivated (triploid) banana (*Musa* sp.) (Buddenhagen, 1960; Buddenhagen *et al.*, 1962). It is believed that this race is native to Central America where it is widely distributed in soils and where it exists in the form of different strains. Until recently, no confirmed report was available of the presence of moko disease in Southern Asia (homeland of the genus *Musa*) and elsewhere outside the Americas (Buddenhagen, 1960). Lately, however, moko has been reported from the Philippines (Rillo, 1979) where it must have been introduced through the smuggling of infected plant material. The damage that this introduction will cause to the local banana industry and that of neighboring countries remains yet to be assessed.

10.4.2 Coffee Rust and the Western Hemisphere

Until 1970 coffee rust (*Hemileia vastatrix*) was not known in coffee producing countries of the western hemisphere. Its accidental introduction in Brazil in 1970 was followed by a rapid spread throughout the country. Now the fungus is present in Argentina, Paraguay, Peru and Bolivia. A separate, possibly intentional introduction of coffee rust was experienced in Nicaragua in 1976 and in El Salvador in 1979. The cost to Nicaragua in an attempt to eradicate the disease was more than 20 million dollars.

10.4.3 Bacterial Blights of Rice in West Africa

Until a few years ago, two of the major pathogens of rice, *Xanthomonas oryzae* and *X. oryzicola* were confined to Asia. Recently they have become established in West Africa and in Brazil where none of the local rice varieties have any resistance against them. Again, the consequence of these pathogen introductions are yet to be fully appraised.

Many more examples could be given of the quarantine risks arising from the international transfer of plant propagation material. However, this is not the purpose of this paper. Here we should instead emphasize the problem and see what could be done to find a suitable solution.

10.5 REMEDIAL ACTION

As indicated in the beginning of this paper, it has been suggested that to find remedy for the errors experienced in international development, we should try to reverse the order of priority assigned to economic need, technology, engineering and to the biology of the natural world. We now propose that to find a remedy for some of the risks and instability introduced into world agriculture in recent years, we should also try to follow a similar approach.

As mentioned before, under the strong pressure to produce more yield per unit of land, we have paid more attention to immediate economic needs, we have adopted the best western, energy-depending technology but, in doing so, we have overlooked some of the most basic biological phenomena of the natural world. In the field of plant protection, this has resulted in an increased number of hazards, some of which have been described in the previous pages. We now suggest to reverse this course of action. This could be achieved in three possible ways.

10.5.1   Identifying Socio-economic Constraints

All too often it has been believed that given a certain technology and provided the necessary financial resources, the process of development could take off without any further impedement. Much disappointment has resulted from this way of thinking. This has been particularly true in agriculture and especially at the level of the peasant farmer. This is the type of farmer who still represents the largest proportion of the people living in the developing world. In most countries, traditional farmers have remained practically untouched by scientific progress and technological innovations. Their problems have been ignored because they are either too complex to understand or too difficult to solve. This has resulted in what Putter (1978) has called the "cognitive dissonance": that is, the practical impossibility of appreciating first the basic culture, actual needs and socio-economic constraints of the peasant farmer and of communicating later the most suitable improvements.

Only recently it was recognized that cropping decisions by traditional farmers are affected by many more factors than those of market-oriented farmers, and that all these factors must be considered at once if changes and improvements are to be made. The work conducted in Guatemala (Hildebrand, 1979) offers one of the most notable examples of how identification of these socio-economic constraints is made and how appropriate advice is given to the farmers. Multidisciplinary teams of social scientists and agricultural technicians working together learn first from the same peasant farmer and from his surroundings the agro-socio-economic conditions that determine his decision on cropping pattern. Following this survey, the team suggests the type of technology the farmer can and is likely to adopt. Where this is lacking, they recommend the development of an appropriate technology. This is generated both at the experimental station and in the farmer's field but it is only validated under the farmer's actual conditions. The level of acceptance by the farmer is finally measured. The success of this approach depends mainly on winning the farmer's confidence, on establishing good lines of communication with him and on proposing changes that he can accept. It is the farmer himself in the end who decides on acceptability of the new technology, which must unquestionably improve his productivity and increase his income.

One of the facts learned from this approach in Guatemala has been the wide variability among different areas of the country that does not permit the adoption of the same technology to contiguous areas otherwise appearing homogeneous as far as farm size and cropping patterns are concerned. This again proves the fallacy of the worldwide trend of developing package technology in far away experiment stations without due consideration of where and how this technology is to be used.

Another interesting point resulting from this work in the Western Highlands of Guatemala concerns the introduction of high yielding wheat varieties. When these were compared with local land races (criollo wheat), it was discovered that although the former could produce higher yields, the efficiency of conversion of nitrogen to wheat grain was much higher in the local wheat. Even in economic terms this higher conversion made the scarce capital invested in fertilizers by the subsistence farmer of Guatemala more productive and fully justified his choice.

These and other examples show how essential it is to understand the socio-economic constraints before being able to generate and transfer appropriate technology.

As for crop parasite control and avoidance of epidemiological hazards, it should be appreciated that subsistence crops, although of low productivity, are generally free from severe parasite attack. In other words, they are resistant. This type of resistance, derived from many years of natural selection, is also stable because stability is probably the most essential requirement for a subsistence system to evolve and survive. It is a great mistake to believe that modern high-yielding cultivars, which often cannot be grown without protection from pesticides, could be blindly adopted by subsistence or semi-subsistence farmers. On the contrary, efforts should be directed to see that, through modern breeding systems and proper selection, the high yields and quality characteristics of modern varieties should be introduced into subsistence cultivars that the farmer is already growing.

So far, we have discussed the problems of the traditional farmer. But what about the market-oriented grower? How does he differ with regard to newly introduced pest and disease hazards? In many respects, he is much more vulnerable to these epidemiological situations because, in addition to other considerations, he has also an investment and a market to protect. The example of the drop in cacao production in Ecuador as a result of plant disease can be cited again as a good example of this type of vulnerability. Even in the case of the market-oriented grower, socio-economic constraints must be fully explored before a new technology is developed and transferred. The responsibility lies with plant protection and social scientists, who are called to see that sufficient preliminary research is carried out, to reveal all the long-term effects of their research on the farmer. Also, government agencies should see that immediate commercial interests of chemical, seed, fertilizer and equipment companies interested in quick returns do not prevail in introducing elements of risk and instability that are often masked within new technology packages.

10.5.2   Analyzing and Managing the Plant Pathosystem

Robinson (1976) defines the plant pathosystem as a sub-system of the agro-ecosystem characterized by the phenomenon of parasitism. In this wide sense, a plant pathosystem comprises all the plant hosts and their parasites interacting in a given place and time framework. According to this same autor, by a comparative analysis of these interactions both in the wild and in the artificial pathosystem, such as agriculture, it would be possible to gain an understanding of the mechanisms which regulate the natural state of balance that exists between host and parasites, a state that is prevalent in nature but absent in most modern farming systems.

Once these mechanisms are understood, it would be possible to manage them in the most effective way in the artificial crop pathosystems to eliminate or reduce risk and instability and to achieve instead a type of crop loss prevention which would be at the same time cheap, complete and permanent. Again, as Commoner (1973) suggested, we must look at the natural environment before we can design a type of technology that meets the needs of the human condition. Robinson's concept of plant pathosystem analysis and management fits well into this kind of approach. In fact, this type of concept has been taken as a basis to guide the FAO International Program on Horizontal Resistance (IPHR) (Robinson *et al.*, 1977). It may be of interest to recall that it was through an analysis of factors which governed the disease epidemic of *Puccinia polysora* of maize in Africa that the above program was launched. This is recalled here because it is believed pertinent to the subject of this paper. Maize was imported into Africa about four centuries ago. Instead, the pathogen *P. polysora* was left behind in Central and South America until the late 1940's. In its area of origin this fungus causes a leaf disease of little or no economic consequence. When *P. polysora*

polysora reached Africa, all the African maizes were highly susceptible to it because they had lost their horizontal resistance (Van der Plank, 1963), due to changes in gene frequency in the host population. The initial damage caused by the disease was so great that there was fear it could cause a famine in subsistence farmers' communities. In fact, the bulk of maize cultivation in Africa is done by peasant farmers. Various attempts were made to find resistance genes by conventional breeding methods but it was soon discovered that vertical resistance was breaking down so quickly that it was of no value. In the meantime, high levels of horizontal resistance were re-established by the same African farmers without the need of introducing any foreign parent material but simply by saving and replanting seed from survivor plants and by allowing these to polycross at random under intense selection for resistance. The entire process of accumulation of resistance took from five to seven years.

Based on the maize-*Polysora* epidemic in Africa, the FAO program is directed at the establishment of host populations with high genetic flexibility. These are subjected to very strong selection pressure against all locally important parasites in order to achieve changes in gene frequencies and shift modes toward maximum resistance. At the same time selection pressures are also conducted for all agronomically important characters including yield and quality. Most important, all selection is conducted in the area of future cultivation and according to methods and cropping patterns used by local farmers.

Already several projects are under way in a number of FAO member countries to solve concrete crop improvement situations. The results so far appear to be most encouraging.

10.5.3   Reducing Crop Vulnerability

Although this type of action is an essential part of pathosystem analysis and management, it is treated separately here in view of its importance in reducing epidemiological hazards. Robinson (1976) described different categories of vulnerability and different methods for their assessment. Here we are only concerned with those prevention measures which should be taken in advance to reduce these risks.

There are essentially two major approaches to reduce epidemiological hazards: better functioning plant quarantines and better preventive actions. Both of these are discussed below.

10.5.3.1   Plant Quarantines   Plant quarantines are laws and measures to prevent the introduction and spread of exotic parasites to susceptible national crops. To serve their purposes, quarantines must be functioning effectively, must be applied by well trained and well informed personnel, and they must operate within the realm of well defined biogeographical regions (Mathys, 1955). Approaches to see that these objectives are fully achieved are dscribed by Kahn (1977). Unfortunately, in most countries, particularly in the developing world, the above requirements are seldom met and plant quarantines are not operating as they should. This is due to the lack of trained personnel and of facilities, excessive number of ports of entry and large interchange of agricultural commodities, excessive bureaucracy and lack of general understanding of what plant quarantines really are and what purpose they serve. This results in a situation of continuous risk often increased by persons who consciously or unconsciously try to circumvent these laws. These include the unscrupulous importer, the smuggling nurseryman or farmer, the ignorant traveller or the well-informed scientist. The latter represents a special category of risk when he deals with exotic parasites (and he wants to compare them with others in his own country) or he wants to test the behavior of certain foreign crops or varieties but he does not take the proper phytosanitary precautions to avoid the introduction or spread of their parasites.

Remedial action to overcome this problem include establishment of well functioning quarantine systems, education of the public (especially the categories mentioned above), reduction of bureaucracy, identification of healthy sources of foreign propagating material and transfer of agricultural commodities with the application of all necessary safety measures. Of particular interest to developing countries is the safe transfer of genetic resources. A thorough analysis of this subject has been recently made (Hewitt *et al.*, 1977).

10.5.3.2 Preparation. Ideally, to avoid the consequences of exotic parasite introductions or the sudden outbreaks of local insect or disease epidemics, there should be reliable methods to anticipate these events and to prepare adequately for them. In the case of exotic parasites, the problem is essentially focused on knowledge that such parasites already exist in some other parts of the world and that, if introduced in a new area, they are epidemiologically competent to cause serious damage. An example of this is the pathogen *Microcyclus ulei* causing South American leaf blight, one of the most destructive diseases of para rubber (*Hevea brasiliensis*). This pathogen is still confined to Tropical America but absent in Africa and in South and South East Asia, which account for 90% of the total world production. The lack of a viable rubber industry in Tropical America is due to this fungus, which occurs in the form of different races known to attack severely the high-yielding oriental rubber clones. Because of the similarity in climatic conditions between the American tropics and the oriental growing regions, there is no doubt that this pathogen has the epidemiological competence to cause great damage if it should ever be introduced into this part of the world. To prepare against *M. ulei*, plans have been made for the strengthening of plant quarantines, for rapid identification and for an emergency eradication campaign. Lastly, a breeding program has been launched to combine high yielding capacities with disease resistance. To this effect, the Rubber Research Institute of Malaysia established a Unit in Trinidad in 1960 to study the biology and epidemiology of the fungus and to screen rubber clones for resistance under local conditions. However, the need for expanding this type of research to other areas in South America has been suggested (Sripathi, 1973).

It should be mentioned that preparation measures, such as those described above, are not always easy to establish, particularly in developing countries where limited budgets, deficient infrastructures and other day-by-day pressing commitments do not allow neither time nor resources to be diverted to preventive action.

For example, although coffee leaf rust (*Hemileia vastatrix*) was discovered in

Brazil in 1970 and in Nicaragua in 1976, very little preventive action was taken in most of the other coffee producing countries of Central and South America, as shown by the recent outbreaks of the disease in Peru, Bolivia and El Salvador.

In the case of indigenous parasites, methods of preventing serious epidemics are essentially based on the development and functioning of good forecasting/ warning systems. This subject has been recently reviewed by Kranz and Theunissen (1979). Again, in developing countries it is difficult for such a system to function although some notable exceptions exist such as the Filipino attempt to establish a viable "Surveillance and Early Warning System" (Weherend, 1977). Efforts should be made in other countries to see whether similar activities could be effectively established.

10.6 CONCLUDING REMARKS

In this brief review we have seen that the common factor which has contributed to epidemiological hazards of many crops in developing countries has been the drive to achieve the greatest possible yields per unit of land in the shortest possible time.

Emphasis on the economic factor has often meant ignoring other important factors (such as yield quality and resistance to parasites) or important considerations such as, for example, the effects of social change. This does not imply any criticism of what has been done: there was an urgency for increasing food production and this was achieved successfully using essentially commercial criteria of large productions and of immediate profits. However, the time has come to reduce the risk that is so inherent in this approach, by looking deeper into the biological world and by finding here answers that can really and more permanently solve the inherent problems of world agriculture. In particular, answers should be found for the farmers of the developing world, especially those belonging to the "silent majority" of traditional farmers. In a commentary that appeared in *Science* a few years ago (Wade, 1974), a statement is reported by Wilkes, an American biologist, who clearly addresses the present situation. The statement says: "We should export a stable technology, not one that depends on fossil fuels and phosphate rock. The price for maintaining high yields by monocultural farming is to constantly change the genetic material by breeding for resistance against the latest pest. But monocultures are not a long-term solution for small field agriculture. The technology we should offer to poor countries is a system with lower yield and greater genetic safety." We feel now that a new technology is needed and that this should be derived from our study of the real biological and sociological world we live in. Fortunately, steps in this direction are being taken.

10.7 REFERENCES

Anon. (1972). Genetic vulnerability of major crops. National Academy of Sciences. Washington, D. C.

Anon. (1974). The use of mercury and alternative compounds as seed dressings. Report of Joint FAO/WHO Meeting. *FAO, Agricultural Studies* 95.

Anon. (1977). Pembangunan Pertanian Pangan. Direktorat Jenderal Pertanian Taneman Pangan. 40 pp. (mimeo).

Buddenhagen, I. W. (1960). Strains of *Pseudomonas solanacearum* in indigenous hosts in banana plantations of Costa Rica and their relationship to bacterial wilt of bananas. *Phytopathology* 50: 660-664.

Buddenhagen, I. W., Sequiera, L., and Kelman, A. (1962). Designation of races in *Pseudomonas solanacearum. Phytopathology* 52: 726 (Abstr.)

Chiarappa, L. (1979). The probable origin of letal yellowing and its co-identity with other lethal diseases of coconut. *5th Technical Working Party Coconut Production, Protection and Processing. AGP/CNP/79/3.*

Commoner, B. (1973). On the meaning of ecological failures in international development. *In* "The Careless Technology Ecology and International Development." Tom Stacey Ltd. pp. 21-29.

Day, P. R. (ed.) (1977). The genetic basis of epidemics in agriculture. *Ann. N.Y. Acad. Sci.* Vol. 287.

Evans, H. C., Edwards, D. F., and Rodriguez, M. (1977). Research on cocoa diseases in Ecuador: past and present. *PANS* 23: 68-80.

Guyot, H. (1953). La lutte contre *Cercospora musae* dans les bananeraies de Guadeloupe. Essais de nébulisation. *Fruits d'Outremer* 8: 524-532.

Gwin, G. H. (1977). Distribution and impact of lethal yellowing in Florida. *Proc. 3rd Meeting ICLY*, Fort Lauderdale, Fla.

Harlan, J. A. (1976). Diseases as a factor in plant evolution. *Ann. Rev. Phytopathol.* 14: 31-51.

Harries, H. C. (1978). Evolution, dissemination and classification of *Cocos nucifera. Bot. Rev.* 44: 265-319.

Hewitt, Wm. B., and Chiarappa, L. (eds.) (1977). Plant health and quarantine in international transfer of genetic resources. CRC Press, Cleveland, Ohio, 346 pp.

Hildebrand, P. E. (1979). Generating technology for traditional farmers. The Guatemala experience. *Abst. IX Intern. Cong. Plant Protection.* Washington, D.C.

Kahn, R. P. (1977). Plant quarantine: principles, methodology, and suggested approaches. *In* "Plant Health and Quarantine in International Transfer of Genetic Resources" (Wm. B. Hewitt and L. Chiarappa, eds.). CRC Press, Cleveland, Ohio.

Kranz, J., and Theunissen, J. (1979). Prognosis and warning in plant protection. Material of a workshop. Deutsche Stiftung für Internationale Entwiklung, Feldafing, Germany. (Mimeogr.) 318 pp.

Mathys, G. (1955). Thoughts on quarantine problems. *EPPO Bull.* 5: 55.

Mochida, O. (1979). Brown plant hoppers reduce rice population. *Indonesian Agr. Res. Dev. J.* 1: 2-7.

Oka, I. N. (1978). Implementing the integrated control program of the brown plant hopper in Indonesia. FAO Panel of Experts on Integrated Control. Sept. 1978.

Oka, I. N. (1979). Feeding populations of people versus populations of insects: the example of Indonesia and the rice brown plant hopper. *Abstr. IX Intern. Cong. Plant Protection*, Washington, D.C.

Okamoto, H. (1965). Chemical control of rice blast in Japan. *In* "The Rice Blast Disease." IRRI–Johns Hopkins Press. 507 pp.

Putter, C. A. J. (1978). Pathosystem management under subsistence farming conditions. *Abstr. 3rd Intern. Cong. Plant Pathol.*

Rillo, A. R. (1979). Bacterial wilt of banana in the Philippines. *FAO Plant Protection Bull.* 27: 105-108.

Robinson, R. A. (1976). Plant pathosystems. Springer-Verlag, Berlin, Heidelberg, New York. 184 pp.

Robinson, R. A., and Chiarappa, L. (1977). The international programme on horizontal resistance. *FAO Plant Protection Bulletin* 25: 197-200.

Saari, E. E., and Wilcoxon, R. D. (1974). Plant disease situation of high yielding dwarf wheats in Asia and Africa. *Ann. Rev. Phytopathol.* 12: 49-68.

Soowal, J. M. (1979). Economic importance of palms to nursery industry. *Proc. 4th Meeting ICLY*, Fort Lauderdale, Fla.

Sripathi Rao, B. (1973). Some observations on South American leaf blight in South America. *Planter*, Kuala Lumpur 49: 2-9.

Stover, R. H. (1974). Pathogenic and morphologic variation in *Mycosphaerella fijiensis (M. musicola)*. *Proc. Am. Phytopathol. Soc.* 1: 128.

Stover, R. H., and Dikson, J. D. (1976). Banana leaf spot caused by *Mycosphaerella musicola* and *Mycosphaerells fijiensis* var. *difformis*: a comparison of the first Central American epidemics. *FAO Plant Protection Bull.* 24: 36-42.

Sumitro, Djojohadikusumo (1977). Science resources and development. Inst. for Econ. and Soc. Res. Educ. and Inform. (LP 3 ES). 179 pp.

Van der Plank, J. E. (1963). Plant diseases: epidemics and control. Academic Press, New York and London. 349 pp.

Wade, N. (1974). Green Revolution (II): Problems of adapting a western technology. *Science* 186: 1186-1192.

Weherend, O., and Doria, R. (1977). Review of the one-year operation of the surveillance and early warning system. *Plant Protection News*, Manila VI: 30-37.

INDEX

A

Abortion of virus-infected pollen, 266, 268–269

Abutilon mosaic virus
hosts for, 242
seed transmission of, 242

Abutilon mulleri, virus in seeds of, 242

Abutilon thompsonii x, virus in seeds of, 242

Aceratagallia calcaris, as disease vector, 140

Aceratagallia curvata, as disease vector, 138

Aceratogallia longula, as disease vector, 138

Aceratagallia obscura, as disease vector, 138

Aceratagallia sanguinolenta, as disease vector, 138

Acholeplasma, aster yellows and, 91

Acholeplasma laidlawii
disease agents in, 62, 63
MLO infection of, 88

Acinopterus angulatus, as disease vector, 132

Acyrthosiphon pisum, as virus vector, 164, 194

Aegilops spp., virus in seeds of, 243

Agallia albidula, as disease vector, 134

Agallia constricta, as disease vector, 134, 138

Agallia quadripunctata, as disease vector, 134, 138

Agalliana ensigera, as disease vector, 134, 135

Agalliana stricticollis, as disease vector, 135

Agalliinae, disease transmission by, 115

Agalliopsis ancistra, as disease vector, 111, 116, 133, 134, 138

Agalliopsis novella, see Agalliopsis ancistra

Ageratum conuzoides
virus control and, 23, 24
virus overwintering in, 6

Agropyron elongatum, virus in seeds of, 243

Alfalfa
Pierce's disease of grape and, 129
potato witches' broom and, 130
as virus host
in pollen and ovule, 263
in seeds, 242, 298

Alfalfa dwarf disease, *see* Pierce's disease of grapes

Alfalfa mosaic virus (AMV)
cost protein role in, 172
ELISA detection of, 50
host range of, 7, 274
in seeds, 242
pollen transmission of, 263
seed transmission of, 242, 274, 298

Alfalfa witches broom disease, vector of, 141

Alisma plantago, MLOs in, 70

Alkaline earth metals, in MLO-infected plants, 78

Alkaline metals, in MLO-infected plants, 78

Alkaline phosphatase, in ELISA method, 37, 41, 42

Alkaloids, accumulation of, in MLO disease, 78

Alkylmercury pesticides, poisonings from, 325

Alliaria officinalis, MLO infections of, 64

Allium cepa, see Onion

Almond
phony peach disease of, 126
yellows disease transfer to, 87

Almond leaf scorch disease, transmission of, 131

341